W9-DCC-994

ORDINARY DIFFERENTIAL EQUATIONS

ORDINARY DIFFERENTIAL EQUATIONS

A computational approach

Charles E. Roberts, Jr.
Indiana State University

Prentice-Hall, Inc., Englewood Cliffs New Jersey 07632

Library of Congress Cataloging in Publication Data

ROBERTS, CHARLES E., (date)
Ordinary differential equations.

Bibliography: p. 365
Includes index.
1. Differential equations—Numerical
solutions. I. Title.
QA372.R75 519.4 78-13023
ISBN 0-13-639757-3

Printed in the United States of America

10 9 8 7 6 5 4 3 2 1

Editorial/production supervision
by Zita de Schauensee
Cover design by Frederick Charles, Ltd.
Manufacturing buyer: Phil Galea

PRENTICE-HALL INTERNATIONAL, INC., *London*
PRENTICE-HALL OF AUSTRALIA PTY. LIMITED, *Sydney*
PRENTICE-HALL OF CANADA, LTD., *Toronto*
PRENTICE-HALL OF INDIA PRIVATE LIMITED, *New Delhi*
PRENTICE-HALL OF JAPAN, INC., *Tokyo*
PRENTICE-HALL OF SOUTHEAST ASIA PTE. LTD., *Singapore*
WHITEHALL BOOKS LIMITED, *Wellington, New Zealand*

To the memory of my father, Charles
To my loving mother, Evelyn
And to those who bring
the greatest joy and pleasure into my life,
my wife and children, Imogene, Eric, and Natalie

CONTENTS

PREFACE

The main purpose of this text is to present ordinary differential equations in a manner that is compatible with the way they are used in many professions. Therefore, three major objectives are:

1. To present several general methods for obtaining an explicit or implicit solution to a specific ordinary differential equation.
2. To present, discuss, and often prove existence, uniqueness, and continuation theorems for initial value problems.
3. To develop, analyze, and use numerical techniques to produce approximate solutions to ordinary differential equations.

To achieve these objectives, a variety of methods which may be used when attempting to solve a particular differential equation are presented. Next the theory for the particular problem is set forth and discussed. Then elementary numerical methods for solving the problem are developed and analyzed. This order of presenting topics is easy to justify, since this is the order in which one normally proceeds when trying to solve any particular problem in differential equations. First, consider methods for obtaining explicit and implicit solutions. Upon failing to find such a solution, consider the applicable theory. If the associated theory guarantees the existence and uniqueness of a solution, then employ some numerical method to obtain an approximate solution. In actual practice, a numerical method will usually be required.

Most elementary differential equations texts accomplish the first objective by dealing almost exclusively with differential equations that can be solved explicitly, implicitly, or by series. They usually introduce little accompanying general theory and often only mention numerical methods in passing. In contrast, this text includes the standard techniques for obtaining explicit, implicit, and series solutions, emphasizes the general theory, stipulates tests that are relatively simple to apply for determining when a solution exists and is unique, and introduces and uses elementary numerical methods. This text is unique in that theory and numerical methods have been integrated with traditional problem-solving techniques. This parallel presentation of theory, analytic methods, and numerical methods prepares the reader to successfully solve real-life problems that involve ordinary differential equations.

Students who enroll in ordinary differential equations courses normally do so for only one or two semesters as an undergraduate. Furthermore, few of these students ever enroll in a numerical analysis course. However, most students who take differential equations find employment in business, industry, or government and will use numerical methods almost exclusively. Therefore, the single most useful and distinguishing feature of this text is the introduction and explanation of elementary numerical methods as an integral part of the text. The text is also highlighted by numerical case studies that illustrate the necessity of considering the theory prior to calculating a numerical solution.

It is assumed that the readers of this text will be undergraduate students majoring in mathematics, applied mathematics, computer science, one of the various fields of engineering, or one of the physical or social sciences. The only assumptions made regarding the reader's mathematical background are that he has studied calculus; that he can add, subtract, and multiply complex numbers; that he has been introduced to matrices; and that he can calculate the determinant of a square matrix. Concepts with which the reader may not already be familiar—such as supremum, domain, closure of a set, Lipschitz condition, linear dependence, etc.—are introduced and explained to the degree necessary for use within the text. This is done at the location where the concept is first used.

The exercises in the first seven chapters which require a numerical solution can be solved with a calculator and require no knowledge of computer programing. Any computer program that does appear within the text is written in FORTRAN, and the purpose and function of most of the program statements are discussed thoroughly. From this discussion an individual who knows no programing language can understand what is transpiring. In the event that one wishes to solve an exercise which requires a numerical solution using a computer, all that normally needs to be done is to keypunch a program which appears within the text, change one or two program statements which specify the function to be integrated or the conditions to be fulfilled, supply one data card which contains information relative to the problem to be solved, and use the job control cards required by the computer installation at which the program is to be run. Consequently, in order to use a computer to generate a numerical solution to a particular exercise, the reader is not required to have any prior knowledge of computers or of any particular computer programing language.

The topics to be presented and the extent to which they are explored (for example, whether theorems are proven or not) depend, of course, upon the background of the students and the intent of the course. For instance, those who do not wish to pursue the more rigorous mathematical route through this text may omit some of the proofs presented in Section 4.2. I would recommend that an introductory three-semester-hour course (or four- or five-quarter-hour course) cover material from Chapter 1 through Section 6.3, and that a second three-semester-hour course cover material from the remainder of the text. This text could also be used to present a three-semester-hour course titled "numerical solutions to ordinary differential equations" to those students who have completed a "traditional" introductory course. The content of this course would consist of material from Chapter 2, Chapter 4, Section 8.4, Section 9.2, and Chapter 10. The dependence relationship of the chapters is depicted in the diagram on the next page. From this diagram it is evident that a large number of courses of study which require different lengths of time to present and which have different contents can easily be designed using this text.

I would like to thank the following professors for their assistance in reviewing the manuscript: Richard A. Alo, Lamar University; Neil E. Berger, University of Illinois; J. M. A. Danby, North Carolina State University; William R. Fuller, Purdue University; John G. Hocking, Michigan State University; H. L. Pearson, Illinois Institute of Technology; Dieter Schmidt, University of Cincinnati; Jon W. Tolle, The University of North Carolina.

I wish to express my appreciation to my colleague Professor Richard Easton for reading various versions of this manuscript and offering many helpful suggestions. I would also like to thank Mrs. Brenda Fischer for her excellent job of typing the manuscript.

Terre Haute, Indiana CHARLES E. ROBERTS, JR.

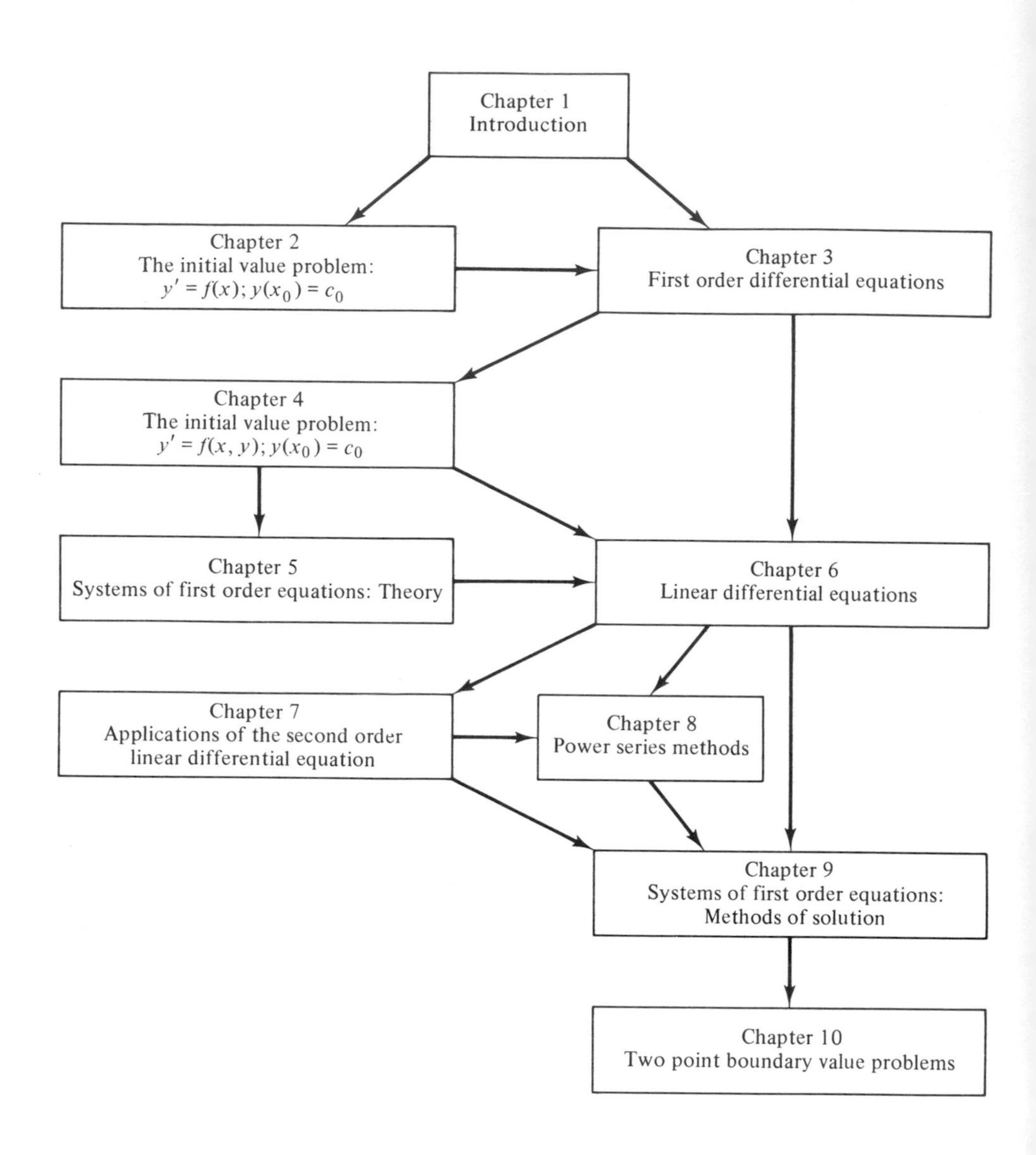

Chapter 1
Introduction
Chapter 2
The initial value problem:
$y' = f(x); y(x_0) = c_0$
Chapter 3
First order differential equations
Chapter 4
The initial value problem:
$y' = f(x, y); y(x_0) = c_0$
Chapter 5
Systems of first order equations: Theory
Chapter 6
Linear differential equations
Chapter 7
Applications of the second order
linear differential equation
Chapter 8
Power series methods
Chapter 9
Systems of first order equations:
Methods of solution
Chapter 10
Two point boundary value problems

1

INTRODUCTION

1.1 HISTORY

The origin of the branch of mathematics known as *differential equations* dates back at least to 1671 and Newton's classification of first order ordinary differential equations into three classes. Sir Isaac Newton (1642–1727) called these equations fluxional equations instead of differential equations. He assumed that the solution of these equations could be expressed as an infinite series, and he successively determined the coefficients in a manner similar to the technique employed today. However, he did not consider the convergence of the series. The notation $\dot{y}$ for the derivative of y with respect to the independent variable was also introduced by Newton.

Gottfried Leibniz (1646–1716) invented the differential notation dy and the symbol $\int$ for integration. To our knowledge he first used these notations in conjunction in 1675, when he wrote:

$$\int y\,dy = \tfrac{1}{2}y^2.$$

In 1676 Leibniz used the term "differential equation" to denote a relationship between two differentials dx and dy. Thus was christened the branch of mathematics that deals with equations which involve differentials or derivatives.

The word *integral* was introduced into mathematics in 1690 by Jacques Bernoulli (1654–1705). In 1691 Leibniz discovered the technique of separation of variables and in 1692 he reduced the linear homogeneous first order differential equation to quadratures. Jean Bernoulli (1667–1748) introduced the concept of an integrating factor in 1694 and the technique of changing the dependent variable. By the end of the seventeenth century the techniques which are usually employed when attempting to solve first order ordinary differential equations were known. As we shall discover, these techniques often prove to be inadequate.

However, early in the development of the study of differential equations, it was believed that elementary functions would be sufficient for the representation of solutions of differential equations arising from geometry and mechanics. Thus, early attempts at solving differential equations were directed toward finding explicit solutions or reducing the solution to a finite number of quadratures. By 1723 at the latest it was recognized that even some first order ordinary differential equations do not have solutions which can be expressed in terms of elementary functions. As a matter of fact, if an ordinary differential equation is written down at random, the probability of being able to write the solution in terms of known functions or their integrals is nearly zero. This emphasizes the necessity for developing methods for obtaining approximate solutions.

In 1739 Léonard Euler (1707–1783) introduced the method of variation of parameters. Jean Bernoulli had unsuccessfully attempted to solve the general linear homogeneous differential equations with constant coefficients. Euler gave a complete discussion of this problem in 1743. He also devised the classical method for solving nonhomogeneous linear differential equations.

No adequate discussion of differential equations as a unified topic existed prior to the lectures developed and presented by Augustin-Louis Cauchy (1789–1857) in the 1820s. In these lectures Cauchy developed the first existence and uniqueness theorems for first order differential equations. Cauchy extended his theory to a system in n first order differential equations in n dependent variables which was equivalent to a single nth order differential equation. Rudolf Lipschitz (1832–1903) generalized Cauchy's existence and uniqueness theorems in 1876. Émile Picard (1856–1941) improved upon the theorems of Cauchy and Lipschitz in 1893 by introducing the method of successive approximations.

1.2 CLASSIFICATION OF DIFFERENTIAL EQUATIONS

By a *differential equation* (DE) we shall mean any equation that involves derivatives or differentials of a function or functions. The *order* of a differential equation is the largest positive integer n for which the nth derivative or differential occurs in the differential equation. If a differential equation is written as a polynomial, then the highest power to which the highest derivative appears in the equation is called the *degree* of the equation.

In the study of differential equations, it is both advantageous and convenient to classify the equations into different categories—much as one classifies chemical compounds into organic and inorganic categories in the study of chemistry. The first and most obvious two categories into which differential equations are classified are those of ordinary differential equations and partial differential equations. This classification is based on the unknown function appearing in the differential equation. If the unknown function depends on only one independent variable, then the differential equation is called an *ordinary differential equation* (ODE). Whereas, if the unknown function depends on two or more independent variables, then the differential equation is called a *partial differential equation* (PDE).

For example,

$$y'' + x(y')^2 + xy = x^3 \tag{1}$$

is a second order ordinary differential equation of degree one;

$$(y''')^2 + yy'y'' + xy' + y^2 = \sin x \tag{2}$$

is a third order ordinary differential equation of degree two;

$$yz_x + x^2z_y = xy \tag{3}$$

is a first order partial differential equation in two independent variables; and

$$u_t = \alpha(u_{xx} + u_{yy} + u_{zz}) \tag{4}$$

is a second order partial differential equation in four independent variables. It should be noted that order is defined for all differential equations, but degree is not defined for some. For instance, $y'' = \sin y$ is a second order differential equation. However, degree is not defined for this equation, since the equation is not a polynomial.

Throughout this text we shall concern ourselves primarily with the solution of ordinary differential equations. However, both ordinary differential equations and partial differential equations are subdivided into two large classes, according to whether they are linear or nonlinear.

The general form of an nth order ordinary differential equation is

$$\phi(x, y, y^{(1)}, \ldots, y^{(n)}) = 0, \tag{5}$$

where $y^{(k)}$ denotes the kth derivative of y with respect to x. An nth order ordinary differential equation is *linear* if ϕ is a linear function in each of the variables $y, y^{(1)}, \ldots, y^{(n)}$. Hence, the general form of a linear nth order ordinary differential equation is

$$a_0(x)y^{(n)} + a_1(x)y^{(n-1)} + \cdots + a_n(x)y = f(x). \tag{6}$$

An nth order ordinary differential equation that cannot be written in the form (6) is called a *nonlinear* nth order ordinary differential equation.

Hence, (1) and (2) are nonlinear ordinary differential equations while

$$x^2y''' + (x^2 - 2)y'' + (\sin x)y' + e^x y = x^2 - 1 \tag{7}$$

is a third order linear ordinary differential equation. Notice that every linear ordinary differential equation is of first degree but not every ordinary differential equation of first degree is linear.

An *explicit solution* of the nth order DE (5) on an interval I is a function $y(x)$ defined on I and satisfying $\phi(x, y(x), y^{(1)}(x), \ldots, y^{(n)}(x)) = 0$ for all x in I. Notice that this definition implies that an explicit solution y has n derivatives on I and, therefore, that $y, y^{(1)}, \ldots, y^{(n-1)}$ are all continuous on I. Generally, the interval I is not specified explicitly, but it is understood to be the largest interval on which y is defined and satisfies (5).

EXAMPLE Consider the differential equation

$$y' + y = 0. \tag{8}$$

The function $y_1(x) = e^{-x}$ is defined and continuous on the interval $(-\infty, \infty)$ and the derivative $y_1'(x) = -e^{-x}$ is defined on $(-\infty, \infty)$. Since

$$y_1'(x) + y_1(x) = -e^{-x} + e^x = 0 \quad \text{for all } x \text{ in } (-\infty, \infty);$$

that is, since $y_1(x) = e^{-x}$ satisfies the differential equation $y' + y = 0$ for all x in $(-\infty, \infty)$, $y_1(x) = e^{-x}$ is an explicit solution of the given differential equation for all real x.

The function

$$y_2(x) = \begin{cases} e^{-x}, & x < 0 \\ 2e^{-x}, & x \geq 0 \end{cases}$$

is not an explicit solution of the given differential equation on the interval $(-\infty, \infty)$, since $y_2(x)$ is not continuous and therefore not differentiable at $x = 0$. However, the function $y_2(x)$ is an explicit solution on any interval that does not contain the point $x = 0$.

A relation $f(x, y) = 0$ is said to be an *implicit solution* of the nth order DE (5) on an interval I if the relation defines at least one function $y_1(x)$ on I such that $y_1(x)$ is an explicit solution of (5) on I. We shall usually refer to both explicit and implicit solutions simply as solutions.

EXAMPLE Consider the differential equation

$$yy' + x = 0. \tag{9}$$

We shall show that the relation

$$f(x, y) = y^2 + x^2 - 16 = 0 \tag{10}$$

is an implicit solution on the interval $(-4, 4)$. The graph of equation (10) is a circle of radius 4 with center at the origin. See Figure 1.1(a). Solving equation (10) for y in terms of x, we get

$$y(x) = \pm\sqrt{16 - x^2}.$$

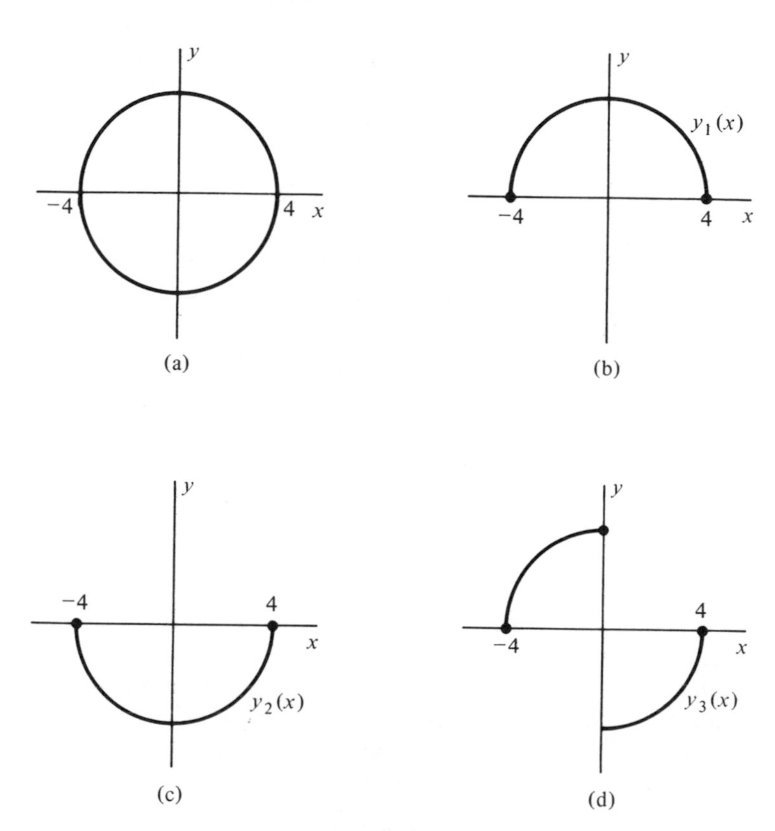

Figure 1.1

The functions

$$y_1(x) = \sqrt{16 - x^2} \tag{11a}$$

and

$$y_2(x) = -\sqrt{16 - x^2} \tag{11b}$$

are both defined and real for x in $[-4, 4]$. The graph of $y_1(x)$ is shown in

Figure 1.1(b) and the graph of $y_2(x)$ is shown in Figure 1.1(c). Differentiating equations (11a) and (11b), we obtain

(12a)
$$y_1'(x) = -\frac{x}{\sqrt{16 - x^2}}$$

and

(12b)
$$y_2'(x) = \frac{x}{\sqrt{16 - x^2}}.$$

Both y_1' and y_2' are defined and real for x in $(-4, 4)$. Substituting y_1 and y_1' into the differential equation (9), we find that

$$\sqrt{16 - x^2}\left(-\frac{x}{\sqrt{16 - x^2}}\right) + x = -x + x = 0.$$

So $y_1(x)$ is an explicit solution of (9) on the interval $(-4, 4)$, and therefore equation (10) is an implicit solution of (9). Likewise, $y_2(x)$ can be shown to be an explicit solution of (9) on the interval $(-4, 4)$. Thus, the implicit solution (10) defines at least two explicit solutions of (9) on the interval $(-4, 4)$. The function

$$y_3(x) = \begin{cases} \sqrt{16 - x^2}, & -4 \le x \le 0 \\ -\sqrt{16 - x^2}, & 0 < x \le 4 \end{cases}$$

shown in Figure 1.1(d) satisfies relation (10), $y^2 + x^2 - 16 = 0$; however, $y_3(x)$ is not an explicit solution of (9) on the interval $(-4, 4)$, since $y_3(x)$ is not continuous and therefore not differentiable at $x = 0$.

In this case it was fairly easy to determine an explicit solution from the implicit solution and to determine the interval on which the solution exists. However, this will not generally be the case. Normally, we will not be able to solve a given relation in x and y explicitly for y. Therefore, we will usually obtain a relation in x and y by some means, verify that this relation formally satisfies the particular differential equation under consideration, and say that the relation is an implicit solution. For example, we will say that the relation

(13)
$$y^3 + 2xy - x^2 = c,$$

where c is a constant, is an implicit solution of the differential equation

(14)
$$y' = \frac{2x - 2y}{3y^2 + 2x}.$$

In order to verify that (13) formally satisfies the differential equation (14), we differentiate (13) with respect to x and solve for y', which gives us (14).

We shall soon discover that, in theory, we will be able to explicitly solve linear differential equations and, in theory, determine the interval on which the solution exists directly from the differential equation itself. However, the best that we will usually be able to accomplish for nonlinear differential equations is to obtain a series or implicit solution. This is one of the primary differences between the kinds of results that we can expect to obtain for linear differential equation versus nonlinear differential equations.

An n-parameter family of functions

$$f(x, y, c_1, c_2, \ldots, c_n) = 0 \tag{15}$$

is called *the general solution* of the nth order DE (5), if (15) is a solution of (5) (explicit or implicit) for every choice of the parameters $c_1, c_2, \ldots, c_n$, and the set of parameters cannot be replaced by another set of parameters with fewer elements and still represent the same set of solutions.

Although the two-parameter family of functions $y = c_1e^{x+c_2}$ satisfies the second order differential equation

$$y'' - y = 0 \tag{16}$$

for every choice of the parameters c_1 and c_2, it is not the general solution. The set of solutions represented by $y = c_1e^{x+c_2}$ can also be represented by the one-parameter family $y = ke^x$, since $y = c_1e^{x+c_2}$ may be rewritten as $y = c_1e^{c_2}e^x = ke^x$. Furthermore, there may be more than one function of the form (15) which is the general solution of (5). For example, $y_1 = c_1e^{-x} + c_2e^x$ and $y_2 = k_1 \sinh x + k_2 \cosh x$ are both general solutions of equation (16). (The reader is asked to verify this fact in Exercise 12 at the end of this section.) So one might well argue that the terminology *a* general solution should be used instead of *the* general solution. However, y_1 and y_2 are just two different representations of the same set of solutions—the general solution. Therefore, we shall follow the customary practice of calling any function of the form (15) which satisfies (5) the general solution. Any solution that is obtained by assigning definite values to the n parameters $c_1, c_2, \ldots, c_n$ of the general solution is called a *particular solution*.

The general solution of the first order differential equation (8) $y' + y = 0$ is the one-parameter family of functions $y = ce^{-x}$. Notice that this is an explicit solution. To verify that this is the general solution, we differentiate and obtain $y' = -ce^{-x}$. Substituting into the differential equation, we see that $y' + y = -ce^{-x} + ce^{-x} = 0$ for all x and all c. So $y = ce^{-x}$ is the general solution on the interval $(-\infty, \infty)$. The function $y_1(x) = e^{-x}$ is the particular solution which is obtained from the general solution by choosing $c = 1$.

The general solution of the first order differential equation (9) $yy' + x = 0$ is

$$y^2 + x^2 = c^2. \tag{17}$$

Notice that this is an implicit solution with one parameter, c. Differentiating

(17), we obtain $2yy' + 2x = 0$ for any c, and dividing by 2 we get equation (9). So we have formally verified that (17) satisfies (9) for any choice of c. Choosing $c = 4$ or $c = -4$, we get the particular implicit solution (10) $y^2 + x^2 - 16 = 0$.

EXAMPLE Show that

$$y = c_1e^{-2x} + c_2e^x + x \tag{18}$$

is the general solution of the second order linear differential equation

$$y'' + y' - 2y = 1 - 2x. \tag{19}$$

Differentiating equation (18) twice, we obtain

$$y' = -2c_1e^{-2x} + c_2e^x + 1 \tag{20}$$

and

$$y'' = 4c_1e^{-2x} + c_2e^x. \tag{21}$$

Substituting equations (18), (20), and (21) into equation (19), we get

$$(4c_1e^{-2x} + c_2e^x) + (-2c_1e^{-2x} + c_2e^x + 1) - 2(c_1e^{-2x} + c_2e^x + x) = 1 - 2x$$

for all constants c_1 and c_2 and all real x. So equation (18) is the general solution of (19) for all real x.

EXAMPLE Verify that

$$y = (x^2 + c)^2 \tag{22}$$

is the general solution of

$$(y')^2 - 16x^2y = 0. \tag{23}$$

Differentiating (22), we get

$$y' = 4x(x^2 + c) \quad \text{for all real } x \text{ and any constant } c. \tag{24}$$

Squaring equation (24) and substituting for the factor $(x^2 + c)^2$ from equation (22), we find that

$$(y')^2 = 16x^2y \quad \text{for all real } x \text{ and any constant } c.$$

So (22) is the general solution of (23) for all real x.

Let us consider for the moment the function $y = x^2 + 1$. The graph of this function is a parabola with axis the y-axis, with vertex at (0, 1), and which opens upward. The derivative of this function is $y' = 2x$. Given the function y we are able to calculate the derivative y'. The inverse problem is: given the function y', how do we obtain the original function y? The differential equation $y' = 2x = f(x, y)$ defines a real value for each point (x, y)

of the xy-plane. The value at the point $(x, y), f(x, y) = 2x$ in this case, represents the slope of the tangent line to the solution of the differential equation which passes through (x, y). A small segment of the tangent line at various points of the xy-plane are shown in Figure 1.2. The one-parameter

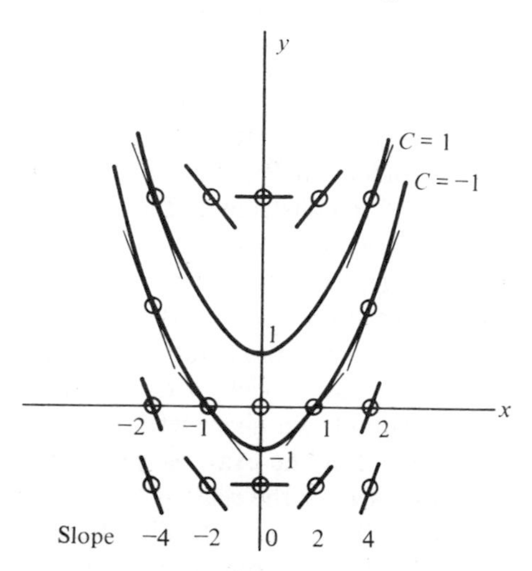

Figure 1.2 Integral curves for the differential equation $y' = 2x$.

family of curves $y(x) = x^2 + C$—parabolas with axis the y-axis, with vertex at $(0, C)$, and which open upward—where C is an arbitrary real number, constitute the set of indefinite integrals of the differential equation $y' = 2x$. The members of this one-parameter family of curves are called the *integral curves* of the differential equation. The curve $y = x^2 + 1$ is a member of this family. Hence, to obtain the original function we must specify, in addition to the differential equation, a point through which the curve is to pass. We might specify, for example, that y is to satisfy $y' = 2x$ and to pass through the point $(1, 2)$—that is, $y(1) = 2$.

Proceeding one step further we see that $y'' = 2$. The set of indefinite integrals of this differential equation is the two-parameter family of curves $y(x) = x^2 + Ax + B$, where A and B are arbitrary real constants. To obtain the original function $y = x^2 + 1$ in this instance, we must specify a combination of conditions that will require us to choose $A = 0$ and $B = 1$. Specifying that (i) $y(1) = 2$ and $y'(1) = 2$ or (ii) $y(1) = 2$ and $y(3) = 10$ will accomplish the desired result. The problem of determining a function that satisfies the differential equation $y'' = 2$ subject to the conditions in (i), called *initial conditions*, is called an *initial value problem*; while the problem of solving the differential equation subject to the conditions in (ii), called *boundary conditions*, is called a *boundary value problem*.

Thus, in the study of ordinary differential equations we are confronted with two large classes of problems—initial value problems and boundary value problems. A precise statement of these two types of problems for nth order ordinary differential equations follows.

An initial value problem (IVP) is a differential equation of the form (5) together with a set of n constraints, the initial conditions (IC), of the form

$$y(x_0) = c_0; \qquad y^{(1)}(x_0) = c_1; \qquad \ldots; \qquad y^{(n-1)}(x_0) = c_{n-1},$$

where $x_0, c_0, c_1, \ldots, c_{n-1}$ are real constants.

A boundary value problem (BVP) is a differential equation of the form (5) together with a set of n constraints, the boundary conditions (BC), specifying values of the function y and/or its derivatives at two or more distinct values of the independent variable x.

Given an algebraic equation such as the polynomial equation $2x^4 - 3x^3 + 3x^2 - 3x + 1 = 0$, we seek the solution set—those values of x which when substituted into the equation yield a true statement. The solution set is often restricted to be a subset of a given set, such as the integers, rationals, reals, or complex numbers. For the example given, the solution set for the integers is $\{1\}$, the solution set for the reals is $\{1, \frac{1}{2}\}$, and the solution set for the complex numbers is $\{1, \frac{1}{2}, +i, -i\}$. The Fundamental Theorem of Algebra states that: "Every polynomial of degree $n \geq 1$ with complex coefficients—$p(x) = a_n x^n + a_{n-1}x^{n-1} + \cdots + a_0$, where $a_n, a_{n-1}, \ldots, a_0$ are complex numbers, $n \geq 1$, and $a_n \neq 0$—has n (not necessarily distinct) roots among the complex numbers." Hence, the Fundamental Theorem of Algebra tells us two things concerning the roots of a polynomial with complex coefficients of degree $n \geq 1$. First, the polynomial has n roots, and second, all the roots can be found in the set of complex numbers.

One would like theorems of this nature for both initial and boundary value problems. That is, one would like to have a Fundamental Theorem for Initial Value Problems and a Fundamental Theorem for Boundary Value Problems which state conditions under which a solution to the problem is guaranteed to exist and which also state conditions under which a solution is guaranteed to be unique. The theory for initial value problems is well established and relatively simple. In Chapter 4 we shall state a Fundamental Theorem for an Initial Value Problem and sketch the proof of the theorem. On the other hand, the theory for boundary value problems is very complex and consequently not as well developed. Therefore, we shall not present any general theory for boundary value problems. The complexities inherent in boundary value problems can be attributed at least partially to the interaction of the boundary conditions with the differential equation. The following example illustrates this interaction.

Consider the relatively simple boundary value problem

$$y'' + y = 0; \qquad y(0) = 0, \quad y(a) = \alpha. \tag{25}$$

The general solution of the differential equation is: $y = A \sin x + B \cos x$, where A and B are arbitrary constants. Imposing the first boundary condition, $y(0) = 0$, results in the equation $0 = B \cos 0$. From which we conclude that $B = 0$. Hence, any solution of the BVP (25) must have the form $y = A \sin x$. We now try to satisfy the second boundary condition $y(a) = \alpha$. If $a \neq n\pi$, where n is an integer, then the BVP (25) has a unique solution, namely, $y = \alpha \sin x/\sin a$. If $a = n\pi$ for some integer n and $\alpha \neq 0$, then there is no solution, since imposing the boundary condition results in the equation —and contradition—

$$\alpha = y(a) = y(n\pi) = A \sin n\pi = 0.$$

If $a = n\pi$ for some integer n and $\alpha = 0$, then there are infinitely many solutions, since any value of A satisfies the equation

$$A \sin n\pi = 0,$$

which results from imposing the second boundary condition.

Because of the inherent complexities, we shall defer the study of the theory of boundary value problems until much later in the text, and then we shall only consider special types of boundary value problems.

EXERCISES

1. For each of the following, state whether the ordinary differential equation is linear or nonlinear, and determine its order and degree.
   (a) $y' = a(x)y + b(x)$
   (b) $y' = a(x)y + b(x)y^n \quad (n \neq 0, n \neq 1)$
   (c) $(y')^2 + xy' = x^3$
   (d) $y'' + k^2y = 0$
   (e) $x\,dx + 2y\,dy = 0$
   (f) $2y\,dx + x\,dy = 0$
   (g) $y(y'')^2 + x(y')^3 + y \sin x = 1$
   (h) $x^2y^{(4)} + y = \tan x$

2. Is $y(x) = 1/x$ a solution of the differential equation $y' = -y^2$
   (a) on the interval $[-1, 1]$? Why?
   (b) on the interval $(0, \infty)$? Why?

3. Is $y(x) = |x|$ a solution of the differential equation $(y')^2 = 1$
   (a) on the interval $[-1, 1]$? Why?
   (b) on the interval $(0, \infty)$? Why?

4. For each of the following differential equations, verify that the given function or functions is an explicit solution and specify the interval or intervals on which the solution exists.
   (a) $y' - y^2 = 1$; $y_1 = \tan x$
   (b) $y' + 3y = 1 + 3x$; $y_1 = x$, $y_2 = 2e^{-3x} + x$

(c) $y'' - 4y = 0$; $y_1 = e^{2x}$, $y_2 = 3 \sinh 2x$
(d) $x^2y'' + xy' - y = 0$; $y_1 = x$, $y_2 = 1/x$
(e) $2x^2y'' + xy' - y = 0$; $y_1 = x$, $y_2 = 1/\sqrt{x}$
(f) $y' = \sin x^2$; $y_1 = \int_0^x \sin t^2\, dt$, $y_2 = -\int_x^\pi \sin t^2\, dt$

5. Verify that $y^2 - x = 1$ is an implicit solution of the differential equation $2yy' = 1$ on the interval $(-1, \infty)$.

6. Verify that $xy^2 + x = 1$ is an implicit solution of the differential equation $2xyy' + y^2 = -1$ on the interval $(0, 1)$.

7. Verify that $x = e^{xy}$ is an implicit solution of the differential equation $y' = (1 - xy)/x^2$ on the interval $(0, \infty)$.

8. Verify that the relation $xy^2 + yx^2 = 1$ formally satisfies the differential equation $y' = -y(2x + y)/x(2y + x)$.

9. Verify that the relation $y = e^{xy}$ formally satisfies the differential equation $y' = y^2/(1 - xy)$.

10. Show that $y = ce^{2x} + xe^{2x}$ is the general solution of the differential equation $y' - 2y = e^{2x}$.

11. Show that $y = c_1 \sin x + c_2 \cos x + x$ is the general solution of the differential equation $y'' + y = x$.

12. (a) Show that $y = c_1e^{-x} + c_2e^x$ is the general solution of the differential equation $y'' - y = 0$.
(b) Show that $y = k_1 \sinh x + k_2 \cosh x$ is also the general solution of the differential equation $y'' - y = 0$.

13. Show that $y = c(x + c)$ is the general solution of the differential equation $(y')^2 + xy' - y = 0$.

14. Given that $y = ce^{x^2}$ is the general solution of the differential equation (*) $y' = 2xy$, solve the initial value problem consisting of the DE(*) and the following initial conditions:
(a) $y(0) = 0$ (b) $y(0) = 2$
(c) $y(1) = e^2$

15. Given that $y = c_1e^x + c_2e^{-x}$ is the general solution of the differential equation (0) $y'' - y = 0$, solve the initial value problem consisting of the DE (0) and the following initial conditions:
(a) $y(0) = 1$, $y'(0) = 0$ (b) $y(0) = 0$, $y'(0) = 1$
Solve the boundary value problem consisting of the DE (0) and the following boundary conditions:
(c) $y(0) = 1$, $y(1) = (e^2 + 1)/2e$ (d) $y(0) = 0$, $y(1) = (e^2 - 1)/2e$

16. Given that $y = c_1 \sin x + c_2 \cos x$ is the general solution of the differential equation $(+)$ $y'' + y = 0$, solve the boundary value problem consisting of the DE(+) and the following boundary conditions:

(a) $y(0) = 0$, $y(\pi/2) = 1$
(b) $y(0) = 0$, $y(\pi) = 0$
(c) $y(0) = 0$, $y(\pi/2) = 0$
(d) $y(0) = 0$, $y(\pi) = 1$

2

THE INITIAL VALUE PROBLEM: $y' = f(x)$; $y(x_0) = c_0$

In this chapter we shall consider the solution of the initial value problem (IVP)

$$y' = f(x); \qquad y(x_0) = c_0. \tag{1}$$

Formally integrating the differential equation $y' = f(x)$ from x_0 to x, we find that

$$y(x) - y(x_0) = \int_{x_0}^{x} dy = \int_{x_0}^{x} f(t)\, dt.$$

Adding $y(x_0)$ to both sides of this equation and substituting the initial condition, we obtain the following integral equation (IE) solution of the IVP (1),

$$y(x) = c_0 + \int_{x_0}^{x} f(t)\, dt. \tag{2}$$

This solution exists whenever the integral appearing in equation (2) exists. For instance, if the function f is continuous or at least piecewise continuous on some finite interval I and $x_0 \in I$, then the solution exists and is unique for all $x \in I$. (The notation $x \in I$ denotes that x is an element of the set I and is read "x is an element of I" or "x is a member of I." We shall also use the notation $x \notin I$ to denote that x is not an element of the set I.) Substituting $x = x_0$ into the integral equation (2), we see that (2) satisfies the initial condition $y(x_0) = c_0$ of the IVP (1). If we assume that f is continuous on a closed interval I and x, $x_0 \in I$, then $y(x)$ of equation (2) is differentiable and in fact $y'(x) = f(x)$, so (2) satisfies the differential equation of the IVP (1). The solution (2) of the IVP (1) also depends continuously on x_0, c_0, and f on any interval I on which the integral exists. That is, a "small" change in x_0, c_0, or f causes a correspondingly "small" change in the solution at any point x for which the solution exists, and the corresponding change occurs in a continuous fashion. In Chapter 4 we shall prove the continuous dependence of the solution (2) on x_0, c_0, and f and provide an expression in x for an upper bound for the absolute value of the difference between the solution of the initial value problem (1) and the initial value problem $y' = g(x)$; $y(x_1) = c_1$. An expression of this kind is useful because differential equations are often used to model physical phenomena. Since a model is seldom, if ever, precise, the modeling function $f(x)$ is only an approximation to the true, but unknown and usually unobtainable, function $g(x)$. The initial values x_0 and c_0 are also usually in error, since these quantities normally represent physical measurements. The corresponding initial values x_1 and c_1 can thus be thought of as representing the true, but unknown, values of the quantities being measured.

EXAMPLE Write the solution of the initial value problem

$$y' = x + \frac{1}{x}; \qquad y(1) = 2 \tag{3}$$

as an integral equation, integrate, and specify the interval I on which the solution exists.

In this case $f(x) = x + 1/x$, which is defined and continuous except for $x = 0$, $x_0 = 1$, and $c_0 = 2$. So the integral equation that is equivalent to the IVP (3) is

$$\begin{aligned} y(x) &= 2 + \int_1^x \left(t + \frac{1}{t}\right) dt \\ &= 2 + \left[\frac{t^2}{2} + \ln|t|\right]\Bigg|_1^x \\ &= \frac{3}{2} + \frac{x^2}{2} + \ln|x|. \end{aligned} \tag{4}$$

Since $f(x)$ is continuous on $(-\infty, 0)$ and $(0, \infty)$, and since the initial condition is specified at $x_0 = 1 \in (0, \infty)$, the solution (4) exists on the interval $I = (0, \infty)$.

EXAMPLE Write the solution of the initial value problem

$$y' = \sec x; \qquad y(0) = 1 \tag{5}$$

as an integral equation, integrate, and specify the interval I on which the solution exists.

The function $f(x) = \sec x$ is defined and continuous on each of the intervals $\left(\frac{(2n+1)\pi}{2}, \frac{(2n+3)\pi}{2}\right)$, where n is any integer. Since $x_0 = 0 \in (-\pi/2, \pi/2)$, the solution to the IVP (5) exists on the interval $I = (-\pi/2, \pi/2)$. The solution of (5) as an integral equation is

$$\begin{aligned} y(x) &= 1 + \int_0^x \sec t\, dt = 1 + [\ln|\sec t + \tan t|]\Big|_0^x \\ &= 1 + \ln|\sec x + \tan x|. \end{aligned}$$

EXAMPLE Write an initial value problem that is equivalent to the integral equation

$$y(x) = 2 + e^{2x} + \int_0^x e^{t^2}\, dt. \tag{6}$$

Since the lower limit of the integral appearing in equation (6) is the constant 0, an appropriate choice for the initial value x_0 is 0. Evaluating (6) at $x_0 = 0$, we obtain the initial condition $y(0) = 3$. Differentiating (6), we see that y satisfies the differential equation $y' = 2e^{2x} + e^{x^2}$. Thus, an initial value problem that is equivalent to (6) is

$$y' = 2e^{2x} + e^{x^2}; \qquad y(0) = 3.$$

EXERCISES

1. Write the solution of each of the following initial value problems as an integral equation and integrate when possible.

(a) $y' = 3x + 1;\ y(1) = 2$ (b) $y' = 2\sin x;\ y(\pi) = 1$

(c) $y' = \cos x^2;\ y(0) = 1$ (d) $y' = x\sin x;\ y(\pi/2) = 1$

2. Write the solution of each of the following initial value problems as an integral equation, integrate, and specify the interval I on which the solution exists.

(a) $y' = \dfrac{1}{x-1};\ y(2) = 1$ (b) $y' = \dfrac{1}{x-1};\ y(0) = 1$

(c) $y' = \dfrac{1}{x^2-1};\ y(2) = 1$ (d) $y' = \dfrac{1}{x^2-1};\ y(0) = 1$

(e) $y' = \tan x;\ y(0) = 0$ (f) $y' = \tan x;\ y(\pi) = 0$

3. Write an initial value problem which is equivalent to each of the following integral equations.

(a) $y(x) = -2 + \int_1^x \sin t^2\, dt$ (b) $y(x) = -2 + \int_x^1 \sin t^2\, dt$

(c) $y(x) = \int_0^x t \tan t\, dt$ (d) $y(x) = 1 - \cos x - \int_x^0 t \sec t\, dt$

2.1 NUMERICAL SOLUTIONS

As we have seen, the solution of the initial value problem

$$y' = f(x); \qquad y(x_0) = c_0 \tag{1}$$

may be written as the integral equation

$$y(x) = c_0 + \int_{x_0}^{x} f(t)\, dt. \tag{2}$$

Many functions $f(t)$ do have an *antiderivative*—that is, a function $F(t)$ such that $F'(t) = f(t)$. In fact, the Fundamental Theorem of Calculus states that if $f(x)$ is continuous on a closed interval I and x_0, $x \in I$, then the function $F(x)$ defined by

$$F(x) = \int_{x_0}^{x} f(t)\, dt$$

is an antiderivative of the function $f(x)$ on the interval I. Moreover, if $f(x)$ is continuous on some closed interval I, if $F(x)$ is any antiderivative of $f(x)$ on I, and if x_0, $x \in I$, then

$$\int_{x_0}^{x} f(t)\, dt = F(x) - F(x_0).$$

So, when $f(x)$ is a continuous function on a closed interval I, when x_0, $x \in I$, and when $F(x)$ is an antiderivative of $f(x)$ on I that can be expressed as an elementary function,* then the solution of the initial value problem (1) is

$$y(x) = c_0 + \int_{x_0}^{x} f(t)\, dt = c_0 + F(x) - F(x_0).$$

---

* Let a be a constant and let $f(x)$ and $g(x)$ be functions. The following operations are called *elementary operations*: $f(x) \pm g(x)$, $f(x) \cdot g(x)$, $f(x)/g(x)$, $(f(x))^a$, $a^{f(x)}$, $\log_a f(x)$, and $T(f(x))$, where T is any trigonometric or inverse trigonometric function. *Elementary functions* are those functions that can be generated using constants, the independent variable, and a finite number of elementary operations.

Many pages in calculus texts and much student effort is devoted to methods for calculating and expressing antiderivatives of continuous functions as elementary functions. Expressions for antiderivatives have been collected and appear in tables of integrals. However, the reader should be aware that there are many relatively simple continuous functions whose antiderivative cannot be expressed as an elementary function. For example, the following list of functions which are defined and continuous on certain intervals do not have antiderivatives that can be expressed as elementary functions:

$$e^{-x^2}, \quad \frac{e^x}{x}, \quad e^x \ln x, \quad \frac{1}{\ln x}, \quad \sin x^2, \quad \frac{\sin x}{x}, \quad \frac{x}{\sin x},$$

$$\frac{\sin^2 x}{x}, \quad \frac{\sin x}{x^2}, \quad x \tan x, \quad \frac{1}{\sqrt{1 - x^3}}.$$

When it is impossible or impractical to express the antiderivative of a function $f(x)$ as an elementary function, then one must use series or numerical integration techniques to calculate the integral of $f(x)$. In some cases even if an antiderivative for $f(x)$ can be expressed as an elementary function, it may still be simpler to approximate the integral of $f(x)$ than to evaluate the antiderivative.

The purpose of this section is to develop a few of the simpler numerical methods for computing approximate solutions to the initial value problem (1) and to give the corresponding formula for estimating the error of the computed solution. We shall not develop the formula for estimating the error of any approximate solution, since such developments usually require the knowledge and use of finite differences—a concept with which the reader is probably not familiar and one that is not presented in this text. Throughout this section the true solution of the initial value problem will be denoted by $y(x)$ and the true solution at x_n by $y(x_n)$. The value of any approximate solution at x_n will be denoted by y_n.

The solution of the simple initial value problem $y' = e^{-x^2}$; $y(0) = 1$ written as an integral equation is

$$y(x) = 1 + \int_0^x e^{-t^2}\, dt.$$

Since there is no elementary function $F(x)$ such that $F'(x) = e^{-x^2}$, the solution $y(x)$ cannot be expressed as an elementary function. Obviously, we cannot compute a numerical solution for each point z in any interval with endpoints 0 and $x \neq 0$, since there are an infinite number of points in the interval. However, we can compute an approximate solution at any particular point z in any interval with endpoints 0 and $x \neq 0$. Consequently, we choose a finite number of points of interest $x_0, x_1, \ldots, x_n$ and compute corresponding approximate values $y_0, y_1, \ldots, y_n$ of the true solution. If we wish to graph a curve that approximates the true solution over the interval with endpoints 0 and x, we draw the polygonal line that joins the points $(x_0, y_0), (x_1, y_1), \ldots, (x_n, y_n)$ or fit a smooth curve through these points. If y_i is a good approximation of $y(x_i)$ and y_{i+1} is a good approximation of $y(x_{i+1})$, then the line segment connecting (x_i, y_i) and (x_{i+1}, y_{i+1}) will usually be a good approximation of the true solution $y(x)$ over the interval with endpoints x_i and x_{i+1}. The questions which naturally arise then are: "What numerical methods are available to provide approximate values $y_0, y_1, \ldots, y_n$ of the true solution at the points $x_0, x_1, \ldots, x_n$?" and "How good are these approximations?"

2.1.1 Riemann Sums

We might try to evaluate the integral $V(x_i) = \int_{x_0}^{x_i} f(t)\,dt$ by appealing directly to the definition of the Riemann integral. The Riemann integral is the integral studied in calculus. This integral was defined about 1850 by the German mathematician Georg Friedrich Bernard Riemann (1826–1866) and later named in his honor. Riemann studied under Carl Friedrich Gauss and made many contributions in the areas of geometry and function theory.

By a *partition P of the interval* $[x_0, x_i]$ we mean any finite set of distinct points, $\{t_0, t_1, \ldots, t_N\}$ ordered so that $x_0 = t_0 < t_1 < \cdots < t_N = x_i$. A partition Q *is a refinement of the partition* P if P is a subset of Q. (A set P is a *subset* of a set Q (denoted by $P \subset Q$) if $x \in P$ implies $x \in Q$.) Let P be any partition of $[x_0, x_i]$ and let $\Delta t_k = t_k - t_{k-1}$ for $k = 1, 2, \ldots, N$. Choose a set $S = \{\tau_1, \tau_2, \ldots, \tau_N\}$ such that $\tau_k \in [t_{k-1}, t_k]$ for $k = 1, 2, \ldots, N$. A *Riemann sum* associated with the integral $V(x_i) = \int_{x_0}^{x_i} f(t)\,dt$ is

$$\sum_{k=1}^{N} f(\tau_k)\,\Delta t_k.$$

Since this sum depends on the function f, the partition P, and the set of points S, we will denote it by $\sum(f, P, S)$. The function f is *Riemann integrable on* $[x_0, x_i]$ if there exists a real number V such that for every $\epsilon > 0$ there exists a partition P of $[x_0, x_i]$ such that if Q is a refinement of P, then

$$\left|\sum(f, Q, S) - V\right| < \epsilon$$

for any choice of the set S.

Suppose that $f(x)$ is known to be integrable on the closed interval $[x_0, x_i]$. Given a desired accuracy, ϵ, for the integral of $f(x)$ on the interval $[x_0, x_i]$, one is not usually able to specify how to choose the partition P of the interval since the partition depends on both $f(x)$ and the interval. So in practice, one normally chooses a partition P of the interval which is *regular*—that is, a partition in which the points are equally spaced, chooses a set S, and forms the Riemann sum $\sum (f, P, S)$. One then chooses (i) a new regular partition P^* which is the refinement of P formed by including in P^* those points halfway between the points of P and (ii) a new set $S^* \supset S$ and forms the Riemann sum $\sum (f, P^*, S^*)$. The numerical value of the two Riemann sums are compared, and if the difference is smaller by one or two orders of magnitude than the desired accuracy, V is approximated by $\sum (f, P^*, S^*)$. If the difference is not small enough, a new Riemann sum $\sum (f, P^{**}, S^{**})$ is formed. The new sum is formed using the partition P^{**}, which is the refinement of P^* formed in the same manner that P^* was formed from P and the set $S^{**} \supset S^*$. This process is repeated until the difference between consecutive sums is less than the desired accuracy by one or two orders of magnitude or until some other prescribed limiting condition is exceeded. For instance, one could specify that the number of points in the partition should not exceed a certain number.

When using this technique, the value chosen for the final difference between consecutive sums, d, should be one tenth or one hundredth the value of the desired accuracy, ϵ, which in turn should be small relative to the value of the integral which is unknown. The computation of a few Riemann sums should lead to an indication of an acceptable range of values for ϵ. For example, if the first few Riemann sums are 4.5, 3.4, 2.6, and 2, then an acceptable value of ϵ would be .1 and $d = .1\epsilon = .01$. Thus, ϵ is small relative to the values of the Riemann sums and d is small relative to the differences between consecutive Riemann sums of 1.1, .8, and .6. If successive Riemann sums are found to be 1.5, 1.3333, 1.2500, 1.2000, 1.1667, 1.1429, 1.1250, 1.1111, 1.1000, and 1.0909, then the value of 1.0909 is selected for the integral since the difference between this value and the value of the previous Riemann sum is the first difference between consecutive sums that is less than .01. The absolute value of the difference between 1.0909 and the value of the integral will not normally be less than d but less than $10d = \epsilon$, the desired accuracy, or perhaps $100d = 10\epsilon$.

If the function $f(t)$ is continuous and $f(t) \geq 0$ for all $t \in [x_0, x_i]$, then

$$V(x_i) = \int_{x_0}^{x_i} f(t)\, dt$$

is equal to the area enclosed by the lines $x = x_0$, $x = x_i$, and $y = 0$, and the curve $y = f(t)$. The N summands of the Riemann sum $\sum (f, P, S)$ can be interpreted as the areas, $f(\tau_j)(t_j - t_{j-1})$, of the N rectangles in Figure 2.1. Hence, this method of approximation is sometimes called the "rectangular rule," and $V(x_i)$ is approximated by

$$I_R = \sum_{j=1}^{N} f(\tau_j)(t_j - t_{j-1}). \tag{7}$$

This formula is also the Riemann approximation of the integral $V(x_i)$ for any integrable function f.

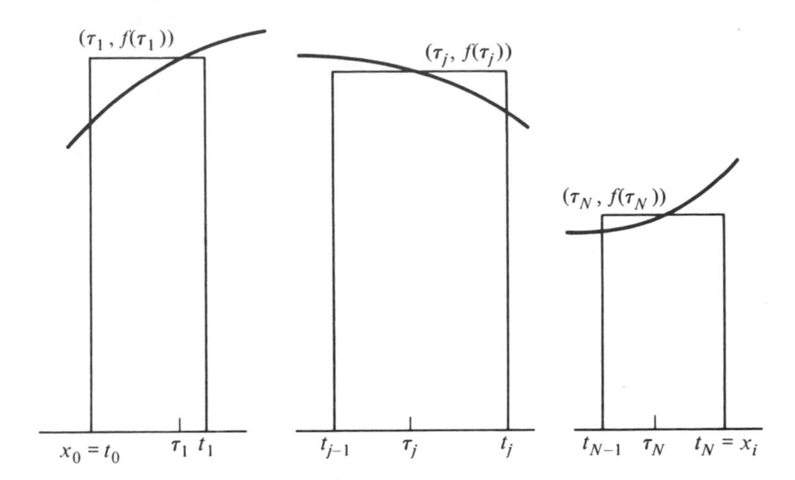

Figure 2.1 Geometric interpretation of Riemann sum.

If a partition P of $[x_0, x_i]$ is a regular partition with N members, then $t_j - t_{j-1} = h$, a constant, and the resulting formula for approximating the integral $V(x_i)$ by Riemann sums is

$$I_R = \sum_{j=1}^{N} f(\tau_j)(t_j - t_{j-1}) = h \sum_{j=1}^{N} f(\tau_j), \tag{8}$$

where $\tau_j \in [t_{j-1}, t_j]$. Formula (8) can be used as an approximation of the integral $V(x_i)$ provided that the function f is integrable on $[x_0, x_i]$. If, in addition, f' is a continuous function on the interval $[x_0, x_i]$, then it can be proven that the error of the Riemann approximation (8) satisfies

$$E_R = \frac{h(x_i - x_0)f'(\xi)}{2} \tag{9}$$

for some ξ where $x_0 < \xi < x_i$. Hence, we have the following upper bound for the error on the interval $[x_0, x_i]$:

$$|E_R| \leq \frac{h(x_i - x_0)}{2} \max_{x_0 \leq x \leq x_i} |f'(x)|. \tag{10}$$

If the members of the set S are chosen to be the left endpoints of each subinterval (i.e., $\tau_j = t_{j-1}$; $j = 1, 2, \ldots, N$), then equation (8) reduces to

$$I_R = h \sum_{j=1}^{N} f(t_{j-1}). \tag{11}$$

If the members of the set S are chosen to be the right endpoints of each subinterval, then equation (8) becomes

$$I_R = h \sum_{j=1}^{N} f(t_j). \tag{12}$$

EXAMPLE Compute a Riemann sum to approximate $I = \int_0^1 e^{-t^2}\,dt = \int_0^1 f(t)\,dt$ by dividing the interval [0, 1] into 10 equal parts and estimate the error.

Table 2.1 contains the values of t_j, t_j^2, and $f(t_j) = e^{-t_j^2}$ at the points of subdivision t_j. If the members of S are chosen to be the left endpoints of the subintervals, then using equation (11) we calculate the following approximation to I:

$$\sum(f, P, S) = h \sum_{j=1}^{N} f(t_{j-1}) = .1 \sum_{j=1}^{10} e^{-t_{j-1}^2}$$

$$= .1(1.000\ 000 + .990\ 050 + \cdots + .444\ 858) = .777\ 817.$$

Table 2.1

j	t_j	t_j^2	$e^{-t_j^2}$	j	t_j	t_j^2	$e^{-t_j^2}$
0	.0	.00	1.000 000	5	.5	.25	.778 801
1	.1	.01	.990 050	6	.6	.36	.697 676
2	.2	.04	.960 789	7	.7	.49	.612 626
3	.3	.09	.913 931	8	.8	.64	.527 292
4	.4	.16	.852 144	9	.9	.81	.444 858
				10	1.0	1.00	.367 879

If the members of S are chosen to be the right endpoints of the subintervals, then using equation (12), we find the following approximation to I:

$$\sum(f, P, S) = h \sum_{j=1}^{N} f(t_j) = .1 \sum_{j=1}^{10} e^{-t_j^2}$$

$$= .1(.990\ 050 + .960\ 789 + \cdots + .367\ 879) = .714\ 605.$$

Since e^{-t^2} is monotone decreasing over the interval [0, 1], it follows that $.714\ 605 \le I \le .777\ 817$. Another estimate of the error can be obtained from equation (10) once we have calculated the maximum of the absolute value of the derivative of $f(t) = e^{-t^2}$ on the interval [0, 1]. Differentiating, we find that

$$f'(t) = -2te^{-t^2}.$$

On the interval [0, 1],

$$|f'(t)| = |-2te^{-t^2}| = 2te^{-t^2} = g(t).$$

Calculating the derivative of $g(t)$, and setting $g'(t) = 0$, we obtain

$$g'(t) = 2(1 - 2t^2)e^{-t^2} = 0.$$

Solving for t, we find $t = 1/\sqrt{2} \in [0, 1]$. So the maximum of $|f'(t)|$ on [0, 1] occurs at $1/\sqrt{2}$ and therefore the error satisfies

$$|E_R| \le \frac{.1(1)}{2} \max_{0 \le x \le 1} |-2xe^{-x^2}| = .05\left(\frac{2}{\sqrt{2}} e^{-1/2}\right) \doteq .0429.$$

2.1.2 Trapezoidal Rule

To approximate the integral $V(x_i) = \int_{x_0}^{x_i} f(t)\,dt$ by using the trapezoidal method, we again partition the interval $[x_0, x_i]$. Let the members t_j of the partition be such that $x_0 = t_0 < t_1 < \cdots < t_j < \cdots < t_N = x_i$. Evaluate the function f at the points t_j $(j = 0, 1, \ldots, N)$ and join the point $(t_{j-1}, f(t_{j-1}))$ to the point $(t_j, f(t_j))$ by a straight-line segment for $j = 1, 2, \ldots, N$. Thus, we approximate the function $f(t)$ over the interval $[x_0, x_i]$ by a polygonal line —see Figure 2.2.

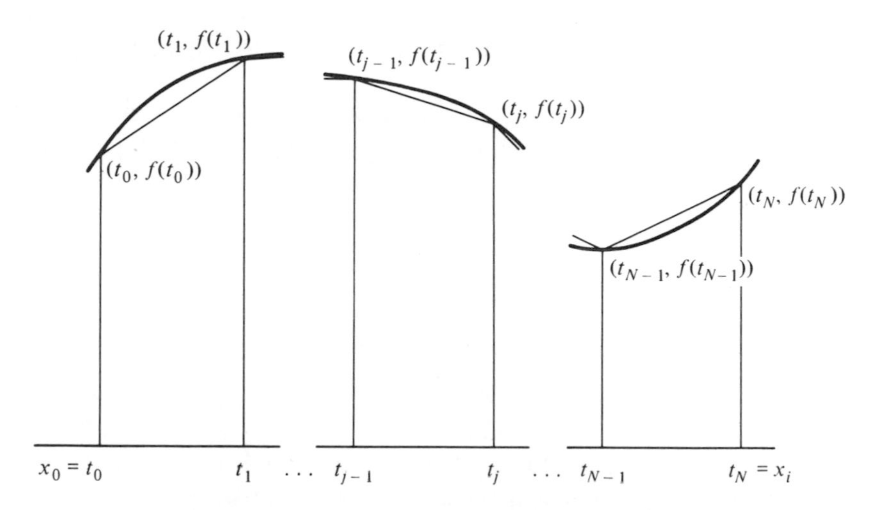

Figure 2.2 Geometric interpretation of trapezoidal rule.

The area of the jth trapezoid of Figure 2.2 is $\frac{1}{2}[f(t_{j-1}) + f(t_j)](t_j - t_{j-1})$. Consequently, the area enclosed by the lines $x = x_0$, $x = x_i$, $y = 0$, and the polygonal line approximating the curve $y = f(t) \ge 0$ is

$$I_T = \frac{1}{2} \sum_{j=1}^{N} [f(t_{j-1}) + f(t_j)](t_j - t_{j-1}). \tag{13}$$

This sum is also the trapezoidal approximation of the integral $V(x_i)$ for any integrable function f. If the partition is a regular partition, then $t_j - t_{j-1} = h$, a constant, and the approximation becomes

$$I_T = \frac{1}{2}h \sum_{j=1}^{N} [f(t_{j-1}) + f(t_j)] = h\left[\frac{1}{2}f(t_0) + \sum_{j=1}^{N-1} f(t_j) + \frac{1}{2}f(t_N)\right]. \tag{14}$$

If $f(t)$ is a linear or constant function, the trapezoidal rule yields the exact value of the integral. That is, the trapezoidal method provides the proper value and slope for f at all points when f is linear or constant. Consequently, we would expect to be able to express the error of this method in terms of the second derivative of f. Formula (14) may be used to approximate the integral $V(x_i)$ provided that the function f is integrable on $[x_0, x_i]$. If, in addition, f has a continuous second derivative on the interval $[x_0, x_i]$, then the error of the trapezoidal rule, $E_T = I_T - V(x_i)$, satisfies

$$(15) \qquad L_T = \min_{x_0 \le t \le x_i} f''(t) \le \frac{12N^2 E_T}{(x_i - x_0)^3} \le \max_{x_0 \le t \le x_i} f''(t) = M_T.$$

EXAMPLE Use the trapezoidal rule with a regular partition and $N = 10$ to approximate $I = \int_0^1 e^{-t^2}\,dt = \int_0^1 f(t)\,dt$ and estimate the error.

Using formula (14) with $N = 10$ and $h = 1/N = 0.1$ and using the function values obtained from Table 2.1, we find the desired approximation to be

$$I_T = h\left[\tfrac{1}{2}f(t_0) + \sum_{j=1}^{9} f(t_j) + \tfrac{1}{2}f(t_{10})\right] = .1\left(\tfrac{1}{2}e^{-t_0^2} + \sum_{j=1}^{9} e^{-t_j^2} + \tfrac{1}{2}e^{-t_{10}^2}\right)$$

$$= .1(.500\ 000 + 6.778\ 167 + .183\ 939\ 5) = .746\ 211.$$

In order to estimate the error, E_T, we must calculate $f''(t)$, L_T, and M_T. We find that

$$f''(t) = 2(2t^2 - 1)e^{-t^2}.$$

The minimum value of $f''(t)$ on the interval $[0, 1]$ occurs at $t = 0$ and the maximum value at $t = 1$. Therefore, $L_T = f''(0) = -2$ and $M_T = f''(1) = .735\ 758$. Using equation (15) we find that E_T satisfies $-2 \le 1200E_T \le .735\ 758$. Hence, $-.001\ 666 \le E_T \le .000\ 614$. Since $E_T = I_T - I$, it follows that

$$.745\ 597 = I_T - .000\ 614 \le I \le I_T + .001\ 666 = .747\ 877.$$

That is, the approximation I_T is accurate to two decimal places.

2.1.3 Simpson's Rule

The method of numerical integration that we shall discuss in this section is named in honor of the English mathematician Thomas Simpson (1710–1761). In 1743 he published the formula for the area under a quadratic curve over a finite interval in his text *Mathematical Dissertations*. However, the rule itself was actually discovered much earlier by the Scottish mathematician James Gregory and published in 1668.

In deriving the rectangular rule the integrand $f(t)$ was approximated by a step function—a piecewise constant function. And in deriving the trapezoidal rule the integrand was approximated by a piecewise linear function. We should expect to obtain a more accurate integration formula if we approximate the integrand by a set of quadratic functions, each defined over a small interval. That is, we should obtain more accurate results if we replace the integrand by a piecewise quadratic function.

In order to derive Simpson's rule for approximating the integral $V(x_i) = \int_{x_0}^{x_i} f(t)\,dt$ we subdivide (partition) the interval $[x_0, x_i]$ into $2N$ subintervals of equal length $h = (x_i - x_0)/2N$. Let the points of subdivision be $t_0 = x_0 < t_1 = x_0 + h < \cdots < t_j = x_0 + jh < \cdots < t_{2N} = x_0 + 2Nh = x_i$. In the first two subintervals—$[t_0, t_1]$ and $[t_1, t_2]$—we approximate the integrand f by a quadratic function $q_1(t)$ of the form $A_1t^2 + B_1t + C_1$, where the constants A_1, B_1, and C_1 are chosen so that $q_1(t)$ passes through $(t_0, f(t_0))$, $(t_1, f(t_1))$, and $(t_2, f(t_2))$. In the next two subintervals we approximate f by another quadratic function q_2 with coefficients A_2, B_2, and C_2 chosen so that q_2 passes through $(t_2, f(t_2))$, $(t_3, f(t_3))$, and $(t_4, f(t_4))$. Thus, we obtain N quadratic functions

$$q_j = A_jt^2 + B_jt + C_j$$

to approximate the integrand f over the interval $[x_0, x_i]$—see Figure 2.3.

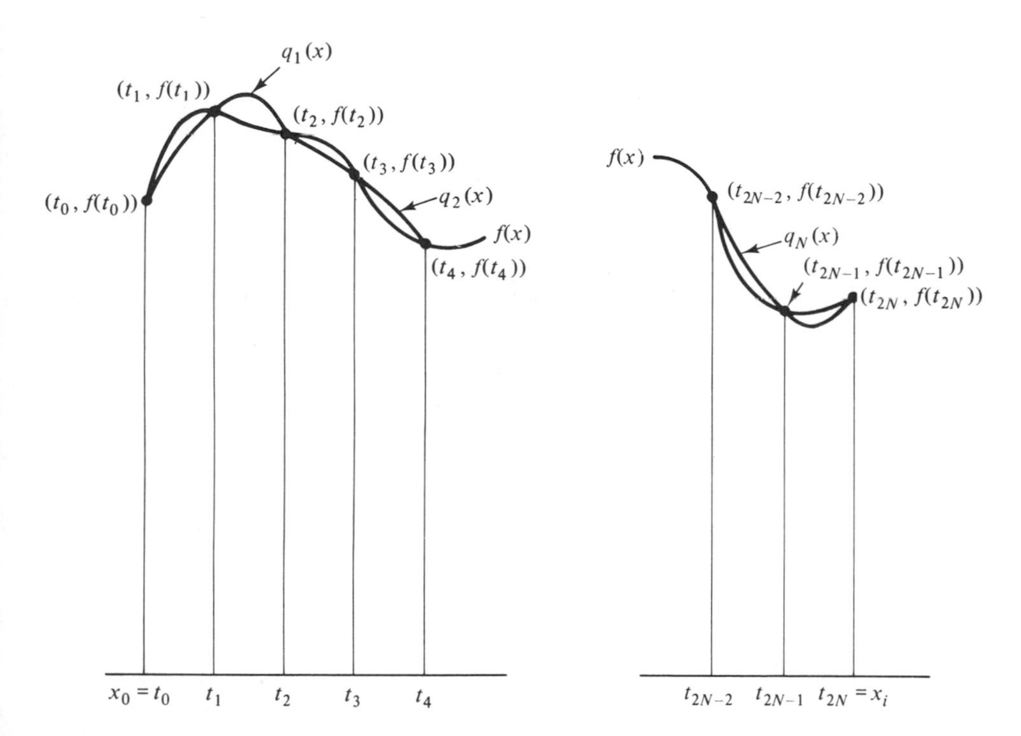

Figure 2.3 Geometric interpretation of Simpson's rule.

For $j = 1, 2, \ldots, N$, let

$$I_j = \int_{t_{2j-2}}^{t_{2j}} q_j(t)\,dt = \int_{t_{2j-2}}^{t_{2j}} (A_j t^2 + B_j t + C_j)\,dt.$$

In order to make the necessary calculations easier, we change the interval of integration from $[t_{2j-2}, t_{2j}]$ to $[0, 2h]$. Then $f(0)$ corresponds to $f(t_{2j-2})$, $f(h)$ to $f(t_{2j-1})$, and $f(2h)$ to $f(t_{2j})$. The function $q_j(t)$ is to pass through $(0, f(0))$, $(h, f(h))$, and $(2h, f(2h))$, so A_j, B_j, and C_j must simultaneously satisfy

$$\begin{aligned} C_j &= f(0) \\ A_j h^2 + B_j h + C_j &= f(h) \\ A_j(2h)^2 + B_j(2h) + C_j &= f(2h). \end{aligned}$$

Substituting $f(0)$ for C_j in the last two equations and rearranging, we see that the unknowns A_j and B_j must satisfy the following system of equations:

$$\begin{aligned} h^2 A_j + hB_j &= f(h) - f(0) \\ 4h^2 A_j + 2hB_j &= f(2h) - f(0). \end{aligned}$$

Solving this system of equations, we find

$$A_j = \frac{f(0) - 2f(h) + f(2h)}{2h^2}$$

and

$$B_j = \frac{-3f(0) + 4f(h) - f(2h)}{2h}.$$

Upon integrating, we find

$$I_j = \int_0^{2h} q_j(t)\,dt = \int_0^{2h} (A_j t^2 + B_j t + C_j)\,dt = A_j \frac{(2h)^3}{3} + B_j \frac{(2h)^2}{2} + 2C_j h$$

$$= \frac{f(0) - 2f(h) + f(2h)}{2h^2} \cdot \frac{8h^3}{3} + \frac{-3f(0) + 4f(h) - f(2h)}{2h} \cdot \frac{4h^2}{2} + 2f(0)h$$

$$= h\left\{\frac{4[f(0) - 2f(h) + f(2h)]}{3} - f(0) + 4f(h) - f(2h)\right\}$$

$$= \frac{h[f(0) + 4f(h) + f(2h)]}{3}.$$

Making use of assumed correspondence, we see that

$$I_j = \int_{t_{2j-2}}^{t_{2j}} q_j(t)\,dt = \frac{h[f(t_{2j-2}) + 4f(t_{2j-1}) + f(t_{2j})]}{3}.$$

The following formula for approximate integration is known as *Simpson's rule*:

(16)

$$I_S = \sum_{j=1}^{N} I_j$$

$$= \frac{h\{[f(t_0) + 4f(t_1) + f(t_2)] + [f(t_2) + 4f(t_3) + f(t_4)] + \cdots + [f(t_{2N-2}) + 4f(t_{2N-1}) + f(t_{2N})]\}}{3}$$

$$= \frac{h[f(t_0) + 4f(t_1) + 2f(t_2) + 4f(t_3) + \cdots + 4f(t_{2N-1}) + f(t_{2N})]}{3}.$$

Simpson's rule can be used to approximate the integral $V(x_i)$ provided that the function f is integrable on $[x_0, x_i]$. If, in addition, f has a continuous fourth derivative on the interval $[x_0, x_i]$, then it can be shown that the error of Simpson's rule, $E_S = I_S - V(x_i)$, satisfies

$$(17) \qquad L_S = \min_{x_0 \le t \le x_i} f^{(4)}(t) \le \frac{180(2N)^4 E_S}{(x_i - x_0)^5} \le \max_{x_0 \le t \le x_i} f^{(4)}(t) = M_S.$$

This error formula indicates that Simpson's rule yields exact results not only for polynomials of degree less than or equal to two but also for third degree polynomials.

EXAMPLE Use Simpson's rule with a regular partition and $2N = 10$ to approximate $I = \int_0^1 e^{-t^2}\,dt = \int_0^1 f(t)\,dt$ and estimate the error.

Since $x_0 = 0$, $x_i = 1$, and $2N = 10$, $h = (x_i - x_0)/2N = 1/10 = .1$. Using formula (16), and the function values obtained from Table 2.1, we find that

$$\begin{aligned} I_S &= .1(f(0) + 4f(.1) + 2f(.2) + \cdots + 4f(.9) + f(1.0))/3 \\ &= .1(1.000\ 000 + 4(.990\ 050) + 2(.960\ 789) + \cdots \\ &\qquad + 4(.444\ 858) + .367\ 879)/3 = .746\ 825. \end{aligned}$$

We can estimate the error of this approximation, E_S, by using (17). The fourth derivative of $f(t) = e^{-t^2}$ is $f^{(4)}(t) = 4(4t^4 - 12t^2 + 3)e^{-t^2}$. By calculating the fifth derivative of f, setting $f^{(5)}(t) = 0$, and solving for t, we find that $f^{(4)}(t)$ has a relative maximum on the interval $[0, 1]$ at $t = 0$, so $M_S = 12$, and $f^{(4)}(t)$ has a relative minimum in the interval $[0, 1]$ at $t_*^2 = .918\ 861$, so $L_S = f^{(4)}(t_*) > -7.419\ 48$. By (17) E_S satisfies $-7.419\ 48 < 1\ 800\ 000 E_S \le 12$, so $-.000\ 004 < E_S \le .000\ 007$. Since $E_S = I_S - I$, it follows that

$$.746\ 818 \le I_S - .000\ 007 \le I < I_S + .000\ 004 \le .746\ 829.$$

This section on numerical integration has been included in order to give the reader some idea of the techniques that are used to approximate integrals of the form $\int_{x_0}^{x_i} f(t)\,dt$. It is by no means meant to be all inclusive. The reader who is interested in more advanced and complex numerical integration schemes and in the derivations of the error estimates presented here should consult texts on numerical analysis, such as, Blum [4], Conte and de Boor [6], Henrici [9], or Lapidus and Seinfeld [14].

EXERCISES

1. Let $I = \int_0^1 e^{-t^2}\,dt$ and suppose that we use a regular partition of the interval [0, 1] with constant subinterval length $h = t_{j+1} - t_j$ to produce numerical approximations to I. How small would h need to be to obtain an approximation to I that was accurate to at least six decimal places, if the approximation were made by (i) a Riemann sum, (ii) the trapezoidal rule, and (iii) Simpson's rule? At least how many subdivisions of the interval [0, 1] would this require in each case?

2. (a) Solve the initial value problem $y' = x \sin x$; $y(0) = 1$.
   (b) Compute $y(1)$ to six decimal places.
   (c) Estimate $y(1)$ using:
      (i) The rectangular rule with $N = 10$. Estimate the error and compare with the true error.
      (ii) The trapezoidal rule with $N = 10$. Estimate the error and compare with the true error.
      (iii) Simpson's rule with $2N = 10$. Estimate the error and compare with the true error.

Table of values

x	$x \sin x$	x	$x \sin x$
.0	.000 000 00	.6	.338 785 48
.1	.009 983 34	.7	.450 952 38
.2	.039 733 87	.8	.573 884 87
.3	.088 656 06	.9	.704 994 22
.4	.155 767 34	1.0	.841 470 98
.5	.239 712 77		

3. The solution of the initial value problem

$$y' = f(x) = \begin{cases} \dfrac{\sin x}{x} & x \neq 0; \\ 1 & x = 0; \end{cases} \qquad y(0) = 3$$

written as an integral equation is $y(x) = 3 + \int_0^x f(t)\,dt$. Estimate $y(1)$ using
   (a) The rectangular rule with $N = 10, 20$. Estimate the error in each case.

(b) The trapezoidal rule with $N = 5, 10, 20$. Estimate the error in each case.

(c) Simpson's rule with $2N = 10, 20$. Estimate the error in each case.

(*Hint:* In calculating the derivatives to be used in the error estimates, use the Maclaurin series expansion for $\sin x$. $\sin x = x - x^3/3! + x^5/5! - \cdots$.)

Table of values

x	$f(x)$	x	$f(x)$
.00	1.000 000 00	.55	.950 340 42
.05	.999 583 39	.60	.941 070 79
.10	.998 334 17	.65	.931 056 01
.15	.996 254 22	.70	.920 310 98
.20	.993 346 65	.75	.908 851 68
.25	.989 615 84	.80	.896 695 11
.30	.985 067 36	.85	.883 859 30
.35	.979 708 02	.90	.870 363 23
.40	.973 545 86	.95	.856 226 85
.45	.966 590 08	1.00	.841 470 98
.50	.958 851 08		

4. Use Simpson's rule and the following table of values to calculate

$$V(x) = \int_0^x t \tan t \, dt$$

for $x = 1$ and $x = 1.4$. Estimate the error. (*Hint:* $\tan 1 \doteq 1.557$, $\sec 1 \doteq 1.851$, $\tan 1.4 \doteq 5.798$, and $\sec 1.4 \doteq 5.883$.)

Table of values

x	$x \tan x$	x	$x \tan x$
.0	.000 000	.8	.823 711
.1	.010 033	.9	1.134 142
.2	.040 542	1.0	1.557 408
.3	.092 801	1.1	2.161 236
.4	.169 117	1.2	3.086 582
.5	.273 151	1.3	4.682 733
.6	.410 482	1.4	8.117 037
.7	.589 602		

5. The Debye function, $D(x)$, of statistical thermodynamics is used in computing the constant-volume specific heat of substances. The function is defined by

$$D(x) = 3x^{-3} \int_0^x \frac{t^3\, dt}{e^t - 1}.$$

Use Simpson's rule and the following table of values for the integrand to calculate $D(.6)$ and $D(1)$.

Table of values

x	$f(x)$	x	$f(x)$
.0	.000 000	.6	.262 736
.1	.009 508	.7	.338 347
.2	.036 133	.8	.417 775
.3	.077 174	.9	.499 451
.4	.130 128	1.0	.581 977
.5	.192 687		

6. Bessel's function of the first kind of order zero, $J_0(x)$, occurs in many physical applications. An integral representation for this function is

$$J_0(x) = \frac{1}{\pi} \int_0^\pi \cos\,(x \sin t)\, dt.$$

Use Simpson's rule and the following table of values for the integrand to compute $J_0(.5)$ and $J_0(3)$. Compare your results with published tabulated values.

Table of values

x	$\cos\,(.5 \sin x)$	$\cos\,(3 \sin x)$
0	1.000 000	1.000 000
$.1\pi$	.988 087	.600 196
$.2\pi$	.957 124	−.191 371
$.3\pi$	.919 296	−.755 393
$.4\pi$	.889 051	−.958 694
$.5\pi$	.877 583	−.989 992
$.6\pi$	.889 051	−.958 694
$.7\pi$	.919 296	−.755 393
$.8\pi$	.957 124	−.191 371
$.9\pi$	.988 087	.600 196
π	1.000 000	1.000 000

2.2 A NUMERICAL CASE STUDY

It is common practice to state problems in mathematical textbooks so that each problem has a definite answer or specific solution. Perhaps this leads some readers to believe that once a problem has been stated, it must be solvable as stated. Certainly this is not necessarily the case. The major problems that face the mathematical analyst are those of defining a physical problem in mathematical terms and interpreting the mathematical solution of the problem in physical terms. These abilities are difficult to convey through textbook examples. Nonetheless, one should be intensely aware that a problem as initially stated is (i) probably not well defined and (ii) probably not solvable. Through a process of redefining the problem and making approximations, one is usually able to effect a solution to the redefined problem.

EXAMPLE Solve the initial value problem $y' = -2(x - .1)^{-3}$; $y(-1) = 1$ on $[-1, 1]$ using (a) the trapezoidal rule with $N = 10$ and $N = 20$ and (b) Simpson's rule with $2N = 10$ and $2N = 20$.

The solution to the given initial value problem written as an integral equation is

$$y(x) = 1 + \int_{-1}^{x} -2(t - .1)^{-3}\, dt. \tag{18}$$

On any interval about $x = -1$ on which the integrand is defined and continuous, the solution of the initial value problem exists. Clearly, the integrand is defined and continuous on any interval that does not contain the point $x = .1$. This point is in the interval $[-1, 1]$, and so we might anticipate some irregularities in the numerical solutions which we calculate when $x \geq .1$.

Integrating (18), one finds the actual solution of the initial value problem to be

$$y(x) = \frac{1}{(x - .1)^2} + \frac{.21}{1.21}.$$

This function is defined and has a continuous derivative for $x \neq .1$. Hence, the solution to the given initial value problem exists on the interval $[-1, .1)$, but may not be extended to the interval $[-1, 1]$, since $\lim_{x \to .1^-} y(x) = +\infty$.

A graph of the actual solution of the initial value problem and the trapezoidal rule with $N = 10$ and $N = 20$ is shown in Figure 2.4. Notice that for $N = 10$, the trapezoidal method produces "solution" values for $x = .2$, .4, .6, .8, and 1.0; whereas for $N = 20$, no solution values are produced for $x \geq .1$, since the value of $-2(x - .1)^{-3}$ cannot be calculated for $x = .1$.

Figure 2.5 is a graph of the "solutions" obtained by using Simpson's rule with $2N = 10$ and $2N = 20$ and the actual solution.

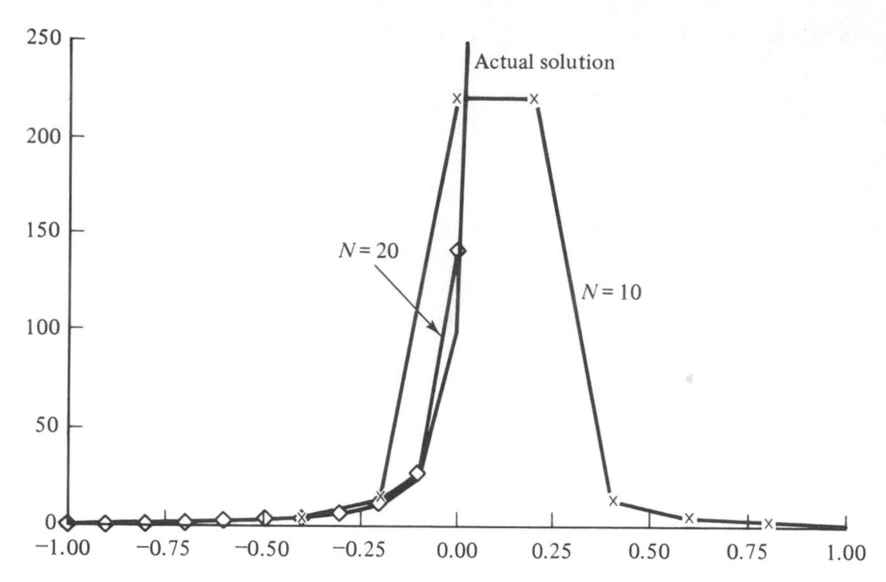

Figure 2.4 Trapezoidal solution to the IVP $y' = -2/(x - .1)^3$; $y(-1) = 1$.

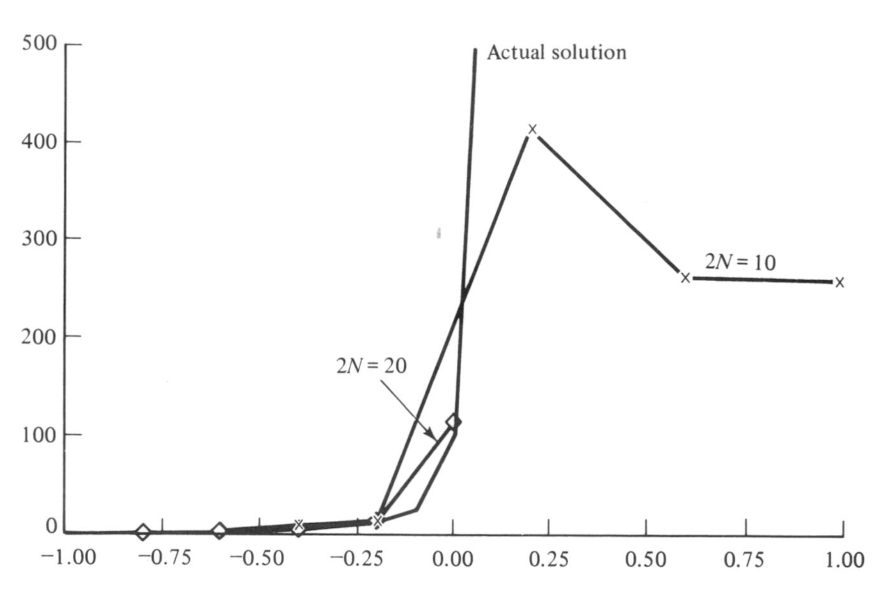

Figure 2.5 Simpson's rule solution to the IVP $y' = -2/(x - .1)^3$; $y(-1) = 1$.

EXERCISE

1. The arc length on [0, 1] of a curve $y(x)$ which is defined and has a continuous first derivative on [0, 1] is given by

$$L = \int_0^1 \sqrt{1 + (y')^2}\, dx.$$

Write computer or programmable calculator programs to solve the initial value problem $L'(x) = \sqrt{1 + (y')^2}$; $L(0) = 0$ on the interval [0, 1] by the following methods:

(a) The trapezoidal rule with $N = 100$, 1000, and 10 000.

(b) Simpson's rule with $2N = 100$, 1000, and 10 000.

Use these programs to attempt to compute an approximate arc length of $y(x)$ on [0, 1] for the functions:

$$\text{(i)} \qquad y(x) = \begin{cases} x^2 \sin \dfrac{1}{x}, & x \neq 0 \\ 0, & x = 0. \end{cases}$$

[Notice that $y'(0) = 0$ but that y' is not continuous at $x = 0$.]

$$\text{(ii)} \qquad y(x) = \begin{cases} x \sin \dfrac{1}{x}, & x \neq 0 \\ 0, & x = 0. \end{cases}$$

[Notice that $y'(0)$ does not exist.]

What, if anything, can you infer from the results?

3

FIRST ORDER DIFFERENTIAL EQUATIONS

In this chapter and the following one, we analyze in detail the initial value problem

$$y' = f(x, y); \qquad y(x_0) = c_0. \tag{1}$$

Sometimes one can solve the IVP (1) by finding the general solution of the differential equation

$$y' = f(x, y). \tag{2}$$

The general solution of (2) is a one-parameter family of curves of the form

$$\phi(x, y, C) = 0. \tag{3}$$

If the general solution (3) of the DE (2) can be found, then the solution of the IVP (1) is obtained from (3) by determining the value of the parameter C so that the initial condition $y(x_0) = c_0$ of the initial value problem is satisfied. In this chapter we shall study some classes of first order differential equation for which a general solution can be obtained, and we shall present some techniques for producing the general solution. Thus, in this chapter we shall concern ourselves with finding the "true" solution of the IVP (1). (We shall use the expression "the 'true' solution" when we wish to emphasize that we are talking about the solution instead of an approximate or numerical solution.) If we are able to write an expression for the "true" solution, then we know that the solution exists and we can determine the properties of the solution from the solution itself. However, most of the time one is not able to write an expression for the "true" solution of the IVP (1) and, therefore, must be satisfied with producing an "approximate" solution and trying to determine the properties of the "true" solution from the initial value problem and the "approximate" solution. Consequently, this chapter is devoted to the methods for finding an expression for the "true" solution of the IVP (1) while Chapter 4 is devoted to the study of the theory for the IVP (1) and to the methods for producing an "approximate" solution to the IVP (1) once it has been established by applying the theory that a solution does indeed exist.

3.1 LINEAR EQUATIONS

First order ordinary differential equations are divided into two large classes—as are all ordinary differential equations—linear and nonlinear. A *linear ordinary differential equation* (LODE) *of first order* is an ordinary differential equation of the form

$$y' = a(x)y + b(x). \tag{4}$$

All other first order ordinary differential equations are called *nonlinear ordinary differential equations of first order*. If $b(x) \equiv 0$, then the DE (4) is called a *homogeneous LODE of first order*. Whereas, if $b(x) \not\equiv 0$, then the DE (4) is said to be a *nonhomogeneous LODE of first order*.

EXAMPLE Find the general solution of the differential equation

$$y' = y + x.$$

Subtracting y from both sides of the given differential equation, we obtain

$$y' - y = x.$$

We then observe that if we multiply this equation by e^{-x}, the left-hand side of the equation is the derivative of the product ye^{-x}. Thus, multiplying by e^{-x}, we get

$$y'e^{-x} - ye^{-x} = \frac{d(ye^{-x})}{dx} = xe^{-x}.$$

Integrating by parts, we find

$$ye^{-x} = \int xe^{-x}\,dx = -xe^{-x} + \int e^{-x}\,dx = -xe^{-x} - e^{-x} + C,$$

where C is an arbitrary constant. Multiplying by e^x, we find the general explicit solution of the given differential equation to be

$$y = -x - 1 + Ce^x.$$

In this example we multiplied the differential equation $y' - y = x$ by the function e^{-x}. The left-hand side of the resulting differential equation was the derivative of the product of two functions—the unknown function y and the function e^{-x}. Any function that we can multiply a given differential equation by and obtain a differential equation in which both sides are the derivatives of some known functions is called an *integrating factor*.

Now let us consider the general first order linear differential equation

$$y' = a(x)y + b(x). \tag{4}$$

Subtracting $a(x)y$, we obtain

$$y' - a(x)y = b(x). \tag{5}$$

The questions that should come to mind are: "Does this differential equation have an integrating factor?" and "If it does, what is it?" Let $u(x)$ be an unknown function. We want to choose $u(x)$ so that when we multiply equation (5) by $u(x)$, the left-hand side of the resulting differential equation is the derivative of the product $yu(x)$. That is, we want to choose $u(x)$ so that

$$\frac{d(yu(x))}{dx} = y'u(x) + u'(x)y = y'u(x) - a(x)u(x)y.$$

Thus, if we can choose $u(x)$ so that

$$u'(x) = -a(x)u(x),$$

$u(x)$ will be an integrating factor. Dividing by $u(x)$ and multiplying by dx, we see that $u(x)$ must satisfy

$$\frac{du(x)}{u(x)} = -a(x)\,dx.$$

Suppose further that $u(x) > 0$ for all x. Integrating, we find that

$$\ln u(x) = -\int^x a(s)\,ds.$$

And exponentiating, we obtain the integrating factor

$$u(x) = \exp\left\{-\int^x a(s)\,ds\right\} > 0 \quad \text{for all } x.$$

In summary, to find the general solution of the general first order linear differential equation

$$y' = a(x)y + b(x) \tag{4}$$

we:

1. Calculate the integrating factor
$$u(x) = \exp\left\{-\int^x a(s)\, ds\right\}.$$
2. Subtract $a(x)y$ from the given differential equation.
3. Multiply by the integrating factor to obtain the differential equation
$$\frac{d(yu(x))}{dx} = b(x)u(x).$$
4. Integrate to get
$$yu(x) = \int^x b(t)u(t)\, dt.$$
5. Divide by $u(x)$ to obtain the general solution,

(6)
$$y = \frac{1}{u(x)}\int^x b(t)u(t)\, dt = \exp\left\{\int^x a(s)\, ds\right\}\int^x b(t)\exp\left\{-\int^t a(s)\, ds\right\} dt.$$

Reviewing the procedure for producing the general solution (6) of the linear differential equation (4), we see that in order to obtain (6) the function $a(x)$ must be integrable—that is, $\int^x a(s)\, ds$ must exist—and the product of the function $b(x)$ and the integrating factor $u(x)$ must be integrable—that is,

$$\int^x b(t)u(t)\, dt = \int^x b(t)\exp\left\{-\int^t a(s)\, ds\right\} dt$$

must exist. If $a(x)$ and $b(x)$ are continuous on some interval I, then all the integrals required to produce the general solution (6) of the linear differential equation (4) exist.

EXAMPLE Solve the initial value problem

$$y' = -2xy + x; \qquad y(0) = 1 \tag{7}$$

In this case $a(x) = -2x$ and $b(x) = x$. Both these functions are defined and continuous for all real x, so the solution to the given linear initial value problem can be calculated using equation (6) and the solution will exist for all real x. An integrating factor for the differential equation

$$y' = -2xy + x \tag{8}$$

is

$$u(x) = \exp\left\{-\int^x a(s)\,ds\right\} = \exp\left\{\int^x 2s\,ds\right\} = e^{x^2}.$$

Adding $2xy$ to equation (8) and multiplying by the integrating factor, we obtain

$$y'e^{x^2} + 2xye^{x^2} = \frac{d(ye^{x^2})}{dx} = xe^{x^2}.$$

Notice that after multiplication by the integrating factor, the left-hand side of the resulting equation is the derivative of the dependent variable times the integrating factor. This will always be the case. Integrating, we find

$$ye^{x^2} = \int^x te^{t^2}\,dt = \tfrac{1}{2}e^{x^2} + C,$$

where C is an arbitrary constant. Dividing by the integrating factor e^{x^2}, we obtain the following general solution to the DE (8):

$$y = \tfrac{1}{2} + Ce^{-x^2}. \tag{9}$$

Imposing the initial condition $y(0) = 1$, we see that the constant C must satisfy

$$1 = \tfrac{1}{2} + C.$$

So $C = \frac{1}{2}$ and the solution of the IVP (7) is

$$y = \frac{1 + e^{-x^2}}{2}.$$

EXAMPLE Solve the initial value problem

$$y' = (\tan x)y + \sin x; \qquad y\left(\frac{\pi}{4}\right) = \sqrt{2}. \tag{10}$$

In this instance the function $a(x) = \tan x$ is defined and continuous on the intervals $I_n = ((2n + 1)\pi/2, (2n + 3)\pi/2)$ where n is an integer and undefined at the endpoints of the intervals. The function $b(x) = \sin x$ is defined and continuous for all real x. So the general solution to the differential equation

$$y' = (\tan x)y + \sin x \tag{11}$$

will exist only on the intervals I_n. Since the initial condition of the IVP (10) is specified at $x = \pi/4 \in (-\pi/2, \pi/2) = I_{-1}$, the solution of the IVP (10) will exist only on the interval $(-\pi/2, \pi/2)$. An integrating factor for the DE (11) is

$$u(x) = \exp\left\{-\int^x a(s)\,ds\right\} = \exp\left\{-\int^x \tan s\,ds\right\} = e^{\ln|\cos x|} = |\cos x|.$$

One often obtains an integrating factor which is the absolute value of some functions, as we did in this instance. The function itself will also always be an integrating factor of the differential equation. This is true because the differential equation is multiplied by the integrating factor and the additional multiplying factor of plus or minus 1 which is contributed by the absolute value of the function does not essentially alter the resulting differential equation. So in the future when we obtain an integrating factor that is the absolute value of a function, we will always multiply the differential equation by the integrating factor which is the function itself.

Subtracting $(\tan x)y$ from the DE (11) and multiplying by the integrating factor $\cos x$, we have

$$(\cos x)y' - (\sin x)y = \frac{d(y \cos x)}{dx} = (\cos x)(\sin x).$$

Integrating, we obtain

$$y(x) \cos x = \int^x (\cos s)(\sin s)\,ds = \tfrac{1}{2}\sin^2 x + C,$$

where C is an arbitrary constant. Thus, the general solution of the DE (11) is

$$y(x) = \tfrac{1}{2}(\sin x)(\tan x) + C(\sec x).$$

The solution to the IVP (10) is the member of this one-parameter family that satisfies the initial condition $y(\pi/4) = \sqrt{2}$. Imposing the initial condition, we see that C must satisfy the equation

$$\sqrt{2} = \left(\frac{1}{2}\right)\left(\frac{1}{\sqrt{2}}\right)(1) + C(\sqrt{2}).$$

Hence, $C = \frac{3}{4}$ and the solution of the IVP (10) is

$$y(x) = \frac{1}{2}(\sin x)(\tan x) + \frac{3(\sec x)}{4}, \qquad x \in (-\pi/2, \pi/2).$$

The linear first order initial value problem corresponding to the differential equation (4) is

$$(12) \qquad y' = a(x)y + b(x); \qquad y(x_0) = c_0.$$

Suppose that $a(x)$ and $b(x)$ are continuous on some interval I containing x_0. Then $\int_{x_0}^t a(s)\,ds$ and the integrating factor

$$E(t) = \exp\left\{-\int_{x_0}^t a(s)\,ds\right\}$$

exist and are continuous for all $t \in I$. Subtracting $a(x)y$ from the differential equation of (12) and multiplying by $E(x)$, we obtain

$$y'E(x) - a(x)E(x)y = \frac{d(yE(x))}{dx} = b(x)E(x).$$

Substituting $z(x) = y(x)E(x)$, the IVP (12) may be written in the following equivalent form:

$$z' = b(x)E(x); \qquad z(x_0) = y(x_0)E(x_0) = y(x_0) = c_0.$$

Of course, the solution of this initial value problem written as an integral equation is

$$z(x) = c_0 + \int_{x_0}^x b(t)E(t)\,dt.$$

Hence, the solution to the IVP (12) is

$$y(x) = \frac{1}{E(x)}\left[c_0 + \int_{x_0}^x b(t)E(t)\,dt\right]. \tag{13}$$

Obviously, being able to write the solution to the IVP (12) in terms of elementary functions depends upon one's ability to calculate explicitly the integral appearing in equation (13).

EXAMPLE Solve the initial value problem

$$y' = -2xy + x; \qquad y(0) = 1 \tag{7}$$

using equation (13).

Since $a(x) = -2x$ and $x_0 = 0$, the integrating factor is

$$E(t) = \exp\left\{-\int_0^t (-2s)\,dt\right\} = \exp\{s^2|_0^t\} = e^{t^2}.$$

Substituting into equation (13) with $b(x) = x$ and $c_0 = 1$, we find the solution to the IVP (7) to be

$$y(x) = \frac{1}{e^{x^2}}\left(1 + \int_0^x te^{t^2}\,dt\right) = e^{-x^2}\left(1 + \left[\frac{1}{2}e^{t^2}\right]\Big|_0^x\right)$$

$$= e^{-x^2}\left(1 + \frac{1}{2}e^{x^2} - \frac{1}{2}\right) = \frac{1 + e^{-x^2}}{2}.$$

EXAMPLE Solve the initial value problem

$$y' = \frac{3}{x}y + 4x^2 + 1; \qquad y(1) = 1. \tag{14}$$

The functions $a(x) = 3/x$ and $b(x) = 4x^2 + 1$ are defined and continuous except for $x = 0$. Since the initial condition is specified at $x_0 = 1 > 0$, we seek a solution $y(x)$ that is valid for $x > 0$. Since $a(x) = 3/x$ and $x_0 = 1$, the integrating factor is

$$E(t) = \exp\left(-\int_1^t \frac{3ds}{s}\right) = e^{-3\ln t} = t^{-3}.$$

Since $b(x) = 4x^2 + 1$,

$$\int_1^x b(t)E(t)\,dt = \int_1^x \frac{4t^2+1}{t^3}\,dt = \left(4\ln t - \frac{1}{2t^2}\right)\Bigg|_1^x = 4\ln x - \frac{1}{2x^2} + \frac{1}{2}.$$

So by equation (13), the solution to the IVP (14) which is valid for $x > 0$ is

$$y(x) = x^3\left(1 + \ln x^4 - \frac{1}{2x^2} + \frac{1}{2}\right) = \frac{3x^3}{2} + x^3\ln x^4 - \frac{x}{2}.$$

Often one will find the differential equation of this example written in the form $xy' = 3y + 4x^3 + x$. The point $x = 0$ is called a *singular point* of the differential equation, since the function $a_0(x) = x$ which multiplies y' is zero for $x = 0$. In general, one might find linear differential equations of the first order written in the form $a_0(x)y' = a_1(x)y + a_2(x)$, where $a_0(x)$, $a_1(x)$, and $a_2(x)$ have no common factor. Using this representation, all points x for which $a_0(x) = 0$ are called *singular points*. The relation of singular points to the solution of ordinary differential equations will be discussed later.

Formula (13) is appealing from the standpoint that it gives a concise representation of the solution to the initial value problem (12). However, one should not memorize the formula but rather remember the steps in deriving the formula. Also, it is often easier, and perhaps more efficient, to find the general solution of the differential equation by employing indefinite integration instead of definite integration and then to determine the value of the parameter that satisfies the initial condition.

Various nonlinear first order differential equations can be transformed into linear first order differential equations by some transformation of the variables. Perhaps most notable among such equations is *Bernoulli's equation*

$$y' = P(x)y + Q(x)y^n, \tag{15}$$

where n is a real constant. This equation is so named in honor of Jacques Bernoulli (1654–1705). For $n = 0$ and $n = 1$, equation (15) is linear and the solution immediate. In 1696 Leibniz (1646–1716) showed that the transformation of variable $z = y^{1-n}$ reduces Bernoulli's equation to a linear first order differential equation. Differentiating the transformation equation with respect to x we obtain $z' = (1 - n)y^{-n}y'$. Multiplying equation (15) by $(1 - n)y^{-n}$ and performing the transformation of variables results in the following linear first order differential equation whose general solution can readily be produced:

$$z' = (1 - n)P(x)z + (1 - n)Q(x).$$

EXAMPLE Find the solution of the differential equation

$$y' = xy - xy^2. \tag{16}$$

This is a Bernoulli equation with $P(x) = x$, $Q(x) = -x$ and $n = 2$. Multiplying equation (16) by $(1 - n)y^{-n} = -y^{-2}$ and adding xy^{-1}, we obtain

$$-y^{-2}y' + xy^{-1} = x. \tag{17}$$

Let $z = y^{1-n} = y^{-1}$. Then $z' = -y^{-2}y'$. Substituting into equation (17), we obtain

$$z' + xz = x. \tag{18}$$

The integrating factor for the linear DE (18) is

$$\exp\left\{\int^x s\,ds\right\} = e^{x^2/2}.$$

Multiplying by the integrating factor, we find

$$\frac{d(ze^{x^2/2})}{dx} = xe^{x^2/2}.$$

Integrating results in the following general solution to the linear DE (18):

$$ze^{x^2/2} = e^{x^2/2} + C.$$

Substituting $z = y^{-1}$ and solving for y results in the following solution to the nonlinear Bernoulli equation (16):

$$y = \frac{e^{x^2/2}}{e^{x^2/2} + C} = (1 + Ce^{-x^2/2})^{-1}.$$

EXERCISES

Solve the initial value problems of Exercises 1–9 by first finding the general solution to the differential equation and then determining the value of the constant of integration that satisfies the initial condition. Specify the interval on which each solution is valid.

1. $y' = 3y;\ y(0) = -1$

2. $y' = -y + 1;\ y(0) = 1$

3. $y' = -y + 1;\ y(0) = 2$

4. $y' = y/x;\ y(1) = 2$

5. $y' = y/x;\ y(-1) = 2$

6. $y' = y/(x - 1) + x^2;\ y(0) = 1$

7. $y' = (\cot x)y + \sin x;\ y(\pi/2) = 0$

8. $y' = xy + 2x;\ y(0) = 1$

9. $y' = xy + 2;\ y(0) = 1$

Solve the initial value problems of Exercises 10–15 by using equation (13). Specify the interval on which each solution is valid.

10. $y' = y + 1;\ y(0) = 1$

11. $y' = xy + x;\ y(1) = 2$

12. $y' = (\cot x)y + \sin x;\ y(\pi/2) = 0$

13. $y' = xy + 1;\ y(1) = 2$

14. $y' = y/x + \sin x^2;\ y(-1) = 1$

15. $y' = 2y/x + e^x;\ y(1) = \frac{1}{2}$

Solve the nonlinear initial value problems of Exercises 16–19 by solving the Bernoulli equation and satisfying the initial condition. Specify the interval on which each solution is valid.

16. $y' = 2x/y;\ y(0) = 2$

17. $y' = -2y + y^2;\ y(0) = 1$

18. $y' = -y/x + y^{1/2};\ y(1) = 1$

19. $y' = y/x + x^2y^3;\ y(-1) = -1$

20. Let $y_1(x)$ be a solution of the homogeneous linear first order differential equation

$$y' = a(x)y.$$

Show that $cy_1(x)$ is also a solution for any constant c.

21. (a) Show that $y(x) \equiv 0$ is *the* solution of the initial value problem

$$y' = a(x)y; \qquad y(x_0) = 0 \tag{0}$$

(b) Suppose that $y_1(x)$ and $y_2(x)$ are two different solutions of the non-homogeneous linear initial value problem

$$y' = a(x)y + b(x); \qquad y(x_0) = c_0. \tag{*}$$

Show that $y_3(x) = y_1(x) - y_2(x)$ satisfies the homogeneous linear initial value problem (0). Therefore, $y_3(x) \equiv 0$ and, consequently, $y_1(x) \equiv y_2(x)$. The results of this exercise prove that the solution of the linear initial value problem (*) is "unique."

3.2 EQUATIONS WRITTEN AS $M(x, y)\,dx + N(x, y)\,dy = 0$

Often it is convenient to write the differential equation $y' = f(x, y)$ in the form

$$y' = -\frac{M(x, y)}{N(x, y)} \tag{19}$$

or in the equivalent form

$$M(x, y)\,dx + N(x, y)\,dy = 0. \tag{20}$$

Note that it is always possible to write the differential equation $y' = f(x, y)$ in the form of equation (19) or (20) by taking $M(x, y) = -f(x, y)$ and $N(x, y) = 1$. However, other representations may be much more useful in effecting a solution to the differential equation.

3.2.1 Separable Equations

If $M(x, y)$ is a function of x alone—that is, $M(x, y) = g(x)$—and if $N(x, y)$ is a function of y alone—$N(x, y) = h(y)$, then the differential equation is said to be separable and may be written as

$$g(x)\,dx + h(y)\,dy = 0.$$

Integrating this differential equation results in the following implicit solution:

$$\int^x g(s)\,ds + \int^y h(t)\,dt = 0. \tag{21}$$

The explicit solution and the determination of the interval of existence of the solution depends upon being able to represent both integrals in equation (21) as elementary functions and upon being able to solve the resulting implicit equation for y.

EXAMPLE Solve the linear differential equation

$$y' = xy + 2x \tag{22}$$

by separating variables.

Writing the derivative y' as the ratio of differentials, dy/dx, and factoring the right-hand side of equation (22), we obtain the equivalent equation

$$\frac{dy}{dx} = x(y + 2).$$

Multiplying by dx, subtracting dy, and dividing by $(y + 2)$, we get

$$x\,dx - \frac{dy}{y + 2} = 0.$$

Integration yields the implicit solution

$$\int^x t\,dt - \int^y \frac{ds}{s + 2} = 0$$

or

$$\frac{x^2}{2} - \ln|y + 2| + C = 0,$$

where C is an arbitrary constant. Solving for $|y + 2|$, we find

$$|y + 2| = e^{x^2/2 + C} = e^C e^{x^2/2}. \tag{23}$$

Both the left-hand side and the right-hand side of equation (23) are positive. Since $|y + 2| = \pm(y + 2)$ and e^C is an arbitrary positive constant, we may remove the absolute value appearing in equation (23) if we replace e^C by an arbitrary constant K which may be positive, negative, or zero. Doing so, we obtain the following explicit general solution of the DE (22):

$$y + 2 = Ke^{x^2/2}$$

or

$$y = -2 + Ke^{x^2/2}.$$

EXAMPLE Find the solution of the initial value problem

$$y' = xe^{y - x^2}; \qquad y(0) = 0. \tag{24}$$

We may rewrite the differential equation of (24) in the form

$$\frac{dy}{dx} = \frac{xe^{-x^2}}{e^{-y}}.$$

From this representation it is evident that the variables separate, resulting in the differential equation

$$-xe^{-x^2}\,dx + e^{-y}\,dy = 0.$$

So the implicit solution of the IVP (24) written as an integral equation is

$$\int_0^x - se^{-s^2}\,dt + \int_0^y e^{-t}\,dt = 0.$$

Hence, the implicit solution of the IVP (24) is

$$\tfrac{1}{2}e^{-s^2}\big|_0^x - e^{-t}\big|_0^y = \tfrac{1}{2}e^{-x^2} - \tfrac{1}{2} - e^{-y} + 1 = 0.$$

Solving this equation explicitly for y, we find the solution of the IVP (24) to be

$$y(x) = -\ln\left(\frac{e^{-x^2} + 1}{2}\right).$$

Notice that the solution exists for all x.

3.2.2 Homogeneous Equations

In this section we identify a particular class of differential equations of the form (20) $M(x, y)\,dx + N(x, y)\,dy = 0$, for which the variables are not immediately separable but which can be reduced to separable equations by a change of variable. This class of equations is called *homogeneous equations.*

A function $f(x, y)$ is said to be *homogeneous of degree n in x and y* if and only if $f(sx, sy) = s^n f(x, y)$. Thus, the function

$$f(x, y) = x^2 + 4xy + 2y^2$$

is homogeneous of degree 2 in x and y, since

$$\begin{aligned} f(sx, sy) &= (sx)^2 + 4(sx)(sy) + 2(sy)^2 \\ &= s^2(x^2 + 4xy + 2y^2) = s^2 f(x, y). \end{aligned}$$

Likewise, the function $f(x, y) = x \ln x - x \ln y + y \sin(x/y)$ is homogeneous of degree 1 in x and y. The function $f(x, y) = x \ln x + x \ln y$, on the other hand, is not homogeneous in x and y.

Theorem 3.1 If $M(x, y)$ and $N(x, y)$ are both homogeneous of degree n in x and y, then the function $f(x, y) = -M(x, y)/N(x, y)$ is homogeneous of degree zero in x and y.

Proof By hypotheses $M(sx, sy) = s^n M(x, y)$ and $N(sx, sy) = s^n N(x, y)$. So

$$f(sx, sy) = \frac{-M(sx, sy)}{N(sx, sy)} = \frac{-s^n M(x, y)}{s^n N(x, y)} = s^0 f(x, y).$$

Theorem 3.2 If $f(x, y)$ is homogeneous of degree zero in x and y, then f is a function of y/x.

Proof Let $y = vx$. Then since f is homogeneous of degree zero in x and y, $f(x, y) = f(x, vx) = x^0 f(1, v) = f(1, y/x) = g(y/x)$.

The application of these two theorems to the solution of the DE (20) $M(x, y)\,dx + N(x, y)\,dy = 0$ is fairly obvious. If M and N are both homogeneous of degree n in x and y, then

$$\frac{dy}{dx} = \frac{-M(x, y)}{N(x, y)} = f(x, y) = g(y/x),$$

since f is homogeneous of degree zero in x and y. If we introduce a new dependent variable, $v = y/x$, then $y = vx$ and

$$\frac{dy}{dx} = v + x\frac{dv}{dx} = g(v).$$

This last differential equation in x and v is separable and may be written as

$$\frac{dx}{x} = \frac{dv}{g(v) - v}. \tag{25}$$

The solution of the original DE (20) is obtained by integrating (25) and replacing v by y/x. Of course, it may be impossible to express the integral of the right-hand side of (25) in terms of elementary functions.

We have shown that if M and N of the DE (20) are both of degree n in x and y, then the substitution $y = vx$ will always transform the DE (20) into a differential equation in v and x in which the variables are separable. Likewise, if M and N of the DE (20) are of degree n in x and y, the substitution $x = wy$ will always transform the DE (20) into a differential equation in w and y in which the variables are separable.

EXAMPLE Solve the differential equation

$$(x^2 + y^2)\,dx + xy\,dy = 0.$$

In this case $M = x^2 + y^2$ and $N = xy$ are both homogeneous of degree 2 in x and y. Hence, if we make the change of variables $y = vx$, we will obtain a differential equation in v and x in which the variables separate. Letting

$y = vx$, we find $dy = v\,dx + x\,dv$. Performing the change of variable, we obtain the differential equation

$$(x^2 + v^2x^2)\,dx + x^2v(v\,dx + x\,dv) = 0.$$

Dividing by x^2 and collecting like terms leads to

$$(1 + 2v^2)\,dx + xv\,dv = 0.$$

Separating the variables yields

$$\frac{dx}{x} = \frac{-v\,dv}{1 + 2v^2}.$$

Integrating, we find

$$\ln|x| = -\tfrac{1}{4}\ln|1 + 2v^2| + \ln|C|,$$

where C is an arbitrary constant. Combining logarithms and exponentiating yields

$$|x|(1 + 2v^2)^{1/4} = |C|.$$

Raising this equation to the fourth power, substituting $v = y/x$, and simplifying results in the following implicit solution of the original differential equation:

$$x^2(x^2 + 2y^2) = C^4.$$

EXAMPLE Solve the differential equation $y^2\,dx - (2x^2 + 3xy)\,dy = 0$.

Since the coefficients $M = y^2$ and $N = -(2x^2 + 3xy)$ are again homogeneous of degree 2, we could make the substitution $y = vx$ and obtain a differential equation in x and v in which the variables separate. However, the simplicity of the coefficient M of dx indicates that instead we should make the change of variable $x = wy$. Calculating differentials, we find $dx = w\,dy + y\,dw$. Substituting into the given differential equation, we obtain

$$y^2(w\,dy + y\,dw) - [2(wy)^2 + 3wy^2]\,dy = 0.$$

Dividing by y^2, separating variables, and expanding by partial fractions leads to

$$\frac{dy}{y} = \frac{1}{2}\frac{dw}{w(w + 1)} = \frac{1}{2}\left(\frac{1}{w} - \frac{1}{w + 1}\right)dw.$$

Integrating, we find

$$\ln|y| = \tfrac{1}{2}\ln|w| - \tfrac{1}{2}\ln|w + 1| + \ln|C|.$$

Combining logarithms and exponentiating results in

$$|y||w + 1|^{1/2} = |C||w|^{1/2}.$$

Substituting x/y for w and multiplying by $|y|^{1/2}$ yields the following implicit solution of the given differential equation:

$$|y||x + y|^{1/2} = |C||x|^{1/2}.$$

Squaring both sides of this equation, we may write the solution as

$$y^2|x + y| = C^2|x|. \tag{26}$$

Since $|x + y| = \pm(x + y)$ and $|x| = \pm x$, depending upon the values of x and y and since C is arbitrary, we may replace $C^2 > 0$ by a new arbitrary constant K (which may be positive, negative, or zero) and represent the solution more simply as

$$y^2(x + y) = Kx.$$

Notice that the arbitrary constant K compensates for all possible combinations of signs that could occur on the factors $x + y$ and x of equation (26).

EXERCISES

In Exercises 1–8, find the solution of the given separable or homogeneous differential equation.

1. $y' = x^2/y$
2. $y' = e^{x+y}$
3. $y' = x^2 \sin y$
4. $(x + y)\,dx + x\,dy = 0$
5. $y\,dx - (x + y)\,dy = 0$
6. $y' = y/x + y^2/x^2$
7. $(x^2y + y)\,dx - dy = 0$
8. $(y^2 + 2xy)\,dx - x^2\,dy = 0$

Find the solution of the following initial value problems.

9. $y\,dx - x^2\,dy = 0$; $y(1) = 1$
10. $(x - y)\,dx + (y - x)\,dy = 0$; $y(0) = 1$
11. $y' = \sin x/\cos y$; $y(\pi) = 0$
12. $y' = (y + x)/x$; $y(-1) = 1$
13. $xe^y\,dx + dy = 0$; $y(0) = 0$
14. $(x + ye^{x/y})\,dy - y\,dx = 0$; $y(0) = -1$

3.2.3 Exact Equations

Before proceeding to the primary topic of this section we shall present a few basic mathematical concepts as they relate to the xy-plane. The reader needs to be familiar with these concepts in order to genuinely understand the theorem regarding exact equations which is to be presented in this section and the fundamental theory of ordinary differential equations which will be presented throughout the remainder of the text. Many readers may already be familiar with some of these concepts, in which case this brief section will largely constitute a review.

A set S of the xy-plane is *open* if every point in S is the center of a circle that is a subset of S. The following subsets of the xy-plane are examples of open sets:

$$\begin{aligned}
D_1 &= \{(x, y) \mid x^2 + y^2 < 4\}\\
D_2 &= \{(x, y) \mid 1 < x^2 + y^2 < 4\}\\
D_3 &= \{(x, y) \mid |x| < 2 \text{ and } |y| < 2\}\\
D_4 &= \{(x, y) \mid x > 0 \text{ and } y > 0\}\\
D_5 &= \{(x, y) \mid y > \epsilon > 0\}\\
D_6 &= \{(x, y) \mid (x - 2)^2 + y^2 < 1 \quad \text{or} \quad (x - 5)^2 + y^2 < 1\}.
\end{aligned}$$

See Figure 3.1.

An open set of the xy-plane is *connected* (*arcwise connected*) if any two points in S can be joined by a polygonal line that lies entirely in S. The sets D_1, D_2, D_3, D_4, and D_5 are all open and connected. However, the set D_6 is open but not connected, since no polygonal line which joins the points (2, 0) and (5, 0) lies entirely in D_6.

A nonempty, open, connected set in the xy-plane is called a *domain*. The sets D_1, D_2, D_3, D_4, and D_5 are domains. The set D_6 is not a domain since it is not connected. The set $D_7 = \{(x, y) \mid x^2 + y^2 \leq 1\}$ is not a domain since it is not an open set.

The DE (20) $M(x, y)\,dx + N(x, y)\,dy = 0$ is said to be *exact in a domain* D if and only if there exists a function $F(x, y)$ such that

$$(27) \qquad dF(x, y) = M(x, y)\,dx + N(x, y)\,dy \quad \text{for all } (x, y) \in D.$$

If there is a function F satisfying equation (27), then obviously the general solution to the DE (20) is $F(x, y) = C$, where C is an arbitrary constant. In some instances it is possible to determine the function F by inspection. For example, the differential equation $y\,dx + x\,dy = 0$ is exact, since $F(x, y) = xy$ satisfies $dF = y\,dx + x\,dy$. Usually, however, it is not possible to find the function F by inspection. It would be convenient if there were a test that we could perform to determine whether or not the DE (20) is exact and, consequently, whether or not there exists a function F satisfying (27). In the event that the DE (20) is found to be exact, it is also desirable to have

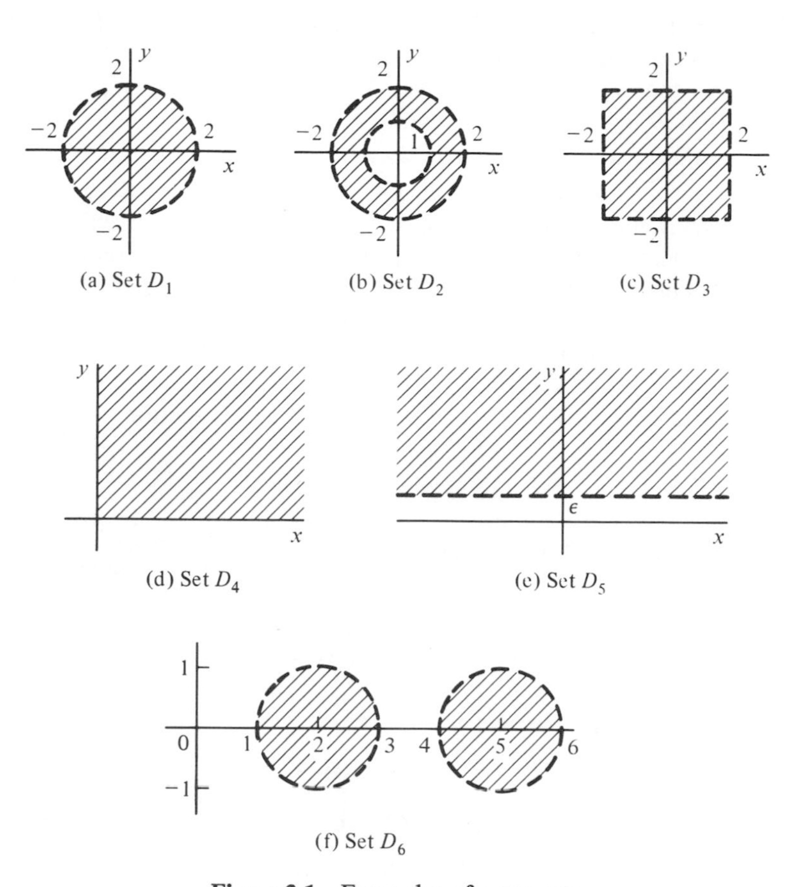

Figure 3.1 Examples of open sets.

a procedure for producing the function F satisfying (27), since $F(x, y) = C$ will then be the solution to the DE (20). The following theorem provides the test for exactness and the proof provides a procedure for producing F when the DE (20) is exact.

Theorem 3.3 If $M(x, y)$, $N(x, y)$, $\partial M(x, y)/\partial y = M_y(x, y)$, and $\partial N(x, y)/\partial x = N_x(x, y)$ are continuous functions of x and y in a simply connected* domain D, then the DE (20) $M(x, y)\,dx + N(x, y)\,dy = 0$ is exact in D if and only if $M_y(x, y) = N_x(x, y)$ for all $(x, y) \in D$.

* Intuitively, a simply connected domain is a domain that has no point, line, or hole removed.

Proof Suppose that the DE (20) is exact in D. Then there exists a function F satisfying $dF = M\,dx + N\,dy = F_x\,dx + F_y\,dy$ in D. Hence, $M = F_x$ and $N = F_y$ in D. Since $M_y = F_{yx}$ and $N_x = F_{xy}$ are continuous in D, $F_{yx} = F_{xy}$ and, therefore, $M_y = N_x$.

Suppose that $M_y = N_x$ in D. Let $\phi(x, y) = \int^x M(t, y)\,dt$. Note that ϕ always exists for $(x, y) \in D$, since M is a continuous function of x and therefore is integrable with respect to x in D. Note also that $\phi_x = M(x, y)$. Taking the partial derivative of this equation with respect to y and invoking the continuity conditions, we obtain $\phi_{yx} = M_y = N_x = \phi_{xy}$. Integrating the last equality with respect to x, we find $N = \phi_y + g(y)$. An arbitrary function $g(y)$ appears in this equation instead of an arbitrary constant of integration, since the partial derivative with respect to x of any function of y alone is zero. Now consider the function

$$F(x, y) = \phi(x, y) + \int^y g(t)\,dt.$$

$$dF = F_x\,dx + F_y\,dy = \phi_x\,dx + [\phi_y + g(y)]\,dy = M\,dx + N\,dy.$$

Thus, the DE (20) is exact in D.

EXAMPLE Solve the differential equation $3y(x^2 - 1)\,dx + (x^3 + 8y - 3x)\,dy = 0$.

Calculating M_y and N_x, we find

$$M_y = 3(x^2 - 1) = N_x.$$

So the differential equation is exact. Hence, there exists a function F such that

$$F_x = M = 3y(x^2 - 1) \tag{28}$$

and

$$F_y = N = x^3 + 8y - 3x. \tag{29}$$

Integrating equation (28) with respect to x, we obtain

$$F = yx^3 - 3xy + f(y).$$

Taking the partial derivative of F with respect to y and invoking equation (29) yields

$$F_y = x^3 - 3x + f'(y) = x^3 + 8y - 3x.$$

Hence, $f'(y) = 8y$ and, therefore, $f(y) = 4y^2 + K$. Consequently, the solution to the given differential equation is

$$F(x, y) = yx^3 - 3xy + 4y^2 + K = C.$$

We also could have solved this differential equation by first integrating equation (29) with respect to y, obtaining

$$F = x^3y + 4y^2 - 3xy + h(x).$$

Taking the partial derivative of F with respect to x and invoking equation (28) yields

$$F_x = 3x^2y - 3y + h'(x) = 3y(x^2 - 1).$$

Therefore, $h'(x) = 0$ and consequently $h(x) = K$, and the solution to the given differential equation is, as before,

$$F(x, y) = yx^3 - 3xy + 4y^2 + K = C.$$

Combining the arbitrary constants K and C, we can write the solution as

$$yx^3 - 3xy + 4y^2 = c.$$

EXAMPLE Solve the differential equation

$$(e^y + y\cos x)\,dx + (xe^y + \sin x + 1)\,dy = 0.$$

Calculating M_y and N_x, we find

$$M_y = e^y + \cos x = N_x.$$

Therefore, the differential equation is exact and there exists a function F such that

$$F_x = M = e^y + y\cos x.$$

Integrating with respect to x, we find

$$F = xe^y + y\sin x + g(y). \tag{30}$$

Taking the partial derivative of F with respect to y and equating to N, we find that $g(y)$ must satisfy

$$F_y = xe^y + \sin x + g'(y) = xe^y + \sin x + 1.$$

So $g'(y) = 1$. Integrating, we get $g(y) = y + K$, where K is an arbitrary constant. Substituting for $g(y)$ in equation (30) and setting $F = C$, we find the implicit solution of the given exact differential equation to be

$$xe^y + y\sin x + y + K = C$$

or, combining constants,

$$xe^y + y\sin x + y = c.$$

The following example illustrates the necessity of the domain D being simply connected.

EXAMPLE Consider the differential equation

$$\frac{-y\,dx + x\,dy}{x^2 + y^2} = 0$$

on the domain $D = \{(x, y) \mid 0 < x^2 + y^2 < 4\}$. On this domain M, N, M_y, and N_x are all defined and continuous and $M_y = N_x = (y^2 - x^2)/(x^2 + y^2)^2$. Hence, one might expect to find a function $F(x, y)$ defined on the domain D such that $F(x, y) = C$, where C is an arbitrary constant, is the general solution of the given differential equation on D. The "function" $F(x\ y) = \arctan(y/x) = C$ appears to be a solution of the differential equation on D. However, the "function" F is not a function in the strictest sense of the word since it is not single-valued. For instance, for $(x, y) = (1, 0)$, $F(1, 0) = \arctan(0) = 2n\pi$, where n is any integer. Since F is not a function, F is not a solution of the differential equation. If we require that $-\pi \leq C < \pi$, then $G(x, y) = \operatorname{Arctan}(y/x) = C$ is a single-valued function on D. However, $G(x, y) = C$ is not a solution to the differential equation on D. G is discontinuous in D on the line segment $L = \{(x, y) \mid y = 0 \text{ and } -2 < x < 0\}$, since $\lim_{(x,y)\to(-a,0)} G(x, y) = \pi$ if the limit is taken along any path in D for which $y > 0$; whereas $\lim_{(x,y)\to(-a,0)} G(x, y) = -\pi$ if the limit is taken along any path in D for which $y < 0$. Since G is not continuous on the line segment L, G is not differentiable on L and, therefore, cannot satisfy the differential equation on L.

3.2.4 Integrating Factors

If the DE (20) $M(x, y)\,dx + N(x, y)\,dy = 0$ is tested for exactness and found not to be exact, then it is natural to attempt to make it exact by multiplying it by an appropriately chosen function $u(x, y)$, called an *integrating factor*. That is, we try to choose a function $u(x, y)$ in such a manner that the differential equation $uM\,dx + uN\,dy = 0$ is exact. From Theorem 3.3 we know that this differential equation will be exact in a simply connected domain D if and only if uM, uN, $\partial(uM)/\partial y$, and $\partial(uN)/\partial x$ are continuous functions of x and y in D and $\partial(uM)/\partial y = \partial(uN)/\partial x$. Calculating the partial derivatives of the last equation, we see that for u to be an integrating factor, it must satisfy the partial differential equation

$$uM_y + Mu_y = uN_x + Nu_x. \tag{31}$$

In general, it is at least as difficult to solve the PDE (31) and determine the integrating factor u as it is to solve the original DE (20). Also, the PDE (31) may not have a unique solution. Hence, (31) may give rise to more than one integrating factor. In order to simplify the situation, we will only try to find integrating factors u which are a function of a single variable—either x or y.

If the integrating factor u which we seek is to be a function of x alone, then $u_y = 0$ and $u_x = du/dx$. In this case the PDE (31) for u reduces to the following ordinary differential equation in u: $uM_y = uN_x + N\,du/dx$. Or, equivalently,

$$\frac{du}{u} = \frac{1}{N}(M_y - N_x)\,dx.$$

If the coefficient of dx on the right-hand side of this equation is a function of x alone, say $f(x)$, then upon integrating and taking the exponential of both sides, we find $u = \exp\left(\int^x f(t)\,dt\right)$.

If the integrating factor u is to be a function of y alone, then $u_x = 0$, $u_y = du/dy$, and the PDE (31) reduces to the ordinary differential equation in u: $uM_y + M\,du/dy = uN_x$. Or, rearranging,

$$\frac{du}{u} = -\frac{1}{M}(M_y - N_x)\,dy.$$

If the coefficient of dy on the right-hand side of this equation is a function of y alone, say $g(y)$, then an integrating factor is $u = \exp\left(\int^y g(t)\,dt\right)$.

Hence, we have derived the following two rules for integrating factors:

If $(1/N)(M_y - N_x) = f(x)$, then $u = \exp\left(\int^x f(t)\,dt\right)$ is an integrating factor for the DE (20) $M\,dx + N\,dy = 0$.

If $-(1/M)(M_y - N_x) = g(y)$, then $u = \exp\left(\int^y g(t)\,dt\right)$ is an integrating factor for the DE (20).

Let us consider again in this context the general linear differential equation of first order $y' = a(x)y + b(x)$. Rewriting this equation in the form $M\,dx + N\,dy = 0$, we obtain $[a(x)y + b(x)]\,dx - dy = 0$. Since $M_y = a(x)$ and $N_x = 0$, this equation is not exact unless $a(x) = 0$. Assuming that $a(x) \not\equiv 0$, we try to find an integrating factor for the differential equation. We find that $(1/N)(M_y - N_x) = -a(x)$, which is a function of x alone. So an integrating factor for the differential equation is $u(x) = \exp\left(-\int^x a(s)\,ds\right)$. Multiplying by the integrating factor, we obtain the following exact differential equation

$$[a(x)y + b(x)]u(x)\,dx - u(x)\,dy = 0. \tag{32}$$

Since this equation is exact, there exists a function $F(x, y)$ such that $F_y = -u(x)$. Integrating with respect to y we see that $F = -u(x)y + g(x)$. Taking the partial derivative of F with respect to x and equating this to the coefficient of dx in the exact DE (32), we find $F_x = a(x)u(x)y + g'(x) = [a(x)y + b(x)]u(x)$. Hence, $g'(x) = b(x)u(x)$. Therefore, $g(x) = \int^x b(t)u(t)\,dt$ and the general solution to the linear first order differential equation is

$$F(x, y) = -u(x)y + \int^x b(t)u(t)\,dt = C,$$

where C is an arbitrary constant. Solving for y, we find the general solution of the linear first order differential equation written as an integral equation is

$$y(x) = \frac{1}{u(x)}\left[\int^x b(t)u(t)\,dt - C\right]. \tag{33}$$

Compare this equation with equation (13). What is the difference between $u(x)$ and $E(x)$?

EXAMPLE Find the general solution to the differential equation $(x^2 + y^2)\,dx + x(x - 2y)\,dy = 0$.

First we check for exactness and find that $M_y = 2y$ and $N_x = 2x - 2y$. Since $M_y \neq N_x$ the given differential equation is not exact, and we proceed by seeking an integrating factor. Since

$$\frac{1}{N}(M_y - N_x) = \frac{2(2y - x)}{x(x - 2y)} = -\frac{2}{x},$$

which is a function of x alone, an integrating factor is

$$u(x) = \exp\left\{\int^x \frac{-2\,dt}{t}\right\} = e^{-2\ln|x|} = e^{\ln x^{-2}} = x^{-2} = \frac{1}{x^2}.$$

Multiplying the given differential equation by the integrating factor $1/x^2$, we obtain the following exact differential equation:

$$\left(1 + \frac{y^2}{x^2}\right)dx + \left(1 - \frac{2y}{x}\right)dy = 0.$$

Applying the method of the previous section for solving exact equations yields the general solution

$$F(x, y) = x - \frac{y^2}{x} + y = C.$$

EXAMPLE Find the general solution of the differential equation

$$(2xy - y^2 + y)\,dx + (3x^2 - 4xy + 3x)\,dy = 0.$$

The differential equation is not exact, since $M_y = 2x - 2y + 1$ and $N_x = 6x - 4y + 3$. $M_y - N_x = -4x + 2y - 2$. Hence,

$$\frac{1}{N}(M_y - N_x) = \frac{-2(2x - y + 1)}{3x^2 - 4xy + 3x},$$

which is not a function of x alone. Thus, there is no integrating factor for the differential equation which is a function of x alone. Since

$$-\frac{1}{M}(M_y - N_x) = \frac{2(2x - y + 1)}{y(2x - y + 1)} = \frac{2}{y},$$

an integrating factor for the differential equation which is a function of y alone is

$$u(y) = \exp\left\{\int^{y} \frac{2\,dt}{t}\right\} = e^{2\ln|y|} = e^{\ln y^2} = y^2.$$

Multiplying the given differential equation by the integrating factor, we obtain an exact differential equation which when solved by the method of the previous section yields the following general solution:

$$F(x, y) = x^2y^3 - xy^4 + xy^3 = C.$$

EXERCISES

Determine if each of the following differential equations is exact or not. If the equation is exact, find the solution. If the equation is not exact, find an integrating factor which is a function of x or y alone and then find the solution.

1. $(y^2 + 4xy + 1)\,dx + 2x(x + y)\,dy = 0$
2. $(x^3 - y)\,dx + x\,dy = 0$
3. $y' = -\sin y/(2y + x\cos y)$
4. $(2x + y)\,dx + x(1 - x - y)\,dy = 0$
5. $(2y - x/y)y' = 1 + \ln y \quad (y > 0)$
6. $x(x + 2y)\,dy + (1 + xy)\,dx = 0$
7. $y' = -(\cos x + y/x)$

For each of the following differential equations, determine if the equation is a linear equation or a nonlinear equation, a Bernoulli equation, a separable equation, a homogeneous equation, an exact equation or if an integrating factor will make the equation exact, and find the general solution to the differential equation.

8. $y' = x/y$
9. $y\,dx - x\,dy = 0$
10. $(x + y)\,dx + (y - x)\,dy = 0$
11. $(x + y)\,dx + (x - y)\,dy = 0$
12. $(x^2 - y)\,dx - (x + y)\,dy = 0$
13. $(x^2 - y)\,dx + x\,dy = 0$
14. $xy(1 - y)\,dx - 2\,dy = 0$
15. $(2y + x)\,dx + (y - 2x)\,dy = 0$

16. $x(1 - y^3)\,dx - 3y^2\,dy = 0$

17. $(e^y \cos x - 2x \sin y)\,dx + (e^y \sin x - x^2 \cos y)\,dy = 0$

18. $y(\ln x - \ln y)\,dx + x(\ln y - \ln x)\,dy = 0$

19. $y(2x - 1)\,dx + x(x + 1)\,dy = 0$

20. $(y + xy^{1/2})\,dx - dy = 0$

3.3 APPLICATIONS OF FIRST ORDER EQUATIONS

In this section we shall consider some typical examples of applications of first order differential equations to problems in biology, economics, physics, and social science. Usually a mathematical model is constructed that approximates an actual problem. This model often involves a differential equation. The solution of the differential equation and model, if one exists, will usually approximate the true solution of the actual problem—unless the model was poorly constructed. It is hoped that after reading this section the reader will have a better appreciation for and understanding of the applications of first order differential equations.

3.3.1 Radioactive Decay

It has been found experimentally that radioactive substances decompose at a rate proportional to the quantity of substance present. If we let $Q(t)$ represent the quantity of substance present at time t, then the statement above may be expressed mathematically by the differential equation

$$\frac{dQ}{dt} = kQ, \tag{34}$$

where k is the constant of proportionality. The general solution of (34) is

$$Q(t) = Ce^{kt}, \tag{35}$$

where C is an arbitrary constant. Suppose that at time t_0 the amount of substance present is Q_0 and that at a later time t_1 the amount of substance present is Q_1. Thus, the solution is to satisfy the conditions

$$Q(t_0) = Q_0 \tag{36a}$$

and

$$Q(t_1) = Q_1. \tag{36b}$$

In order for (35) to satisfy (36a) we must have $Q_0 = Ce^{kt_0}$. Hence, $C = Q_0e^{-kt_0}$ and the solution has the form

$$Q(t) = Q_0e^{k(t - t_0)}. \tag{37}$$

For (37) to satisfy (36b), we must have $Q_1 = Q_0 e^{k(t_1 - t_0)}$. Consequently, the solution of (34) subject to (36a) and (36b) is (37), where

$$k = \frac{\ln Q_1 - \ln Q_0}{t_1 - t_0}. \tag{38}$$

Since $Q_1 < Q_0$, $\ln Q_1 < \ln Q_0$ and therefore $k < 0$.

The rate of decay of a radioactive substance is often expressed in terms of *half-life*; that is, the time required for any given quantity of the substance to be reduced by a factor of one-half. If in equation (38) we let $Q_1 = \frac{1}{2}Q_0$, then the half-life, $T = t_1 - t_0$, satisfies

$$kT = -\ln 2. \tag{39}$$

Consequently, if either k or T is known or can be determined experimentally, the other variable can be determined from equation (39).

The half-life of uranium 238 is 4.5 billion years, the half-life of potassium 40 is 1.4 billion years, and the half-life of rubidium 87 is 60 billion years. By checking the ratio of elements such as these to the elements into which they decay radioactively, geologists and archaeologists can reliably estimate dates of significant events that occurred millions and even billions of years ago. However, since they decay so slowly, radioactive elements with half-lives of billions of years are not suitable for dating events that took place relatively recently.

In the late 1940s and early 1950s the American chemist Willard F. Libby (1908–) developed the technique of *radiocarbon dating*, which can be used to estimate the dates of events that occurred up to 50,000 years ago. Libby was awarded the 1960 Nobel Prize in Chemistry for this achievement. The technique is based upon a phenomenon that involves the radioactive isotope carbon 14, ^{14}C, which is called *radiocarbon* and has a half-life of 5568 years. Radioactive carbon is constantly being produced in the earth's upper atmosphere by incoming cosmic rays. These rays produce neutrons, which in turn collide with nitrogen 14 to yield carbon 14. The radioactive carbon is oxidized and forms radioactive carbon dioxide, which circulates in the earth's atmosphere. Plants that "breathe" carbon dioxide also breathe radioactive carbon dioxide and through their life processes absorb radiocarbon, ^{14}C, in their tissue. Likewise, animals that eat these plants absorb radiocarbon in their tissue. The rate of absorption of radiocarbon by living tissue is in equilibrium with the rate of disintegration. However, when a plant or animal dies, it ceases to absorb radiocarbon and only the process of disintegration continues. The age of a substance of organic origin can be estimated by measuring the radioactivity of carbon 14 of a sample of that substance. For example, a piece of charcoal that has one-half the radioactivity of a living tree died approximately 5568 years ago, and a piece of charcoal that has one-fourth the radioactivity of a living tree died approximately 11 136 years ago.

Solving equation (39), $kT = -\ln 2$ for the decay constant, k, and substituting 5568 for the half-life, T, we find the decay constant for radioactive carbon, ^{14}C, to be

$$k = -\frac{\ln 2}{5568 \text{ years}} = -.000\ 124\ 49/\text{year}$$

From equation (37) we see that the amount, $Q(t)$, of ^{14}C present at time $t \geq t_0$ in some organic substance is

$$Q(t) = Q_0 e^{k(t-t_0)},$$

where Q_0 is the amount that was present at the time t_0 that the substance died. Differentiating, we find that the rate, $R(t)$, of disintegration of ^{14}C at any time $t \geq t_0$ is

$$R(t) = kQ_0 e^{k(t-t_0)}.$$

At time $t = t_0$ the rate of disintegration is

$$R(t_0) = kQ_0.$$

So the ratio of the rates of disintegration is

$$\frac{R(t)}{R(t_0)} = e^{k(t-t_0)}.$$

Solving for the time since the death of the substance, $t - t_0$, we find

$$t - t_0 = \frac{1}{k} \ln \left(\frac{R(t)}{R(t_0)} \right).$$

Assuming that for any particular living substance the rate of disintegration of ^{14}C is a constant (that is, the rate of disintegration is the same now as it was in the past), the time that a particular sample of the same substance died can be calculated from the equation above. For example, in 1950 the rate of radioactive disintegration of ^{14}C from a piece of charcoal found in the Lascaux cave in France was .97 disintegration per minute per gram. Tissue of living wood has a disintegration rate of 6.68 disintegrations per minute per gram. So the tree from which the charcoal came died

$$t - t_0 = \frac{1}{-.000\ 124\ 49} \ln \left(\frac{.97}{6.68} \right) = 15\ 500 \text{ years before 1950.}$$

We have only discussed radioactive substances with relatively long half-lives. However, the reader should be aware that there are radioactive substances with half-lives on the order of a few years, a year, a day, a second, and even very small fractions of a second.

3.3.2 Population Growth

Suppose that a certain colony of bacteria increases at a rate that is proportional to the number of bacteria in the colony. If the population quadruples in 2 hours, find the size of the colony at the end of 8 hours.

Let $n(t)$ be the number of bacteria in the colony at time t. Stated mathematically, this problem is: "Solve the differential equation

$$\frac{dn}{dt} = kn \tag{40}$$

subject to the conditions

$$n(0) = n_0, \qquad n(2) = 4n_0.\text{"} \tag{41}$$

From (40) and the first condition in (41) we get $n(t) = n_0e^{kt}$. The second condition of (41) results in $4 = e^{2k}$. Solving for e^k, we find that $e^k = 2$. Hence, the solution to (40) subject to (41) is $n(t) = n_0 2^t$. Therefore, the size of the colony at the end of 8 hours is $n(8) = n_0 2^8 = 256n_0$.

The population growth model

$$\frac{dn}{dt} = kn; \qquad n(t_0) = n_0, \tag{42}$$

where k is a positive constant has the solution

$$n(t) = n_0e^{k(t-t_0)}. \tag{43}$$

This model is extremely simple and its solution predicts exponential growth. Yet, somewhat surprisingly, this model is often adequate as long as the population has enough room in which to grow and an abundance of food and other natural resources to support its growth. For instance, the estimated world population in 1960 was 3 billion (3×10^9) people, and the estimated growth rate was 1.8% per year. That is, $n(1960) = 3 \times 10^9$ and $k = .018$. So an equation for predicting the world's population based on the model of equation (42) with the stated initial condition and growth rate of 1960 is

$$n(t) = 3 \times 10^9 \times e^{.018(t-1960)}.$$

As the following table shows, this equation represents the world's population growth for the period 1900 to 1970 fairly well in spite of wars, famine, epidemics, and other natural disasters.

World population (billions) 1900–1970

	1900	*1930*	*1950*	*1970*
Actual	1.590	2.015	2.509	3.592
Predicted	1.019	1.748	2.506	3.592

When a population's growth becomes limited by the lack of space, food, or a vital natural resource, the model of equation (42) is no longer valid and the equation must be modified in some way to provide for a reduced rate of growth. In 1837 the Dutch mathematician Pierre-François Verhulst (1804–1849) proposed the following population growth model:

$$\frac{dn}{dt} = kn - \epsilon n^2; \qquad n(t_0) = n_0, \tag{44}$$

where k and ϵ are positive constants and ϵ is small relative to k. This model is known as the *logistic law model.* When the population n is small, the term ϵn^2 will be very small compared to kn, and so the population will grow at nearly an exponential rate. However, when the population n becomes large, the term ϵn^2 approaches the term kn in size and the rate of population growth diminishes.

Separating variables and employing partial fractions, we find that the differential equation of (42) may be rewritten as

$$\frac{dn}{kn - \epsilon n^2} = \frac{1}{k}\left(\frac{1}{n} + \frac{\epsilon}{k - \epsilon n}\right) dn = dt.$$

Integration successively yields

$$\frac{1}{k}\int_{n_0}^{n} \left(\frac{1}{\eta} + \frac{\epsilon}{k - \epsilon\eta}\right) d\eta = \int_{t_0}^{t} d\tau$$

$$\frac{1}{k}(\ln|\eta| - \ln|k - \epsilon\eta|)\Big|_{n_0}^{n} = \frac{1}{k}\ln\left|\frac{\eta}{k - \epsilon\eta}\right|\Big|_{n_0}^{n} = \tau\Big|_{t_0}^{t}$$

$$\frac{1}{k}\ln\left|\frac{n(k - \epsilon n_0)}{n_0(k - \epsilon n)}\right| = t - t_0. \tag{45}$$

For population growth we must have for all time t, $dn/dt = kn - \epsilon n^2 > 0$ or $k - \epsilon n > 0$, since $n > 0$. Thus, the factors n, n_0, $k - \epsilon n$, and $k - \epsilon n_0$ appearing in equation (45) are all positive, and we may therefore omit the absolute value appearing in the equation. Multiplying equation (45) by k and taking exponentials of both sides of the resulting equation, we get

$$\frac{n(k - \epsilon n_0)}{n_0(k - \epsilon n)} = e^{k(t - t_0)}.$$

Solving for n, we find

$$n(t) = \frac{n_0 k e^{k(t-t_0)}}{k - \epsilon n_0 + \epsilon n_0 e^{k(t-t_0)}} = \frac{n_0 k}{\epsilon n_0 + (k - \epsilon n_0)e^{-k(t-t_0)}}. \tag{46}$$

Notice that as $t \to \infty$, $n(t) \to n_0 k/\epsilon n_0 = k/\epsilon$. So, regardless of the initial population size, the population approaches the limiting value of k/ϵ. Therefore, the constants k and ϵ are called the *vital coefficients* of a population. The graph of equation (46) is an S-shaped curve called the *logistic curve.*

Experts have estimated that the earth's human population has a vital coefficient $k = .029$. Given that the population of the world in 1960 was 3 billion people and the growth rate, $(dn/dt)/n$, was 1.8% per year, we can determine the vital coefficient ϵ in the following manner. Dividing the differential equation of (44) by n, we find that the growth rate expressed as a percentage satisfies

$$\frac{1}{n}\frac{dn}{dt} = k - \epsilon n.$$

Substituting the estimated value for k and the known 1960 values given above for n and $(dn/dt)/n$, we find that the vital coefficient ϵ must satisfy

$$.018 = .029 - \epsilon(3 \times 10^9).$$

Solving this equation for ϵ yields $\epsilon = 3.667 \times 10^{-12}$. Using these vital coefficients, the logistic law model predicts a limiting value of the human population of the earth of $n = k/\epsilon = .029/(3.667 \times 10^{-12}) = 7.91 \times 10^9$ people.

3.3.3. Newton's Law of Cooling

It has been shown experimentally that under certain conditions the temperature of an object can be predicted by using Newton's law of cooling, which states: "The rate of change of the temperature of a body is proportional to the difference between the temperature of the body and temperature of the surrounding medium." Hence, if $T(t)$ is the temperature of the body at time t and A is the temperature of the surrounding medium, then according to Newton's law of cooling, T satisfies the differential equation

$$\frac{dT}{dt} = k(T - A). \tag{47}$$

Separating variables, integrating, and exponentiating, we obtain the following general solution to (47):

$$T(t) = Ce^{kt} + A. \tag{48}$$

EXAMPLE A cup of coffee whose temperature is 190°F is poured in a room whose temperature is 65°F. Two minutes later the temperature of the coffee is 175°F. How long after the coffee is poured does it reach a temperature of 150°F?

For this problem $A = 65°\text{F}$ and (48) must satisfy the conditions $T(0) = 190°\text{F}$ and $T(2) = 175°\text{F}$. From the first condition we find

$$C = T(0) - A = 190°\text{F} - 65°\text{F} = 125°\text{F}.$$

Satisfying the second condition requires that

$$175°\text{F} = 125°\text{F}e^{2k} + 65°\text{F}.$$

Hence, $e^{2k} = \frac{110}{125} = \frac{22}{25}$ and the temperature of the coffee as a function of the time after being poured is

$$T(t) = (\tfrac{22}{25})^{t/2}125°\text{F} + 65°\text{F}.$$

The coffee reaches 150°F when t satisfies

$$150°\text{F} = (\tfrac{22}{25})^{t/2}125°\text{F} + 65°\text{F},$$

or $t = 2 \ln (\frac{85}{125})/\ln (\frac{22}{25}) = 6.034$ min.

3.3.4 Mixture Problems

Suppose that at time $t = 0$ a quantity Q_0 of a substance is present in a container. Assume that at time $t = 0$ a fluid containing a concentration c_α of substance is allowed to enter the container at a constant rate α and that the mixture in the container is kept at a uniform concentration throughout by a mixing device. Also assume that at time $t = 0$ the mixture in the container with concentration c_β is allowed to escape at a constant rate β. The problem is to determine the amount Q of the substance in the container at any time. Since the rate of change of the amount of substance in the container dQ/dt equals the rate at which a fluid enters the container times the concentration of the substance in the entering fluid minus the rate at which a fluid leaves the container times the concentration of the substance in the container, Q must satisfy the initial value problem

$$\frac{dQ}{dt} = \alpha c_\alpha - \beta c_\beta; \qquad Q(0) = Q_0. \tag{49}$$

EXAMPLE A 100-gal tank contains a 25% dye solution. A 40% dye solution is allowed to enter the tank at the rate of 10 gallons per minute and the resulting uniform mixture is removed from the tank at the same rate. Derive an expression for the amount of dye in the tank as a function of time.

The initial condition is $Q(0) = Q_0 = .25 \times 100$ gal $= 25$ gal of dye in the tank. We easily find that $\alpha = 10$ gal/min, $c_\alpha = .40$, $\beta = 10$ gal/min, and $c_\beta = Q(t)/100$ gal. Hence, Q must satisfy the initial value problem

$$\frac{dQ}{dt} = (10 \text{ gal/min})(.40) - (10 \text{ gal/min})(Q/100); \qquad Q_0 = 25 \text{ gal}.$$

Separating variables results in

$$\frac{dQ}{40 - Q} = \frac{dt}{10}.$$

Integration yields

$$-\ln|40 - Q| + \ln|C| = \tfrac{1}{10}t.$$

Hence, the general solution is

$$\left|\frac{C}{40 - Q}\right| = e^{t/10}.$$

Since the constant C is arbitrary, it may always be chosen so that the expression appearing within the absolute value sign is positive. Assuming that C is so chosen and solving for Q, we find

$$Q = (40 - Ce^{-t/10}) \text{ gal}.$$

When $t = 0$, $Q(0) = Q_0 = 25 = 40 - C$.

Consequently, $C = 15$ and the solution of the initial value problem is

$$Q = (40 - 15e^{-t/10}) \text{ gal}.$$

The second term on the right becomes small as t becomes large; hence, the limiting value of the concentration of the dye solution is 40%—a concentration equal to the concentration of the incoming fluid. That is, as time passes, less and less of the original dye remains in the tank.

EXAMPLE A 400-gal-capacity tank initially contains a salt solution consisting of 150 gal of water and 25 lb of salt. A salt solution containing 2 lb of salt per gallon is allowed to enter the tank at the rate of 10 gal/min and the resulting mixture is removed at the rate of 5 gal/minute. Find an expression for the number of pounds of salt in the tank as a function of time and find the amount of salt in the tank at the instant the tank starts to overflow.

Initially, the tank contains 25 lb of salt, so $Q_0 = 25$ lb. Also, $\alpha =$ 10 gal/min, $c_\alpha = 2$ lb/gal, and $\beta = 5$ gal/min. Since the rate at which fluid is flowing into the tank is greater than the rate at which the fluid is flowing out of the tank, the tank is filling at a rate $\gamma = \alpha - \beta = 10 - 5 = 5$ gal/min. The number of gallons N of fluid in the tank at time t is $N = 150 + 5t$, since the tank initially contained 150 gal. Therefore, the concentration of salt in the tank at any time is $c_\beta = Q/(150 + 5t)$. Hence, Q satisfies the initial value problem

$$\frac{dQ}{dt} = (10 \text{ gal/min})(2 \text{ lb/gal}) - (5 \text{ gal/min})\,\frac{Q}{(150 + 5t)\text{ gal}}; \qquad Q_0 = 25 \text{ lb}.$$

The differential equation may be rewritten as

$$\frac{dQ}{dt} + \frac{1}{30 + t}\,Q = 20,$$

which is a linear differential equation in Q and thus has as an integrating factor the function

$$v(t) = \exp\left\{\int^t \frac{ds}{30+s}\right\} = e^{\ln(30+t)} = 30 + t.$$

Hence,

$$d(Qv) = 20v\,dt.$$

Integration yields

$$Q(t)(30+t) = \int^t 20(30+t)\,dt$$

or

$$Q(t) = \frac{1}{30+t}(10t^2 + 600t + C).$$

In order to satisfy the initial condition, $Q_0 = Q(0) = 25$, the constant C must satisfy the equation $25 = C/30$. Thus, $C = 750$ and the amount of salt in the tank at time t is

$$Q(t) = \frac{10t^2 + 600t + 750}{30+t}\ \text{lb}.$$

Since the maximum capacity of the tank is 400 gal, the tank overflows when t satisfies $400 = 150 + 5t$. Hence, the tank overflows at $t = 50$ min. At that time the amount of salt in the tank is

$$Q(50) = 55\ 750/80 = 696.875\ \text{lb}.$$

3.3.5 Orthogonal Trajectories

Suppose that we are given a one-parameter family of plane curves

$$f(x, y, c) = 0. \tag{50}$$

A problem that often arises in fields such as electrostatics, hydrodynamics, and thermodynamics is to find a second one-parameter family of curves

$$g(x, y, k) = 0 \tag{51}$$

with the property that each member of this family is perpendicular (orthogonal) to each member of the other family at each point of intersection. The families (50) and (51) are called *orthogonal trajectories*. For example, in electrostatic and gravitational problems, one family of curves is called "equipotential curves" and the other "lines of force"; in hydrodynamics, one family is called "velocity potential curves" and the other "flow lines" or "stream lines"; and in thermodynamics one family is called "isothermal lines" and the other "heat flow lines."

In order to find (51) given (50), we differentiate (50) with respect to x and generally obtain a differential equation of the form

$$h(x, y, dy/dx, c) = 0. \tag{52}$$

Eliminating the parameter, c, from equations (50) and (52) yields a differential equation of the form

$$F(x, y, dy/dx) = 0. \tag{53}$$

The family of curves (50) satisfies this differential equation. If (x, y) is a point on one of the curves, C_1, in the family (50), then the slope of the tangent line to the curve at (x, y), dy/dx, is given implicitly by equation (53). Since we are to find a curve, C_2, passing through (x, y) perpendicular to C_1, the slope of C_2 at (x, y) must be the negative reciprocal of the slope of C_1 at (x, y). Consequently, the family of curves (51) orthogonal to the family (50) must satisfy the differential equation

$$F(x, y, -dx/dy) = 0. \tag{54}$$

Summarizing: To find a family of curves orthogonal to a given family, one first finds the differential equation which the given family satisfies, then replaces dy/dx by $-dx/dy$, and integrates.

EXAMPLE The equipotential curves of a certain electrostatic field are closely approximated by the ellipses

$$x^2 - 2cx + 2y^2 = 0. \tag{55}$$

Find the lines of force.

Differentiating (55) with respect to x and dividing by 2, we obtain

$$x - c + 2y\frac{dy}{dx} = 0. \tag{56}$$

Solving (56) for c and substituting into (55) results in the differential equation

$$-x^2 - 4xy\frac{dy}{dx} + 2y^2 = 0, \tag{57}$$

which the family (55) satisfies. Replacing dy/dx in equation (57) by $-dx/dy$, we see that the family of orthogonal curves satisfies the differential equation

$$-x^2 + 4xy\frac{dx}{dy} + 2y^2 = 0.$$

Rewriting this equation in the form $M\,dx + N\,dy = 0$, we get

$$4xy\,dx + (2y^2 - x^2)\,dy = 0.$$

This equation is not exact; but since $-(1/M)(M_y - N_x) = -\frac{3}{2}y$, it can be made exact by multiplying by the integrating factor,

$$v(y) = \exp\left\{\int^y -\frac{3dt}{2t}\right\} = e^{-(3/2)\ln|y|} = |y|^{-3/2}.$$

Multiplying by the integrating factor $y^{-3/2}$ and rearranging results in

$$4xy^{-1/2}\,dx - x^2y^{-3/2}\,dy + 2y^{1/2}\,dy = 0$$

or

$$2d(x^2y^{-1/2}) + \tfrac{4}{3}d(y^{3/2}) = 0.$$

Integrating, multiplying by $y^{1/2}$, and rearranging, we find that the family of orthogonal curves—lines of force—are given by the equation

$$x^2 + \tfrac{2}{3}y^2 + ky^{1/2} = 0.$$

EXAMPLE Two thin refrigerating rods which are kept at 0°C are inserted at right angles to each other into a thin slab of material. The temperature at any point in the slab is given by the equation

$$T(x, y) = axy,$$

where the origin is the point at which the refrigerating rods cross. Find the lines of heat flow.

Setting $T(x, y) = K$, a constant, we find that the isothermal curves satisfy $xy = K/a = c$. Differentiating with respect to x, we find that the isothermal curves satisfy the differential equation $y + x\,dy/dx = 0$. Substituting $-dx/dy$ for dy/dx, we see that the lines of heat flow satisfy $y - x\,dx/dy = 0$. Separating variables and integrating, we obtain the following equation for the lines of heat flow: $y^2 = x^2 + k$.

3.3.6 Falling Bodies

Newton's second law of motion states: "The rate of change of momentum of a body is proportional to the force acting on the body and is in the direction of the force." Let m be the mass of a body and let v be its velocity. The momentum of the body is mv. Thus, Newton's second law of motion may be expressed mathematically as

$$\frac{d(mv)}{dt} = kF, \tag{58}$$

where k is a constant of proportionality and F is the magnitude of the force acting on the body. If we make the assumption that the mass of the body is a constant and that the units of mass, velocity, and force are chosen so that $k = 1$, then Newton's second law becomes

$$m\frac{dv}{dt} = F. \tag{59}$$

Consider an origin located somewhere above the earth's surface. Let the positive axis of a one-dimensional coordinate system extend from the origin through the gravitational center of the earth. Suppose that at time $t = 0$ a body with mass m is initially located at y_0 and is traveling with a velocity v_0. See Figure 3.2. Assuming that the body is falling freely in a vacuum and that

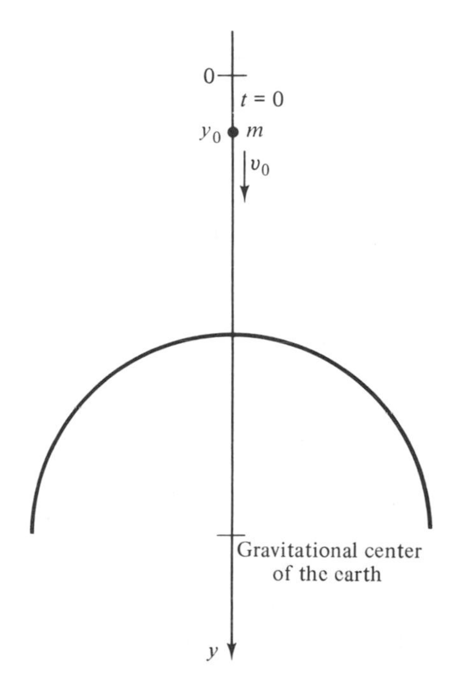

Figure 3.2 Falling body.

it is close enough to the earth's surface so that the only significant force acting on the body is the earth's gravitational attraction, then $F = mg$, where g is the gravitational attraction—approximately 32 feet/second^2 in the English system of measurement and 9.8 meters/second^2 in the metric system. Hence, the velocity of the body satisfies the initial value problem $dv/dt = g$; $v(0) = v_0$. Integrating and satisfying the initial condition, we easily find the solution of this initial value problem to be $v(t) = gt + v_0$. Since $v = dy/dt$, the position of the body satisfies the initial value problem $dy/dt = gt + v_0$; $y(0) = y_0$. Integrating and satisfying the initial condition yields the solution

$y(t) = \frac{1}{2}gt^2 + v_0t + y_0$. If the body is at rest at the origin when $t = 0$, then $y_0 = 0$ and $v_0 = 0$ and consequently $v(t) = gt$ and $y(t) = \frac{1}{2}gt^2$. Eliminating t from the last two equations, we find that $v = \sqrt{2gy}$. The last equation gives the velocity of a freely falling body as a function of the distance that it has fallen from its position of rest.

If we assume that the body is falling freely in air instead of in a vacuum, then we must make some assumption regarding the retarding force due to air resistance. Let us assume that the retarding force is proportional to the velocity of the body. Then the initial value problem for the velocity becomes $dv/dt = g - cv$; $v(0) = v_0$, where $c > 0$ is the constant of proportionality. Separating variables results in

$$\frac{dv}{g - cv} = dt.$$

Integrating, we find

$$-\frac{1}{c}\ln|g - cv| = t + A,$$

where A is an arbitrary constant. Multiplying by $-c$ and exponentiating, we obtain

$$|g - cv| = e^{-c(t+A)} = e^{-cA}e^{-ct} = Be^{-ct}, \tag{60}$$

where B is an arbitrary positive constant. By letting B be an arbitrary constant (positive, negative, or zero), we can remove the absolute value appearing in equation (60). When $t = 0$, $v = v_0$, so $B = g - cv_0$ and therefore

$$g - cv = (g - cv_0)e^{-ct}.$$

Solving for the velocity v, we find

$$v(t) = \frac{g}{c}(1 - e^{-ct}) + v_0e^{-ct}.$$

Since $c > 0$, as $t \to \infty$, $v \to g/c$, the *terminal velocity*—the maximum velocity that the falling body can attain. In this case the distance from the origin satisfies the initial value problem

$$\frac{dy}{dt} = \frac{g}{c}(1 - e^{-ct}) + v_0e^{-ct}; \qquad y(0) = y_0.$$

Integrating and imposing the initial condition, we find

$$y(t) = \frac{g}{c}t + \frac{g}{c^2}(e^{-ct} - 1) + \frac{v_0}{c}(1 - e^{-ct}) + y_0.$$

3.3.7 Curves of Pursuit

Interesting problems in differential equations often arise when one tries to determine the path that one object must take in order to pursue and perhaps capture another object subject to certain constraints on either or both objects.

EXAMPLE A rabbit starts at $(0, a)$ and runs along the y-axis in the positive direction with constant velocity v_R. A dog starts at $(b, 0)$ and pursues the rabbit with velocity v_D. The dog runs so that he is always pointed toward the rabbit. The problem is to determine the path taken by the dog. See Figure 3.3.

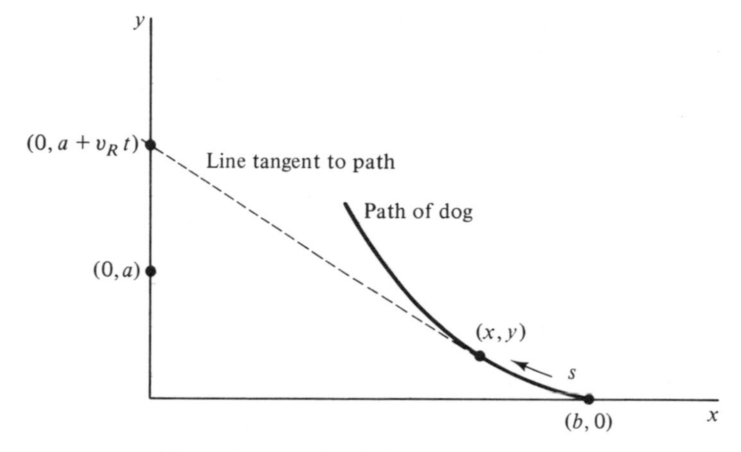

Figure 3.3 Path of dog pursuing rabbit.

At time t the rabbit will be at $(0, a + v_R t)$ and the dog will be at (x, y). Since the line between these two points is to be tangent to the path of the dog, we have

$$\frac{dy}{dx} = \frac{y - a - v_R t}{x}$$

or

$$xy' = y - a - v_R t. \tag{61}$$

The arc length s of the path of the dog satisfies the differential equation

$$ds = -\sqrt{(dx)^2 + (dy)^2}.$$

The minus sign is due to the fact that as s increases, x decreases. Since

$$v_D = \frac{ds}{dt} = -\sqrt{\left(\frac{dx}{dt}\right)^2 + \left(\frac{dy}{dt}\right)^2} = -\sqrt{1 + \left(\frac{dy}{dx}\right)^2}\,\frac{dx}{dt},$$

$$\frac{dt}{dx} = \frac{-1}{v_D}\sqrt{1 + (y')^2}. \tag{62}$$

Differentiating (61) with respect to x, we find $xy'' = -v_R\, dt/dx$. Solving for dt/dx yields

$$\frac{dt}{dx} = \frac{-xy''}{v_R}. \tag{63}$$

Equating (62) and (63) results in

$$\frac{1}{v_D}\sqrt{1 + (y')^2} = \frac{xy''}{v_R}. \tag{64}$$

Let $p = y'$. Then $p' = y''$ and when $t = 0$, $(x, y) = (b, 0)$ and $p = -a/b$. Hence, p satisfies the initial value problem

$$p' = \frac{v_R}{xv_D}\sqrt{1 + p^2}; \qquad p(b) = -\frac{a}{b}. \tag{65}$$

Separating variables and letting $c = v_R/v_D$, we get

$$\frac{dp}{\sqrt{1 + p^2}} = \frac{c\,dx}{x}.$$

Integration and exponentiation results in

$$p + \sqrt{1 + p^2} = Kx^c. \tag{66}$$

In order to satisfy the initial condition, we must have

$$K = b^{-c-1}(-a + \sqrt{a^2 + b^2}).$$

Subtracting p from (66) and squaring, we obtain

$$1 = K^2x^{2c} - 2pKx^c.$$

Hence,

$$p = \frac{dy}{dx} = \frac{1}{2}\left(Kx^c - \frac{1}{K}x^{-c}\right). \tag{67}$$

There are two cases to consider: (i) $0 < c < 1$, and (ii) $c \geq 1$. If $0 < c < 1$, then $v_R < v_D$ and the dog catches the rabbit. If $c \geq 1$, then $v_R \geq v_D$ and the dog cannot catch the rabbit. For $0 < c < 1$, the solution of (67) subject to $y(b) = 0$ is

$$y = \frac{1}{2}\left[\frac{K}{c + 1}(x^{c+1} - b^{c+1}) + \frac{1}{K(c - 1)}(x^{1-c} - b^{1-c})\right] \tag{68}$$

and the dog catches the rabbit at $(0, y_*)$, where

$$y_* = \frac{[a\sqrt{a^2 + b^2} - (a^2 + b^2)]c + a^2 + a\sqrt{a^2 + b^2}}{(c^2 - 1)(-a + a^2 + b^2)}$$

and when $t = (y_* - a)/v_R$.

EXERCISES

In Exercises 1–7, assume that the rate of decay or growth is proportional to the amount present.

1. If 5% of a radioactive substance decomposes in 50 years, what percentage will be present at the end of 500 years? 1000 years? What is the half-life of this substance?

2. If the half-life of a radioactive substance is 1800 years, what percentage is present at the end of 100 years? In how many years does only 10% of the substance remain?

3. If 100 grams of a radioactive substance is present 1 year after the substance was produced and 75 grams is present 2 years after the substance was produced, how much was produced and what is the half-life of the substance?

4. In 1977 the rate of carbon 14 radioactivity of a piece of charcoal found at Stonehenge in southern England was 4.16 disintegrations per minute per gram. Given that the rate of carbon 14 radioactivity of a living tree is 6.68 disintegrations per minute per gram and assuming that the tree which was burned to produce the charcoal was cut during the construction of Stonehenge, estimate the date of the construction of Stonehenge.

5. An amount of money, A, which is invested at an annual rate of interest, k, and *compounded continuously* satisfies the differential equation $dA/dt = kA$. How long will it take an investment to double if the annual rate of interest is 4%? 5%? 6%?

6. If the population of a city doubled in the past 25 years and the present population is 100 000, when will the city have a population of 500 000?

7. If the population of the earth in 1960 were 3 billion (3×10^9) and the population in 1970 were 4 billion and if the earth can only support a maximum population of 10 billion, in what year will the limit be reached? In what year will the limit be reached if the maximum population that the earth can support is 20 billion?

8. Assume that a college's enrollment satisfies the logistic law model. If the enrollment 5 years ago was 10 000, if the enrollment now is 15 000, and if the college's maximum enrollment is 25 000, when will the enrollment reach 18 000? 21 000? 24 000?

9. The logistic law model with $k = .03134$ and $\epsilon = 1.589 \times 10^{-10}$ provides a model for the population of the United States. Based on this model, what is the limiting population of the United States? Assume that in 1800 the population of the United States was 5.3 million (5.3×10^6).

Calculate the population of the United States in 1850, 1900, 1950, and 1970 using the logistic law model and compare with the actual population in those years which was 23.2 million in 1850, 76 million in 1900, 151.3 million in 1950, and 204.9 million in 1970.

10. A spherical drop of liquid evaporates at a rate that is proportional to its surface area. Find the radius of the drop as a function of time.

11. One morning it started snowing at a heavy and constant rate. A snowplow started out at 8 A.M. By 9 A.M. the snowplow had gone 2 miles. By 10 A.M. the snowplow had gone 3 miles. Assuming that the snowplow removes a constant volume of snow per hour, determine the time at which it started snowing.

12. A thermometer reading 70°F is taken outside where the temperature is 10°F. Five minutes later the thermometer reads 40°F. How long after being taken outside is the thermometer reading within one-half a degree of the outside temperature?

13. A thermometer reading 80°F is taken outside. Five minutes later the thermometer reads 60°F. After another 5 minutes the thermometer reads 50°F. What is the temperature outside?

14. At 1:00 P.M. a thermometer reading 10°F is removed from a freezer and placed in a room whose temperature is 65°F. At 1:05 P.M. the thermometer reads 25°F. Later the thermometer is placed back in the freezer. At 1:30 P.M. the thermometer reads 32°F. When was the thermometer returned to the freezer and what was the thermometer reading at that time?

15. A invited B for morning coffee. A poured two cups of coffee. B added enough cream to lower the temperature of his coffee 1°F. After 5 minutes A added enough cream to his coffee to lower the temperature 1°F and both A and B began to drink their coffee. Who had the cooler coffee?

16. A tank contains 200 gal of fresh water. A brine solution consisting of 2 lb of salt per gallon is allowed to enter the tank at a rate of 5 gal/min. Assume that the mixture in the tank is kept at a uniform concentration and that the mixture flows out at the rate at which it enters. How many pounds of salt is in the tank at the end of 30 min? What is the limiting value of the number of pounds of salt in the tank?

17. A 300-gal-capacity tank contains a solution of 200 gal of water and 50 lb of salt. A solution containing 3 lb of salt per gallon is allowed to flow into the tank at the rate of 4 gal/min. The mixture flows from the tank at the rate of 2 gal/min. How many pounds of salt are in the tank at the end of 30 min? 60 min? (*Hint:* When does the tank start to overflow?)

18. A 100-gal tank initially contains pure water. A 30% solution of dye flows into the tank at the rate of 5 gal/minute and the resulting mixture flows out at the same rate. After 15 min the process is stopped and fresh water flows into the tank at the rate of 5 gal/min and the mixture flows out at the same rate. Find the concentration of dye in the tank at the end of 30 min.

19. A 100-gal tank is initially filled with a 40% dye solution. A 20% dye solution flows into the tank at the rate of 5 gal/min. The mixture flows out of this tank at the same rate into another 100-gal tank which was initially filled with pure water. The resulting mixture flows out of the second tank at the rate of 5 gal/min. Derive an expression for the amount of dye in the second tank. What is the concentration of dye in the second tank at the end of 30 min?

20. A point source is radiating heat equally in every direction in two-dimensional space. Therefore, the equation of the lines of heat flow are $y = cx$ when the origin is the located at the heat source. Find the equation of the isothermal lines.

21. The temperature in a thin slab is given by $T(x, y) = ae^{-y} \cos x$. Find the equation of the lines of heat flow.

22. Find the family of curves orthogonal to each of the following families.

(a) $y^2 = 2cx$ (b) $y = ce^x$

(c) $y^2 - 2x^2 = c$ (d) $cx^2 + y^2 = 1$

23. Find the equation for the velocity of a falling body if the air resistance is proportional to the square of the velocity. What is the terminal velocity in this case?

24. A boat is moving through the water with velocity v_0. At time $t = 0$ the boat stops its engines and begins to coast. Find an equation for the velocity of the boat, find an equation for the distance of the boat from the point at which the engines were stopped, determine how far the boat travels, and determine how soon after stopping its engines the boat comes to rest, if the only significant force acting on the boat is the retarding force due to friction, and this force is proportional to (1) the velocity of the boat, (2) the square of the velocity of the boat, and (3) the square root of the velocity of the boat.

25. A boy has a string of length a attached to a boat. The boy places the boat at the edge of a rectangular pool, $(a, 0)$, and moves to a corner of the pool, $(0, 0)$. Then the boy walks along the other edge of the pool—the y-axis. The boat glides through the water after him. See Figure 3.4. Find the equation of the path taken by the boat. The curve that is the solution of this problem is called a *tractrix* (Latin *tractum*, drag).

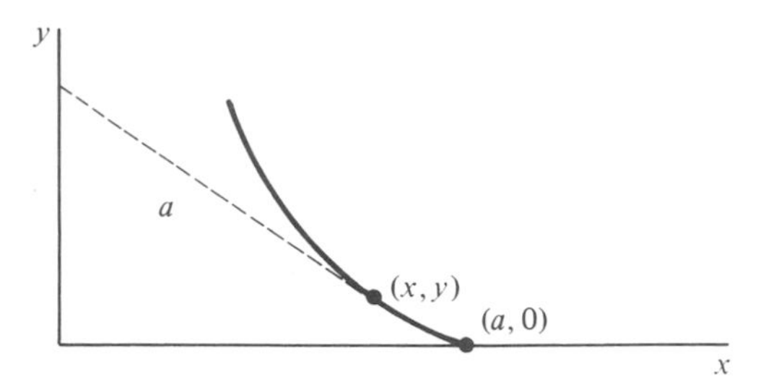

Figure 3.4 The tractrix.

26. The y-axis and the line $x = a > 0$ are the banks of a river. The river flows in the negative y direction with a speed s_R. A boat whose speed in still water is s_B is launched from $(a, 0)$. The boat is steered so that it is always headed toward the origin. See Figure 3.5. Find the path of the boat. Under what conditions will the boat land on the opposite bank? And where will it land?

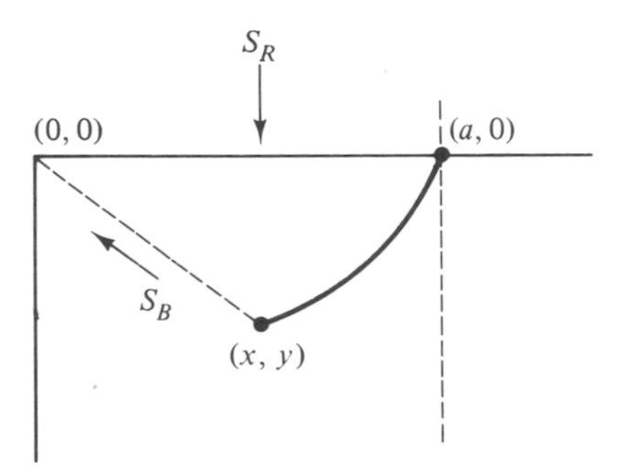

Figure 3.5 Path of boat in river.

27. A tanker is to refuel a submarine. The submarine surfaces and spots the tanker 3 miles away. A dense fog immediately settles over the ocean. The speed of the submarine is twice the speed of the tanker. It is known that the tanker is on a straight course of unknown direction. What course should the submarine take in order to overtake the tanker? (*Hint:* The submarine should proceed 2 miles toward the point at which the tanker was sighted. The point at which the tanker was sighted should be designated as the origin of a polar coordinate system, and the initial line of travel of the submarine should be designated as the coordinate axis.)

4

THE INITIAL VALUE PROBLEM: $y' = f(x, y)$; $y(x_0) = c_0$

The initial value problem

$$y' = f(x, y); \qquad y(x_0) = c_0 \tag{1}$$

is said to be *well-posed* (or *well-set*) on a domain D of the xy-plane if and only if for every $(x_0, c_0) \in D$, there is one and only one function $y(x)$ passing through (x_0, c_0) and satisfying $y' = f(x, y)$ in D and this solution varies continuously with x_0, c_0, and f. Thus, in order to show that the IVP (1) is well-posed in a domain D, one must be able to prove that for each $(x_0, c_0) \in D$: (a) there exists a solution of the differential equation passing through the point (x_0, c_0), (b) there is only one solution of the differential equation passing through each point in D, and (c) the solution of the differential equation depends continuously on the initial values x_0 and c_0 and the function f. Consequently, one is interested in proving three general types of theorems for the IVP (1): (i) existence theorems, (ii) uniqueness theorems, and (iii) continuity theorems. In this chapter we shall prove these types of theorems, illustrate their use and importance, and discuss numerical methods for obtaining "approximate" solutions to the IVP (1). We discussed in Chapter 3 the various techniques that can be employed when trying to obtain the "true" solution of the IVP (1).

First we establish the following relation between a large class of initial value problems of the form (1) and an integral equation.

Theorem 4.1 Let $f(x, y)$ be a continuous function of x and y. A function $\bar{y}(x)$ that is continuous on an open interval I which contains x_0 is a solution of the IVP (1) if and only if $\bar{y}(x)$ is a solution of the integral equation (IE)

$$y(x) = c_0 + \int_{x_0}^{x} f(t, y(t))\, dt. \tag{2}$$

Proof Suppose that $\bar{y}(x)$ satisfies the IVP (1) on an interval I containing x_0. Since $\bar{y}(x)$ satisfies the differential equation $\bar{y}'(x) = f(x, \bar{y}(x))$ on I, $\bar{y}(x)$ is a continuous function of x on I. And since $f(x, y)$ is a continuous function of x and y, $f(x, \bar{y}(x))$ is a continuous function of x on I. Consequently, $f(x, \bar{y}(x))$ is integrable on I. Integrating the differential equation $\bar{y}'(x) = f(x, \bar{y}(x))$, we obtain

$$\bar{y}(x) - \bar{y}(x_0) = \int_{x_0}^{x} \bar{y}'(t)\, dt = \int_{x_0}^{x} f(t, \bar{y}(t))\, dt$$

for $x \in I$. Imposing the initial condition $\bar{y}(x_0) = c_0$, we see that if $\bar{y}$ is a solution of the IVP (1) for $x \in I$, then $\bar{y}$ satisfies the IE (2) for $x \in I$.

Now suppose that $\bar{y}(x)$ is a continuous function of x on some interval I containing x_0 and that $\bar{y}$ satisfies the IE (2). Substituting $x = x_0$ into the IE (2), we see that $\bar{y}$ satisfies the initial condition $\bar{y}(x_0) = c_0$. Differentiating the IE (2), we find $\bar{y}$ satisfies the differential equation $\bar{y}'(x) = f(x, \bar{y}(x))$ for $x \in I$. So if $\bar{y}$ satisfies the IE (2) for $x \in I$, $\bar{y}$ satisfies the IVP (1) for $x \in I$.

EXAMPLE Write an integral equation which is equivalent to the initial value problem

$$y' = 3xy^2; \qquad y(1) = 2.$$

In this case $f(x, y) = 3xy^2$, $x_0 = 1$, and $c_0 = 2$. So an equivalent integral equation is

$$y(x) = 2 + 3\int_{1}^{x} ty^2(t)\, dt.$$

EXAMPLE Write an initial value problem that is equivalent to the integral equation

$$y(x) = 1 - 2e^{3x} - \int_{0}^{x} e^t y(t)\, dt.$$

Since the lower limit of integration is the constant 0, it is appropriate to choose the initial value $x_0 = 0$. Evaluating the given integral equation at zero, we find an appropriate initial condition to be $y(0) = -1$. Formally differentiating the given initial value problem, we find that y satisfies the differential equation

$$y' = -6e^{3x} - e^x y.$$

So an equivalent initial value problem is

$$y' = -6e^{3x} - e^x y; \qquad y(0) = -1.$$

Earlier in the text we found that if $f(x)$ was a piecewise continuous function, then the initial value problem $y' = f(x);\ y(x_0) = c_0$ was equivalent to the integral equation

$$y(x) = c_0 + \int_{x_0}^{x} f(t)\, dt.$$

This equivalence reduced the problem of solving the initial value problem to the problem of calculating the integral of the known function f. So in this case the equivalence of the initial value problem and the integral equation was of practical importance. However, as we shall soon discover, the equivalence of the initial value problem

$$y' = f(x, y); \qquad y(x_0) = c_0 \tag{1}$$

and the integral equation

$$y(x) = c_0 + \int_{x_0}^{x} f(t, y(t))\, dt \tag{2}$$

is of more theoretical value than practical value, since the unknown function y appears not only on the left-hand side of (2) but also on the right-hand side of (2) in the integrand. So in this case the solution of the initial value problem does not reduce to the evaluation of the integral of a known function as it did in the previous case.

EXERCISES

1. Write the following initial value problems as integral equations.

(a) $y' = x;\ y(0) = 1$ (b) $y' = y;\ y(0) = 1$

(c) $y' = 2xy + 1;\ y(1) = -1$ (d) $y' = 1 + y^2;\ y(0) = 0$

2. Write the following integral equations as initial value problems.

(a) $y(x) = -2 + \int_{1}^{x} [ty(t) + y^2(t)]\, dt$

(b) $y(x) = 2\int_{-1}^{x} y^{1/2}(t)\,dt$

(c) $y(x) = 1 + x^2 + \int_{x}^{1} ty^2(t)\,dt$

(d) $y(x) = 2 - \int_{0}^{x} t^2y(t)\,dt$

4.1 EXISTENCE, UNIQUENESS, AND CONTINUATION OF SOLUTIONS

A set S of the xy-plane is *bounded* if there exists a positive constant M such that $x^2 + y^2 \leq M$ for all $(x, y) \in S$. Thus, a bounded set of the xy-plane is a set that can be enclosed in a circle of radius M with center at the origin. The set $A = \{(x, y) \mid (x - 1)^2 + (y - 1)^2 < 4\}$ which is a circle with center (1, 1) and radius 2 is a bounded set. Any value of $M \geq 2 + \sqrt{2}$ will fulfil the requirement of the definition of bounded for the set A. The set $B = \{(x, y) \mid x > 0 \text{ and } y > 0\}$ which is the first quadrant of the xy-plane is not bounded.

It can be shown that if the function $f(x, y)$ is a continuous function of x and y on some bounded domain D of the xy-plane, then for every $(x_0, c_0) \in D$ there exists a solution to the initial value problem

$$y' = f(x, y); \qquad y(x_0) = c_0 \tag{1}$$

on some interval I containing x_0. For proof, see Birkhoff and Rota [3, pp. 124–126]. The following example, however, shows that continuity of the function $f(x, y)$ on a domain D does not guarantee the uniqueness of a solution of the IVP (1) on D.

EXAMPLE The initial value problem

$$y' = 3y^{2/3}; \qquad y(x_0) = 0 \tag{3}$$

has an infinite number of solutions of the form

$$y_{ab}(x) = \begin{cases} (x - a)^3, & x < a \\ 0, & a \leq x, x_0 \leq b. \\ (x - b)^3, & x > b \end{cases}$$

See Figure 4.1. The reader should verify that each function $y_{ab}(x)$ satisfies the IVP (3).

The function $f(x, y) = 3y^{2/3}$ is a continuous function of x and y on the entire xy-plane. So by the existence theorem mentioned above there exists a solution to the IVP (3) on some interval I containing x_0 for any point (x_0, y_0) of the xy-plane. However, as this example shows, a solution through the

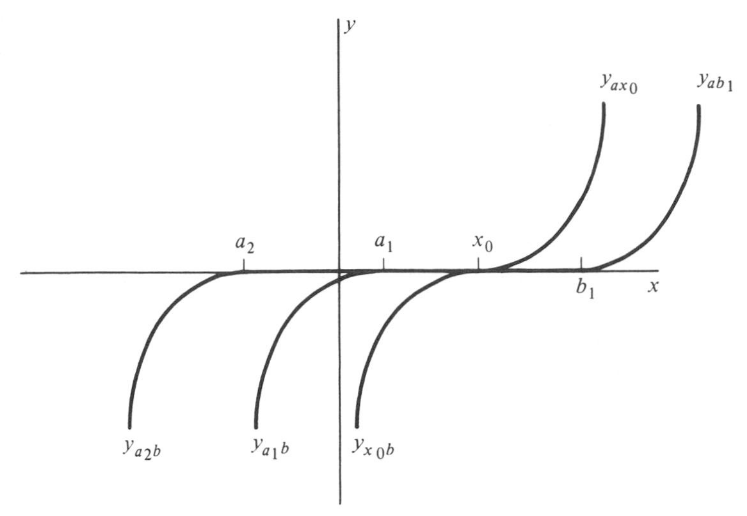

Figure 4.1 Solutions to the IVP $y' = 3y^{2/3}$; $y(x_0) = 0$.

point $(x_0, 0)$ is not unique. So in order to ensure the uniqueness of the solution of the IVP (1), some condition which is stronger than continuity must be satisfied by the function f.

The German mathematician Rudolf Lipschitz (1832–1903), who was a student of Dirichlet, stated a condition on a function $f(x, y)$ which is stronger than continuity in y and weaker than differentiability with respect to y. This condition simplified Cauchy's theory regarding the existence and uniqueness of solutions to differential equations. Quite naturally the condition is known as a *Lipschitz condition.*

A real-valued function $f(x, y)$ *satisfies a Lipschitz condition with respect to y on a domain D* if there exists a constant $M > 0$ such that

$$|f(x, y_1) - f(x, y_2)| \leq M|y_1 - y_2|$$

for all $(x, y_1), (x, y_2) \in D$.

It follows from this definition that if f satisfies a Lipschitz condition with respect to y on a domain D, then for arbitrary fixed x, $f(x, y)$ is a continuous function of y for $(x, y) \in D$. It should be noted that the Lipschitz constant M depends on both the function f and the domain D. That is, as D varies, the same function f may satisfy a Lipschitz condition with respect to y on D for different Lipschitz constants, or f may not satisfy a Lipschitz condition with respect to y on D at all. For example, $f(x, y) = 3y^{2/3}$ satisfies a Lipschitz condition with respect to y on any domain $D = \{(x, y) \mid |y| > \epsilon > 0\}$ and an appropriate Lipschitz constant is $M = 2\epsilon^{-1/3}$. The function f does not satisfy a Lipschitz condition with respect to y on any domain that contains any point of the line $y = 0$.

The following theorem, which concerns the existence and uniqueness of solutions of the IVP (1), is sometimes called the *Fundamental Existence (and Uniqueness) Theorem of Ordinary Differential Equations* or *Picard's Theorem.* The latter name being assigned to the theorem in honor of the French mathematician Émile Picard (1856–1941), whose method of successive approximations is employed in the proof of the theorem.

Theorem 4.2 Let $f(x, y)$ be continuous on some bounded domain D of the xy-plane and satisfy a Lipschitz condition with respect to y in D. Let $(x_0, c_0) \in D$. Then there exists a unique solution $\bar{y}$ to the IVP (1) on some closed interval I with center x_0.

Sketch of Proof It follows from Theorem 4.1 that since f is assumed to be continuous on D, the initial value problem

$$y' = f(x, y); \qquad y(x_0) = c_0 \tag{1}$$

is equivalent to the integral equation

$$y(x) = c_0 + \int_{x_0}^{x} f(t, y(t))\, dt \tag{2}$$

as long as $(x, y(x))$ remains in D. Based on topological concepts, the function f, and the domain D, one shows how to choose a closed interval I with center x_0 so that $(x, y(x))$ remains in D for $x \in I$. (If, for example, the rectangle $R = \{(x, y) \mid |x - x_0| \leq a \text{ and } |y - y_0| \leq b\}$ is a subset of D and if f is bounded by K on R, then I contains $J = [x_0 - \alpha, x_0 + \alpha]$ where $\alpha = \min(a, b/K)$.) After showing how to choose I, one then shows that the IE (2) and consequently the IVP (1) has a unique solution for $x \in I$. This requires using Picard's method of successive approximations. A description of this technique follows.

For $x \in I$ one recursively defines the sequence of function $\langle y_n(x) \rangle_{n=0}^{\infty}$ as follows:

$$y_0(x) = c_0$$

$$y_n(x) = c_0 + \int_{x_0}^{x} f(t, y_{n-1}(t))\, dt, \qquad n = 1, 2, \ldots. \tag{4}$$

In order to theoretically produce the unique solution of the IVP (1), one shows that (i) the sequence of functions $\langle y_n \rangle$ is well-defined on I, (ii) the sequence of functions $\langle y_n \rangle$ converges uniformly to a function $\bar{y}(x)$ which is defined and continuous on I, (iii) the function $\bar{y}(x)$ satisfies the IE (2) and consequently the IVP (1) on I, and (iv) the solution $\bar{y}(x)$ of the IVP (1) is unique on I. Steps (ii) and (iv) require the Lipschitz condition hypothesis. See Greenspan [7] for the complete details of this proof.

The proof of Theorem 4.2 is constructive in the sense that an algorithm is defined within the proof of the theorem which, at least theoretically, produces the entity described in the conclusion of the theorem—the solution of the initial value problem, in this case. Although the construction of the solution to an initial value problem of the form (1) which satisfies the hypotheses of Theorem 4.2 can theoretically be accomplished by Picard's method of successive approximations, the actual construction of such a sequence of functions, $\langle y_n \rangle_{n=0}^{\infty}$, is seldom practical. Theorem 4.2 is called the Fundamental Existence and Uniqueness Theorem; however, it is not the "best" theorem of this kind. The first example of this section and the following remark indicate that some condition in addition to f being continuous on D is required for a solution of the IVP (1) to be unique. Theorem 4.2 has most probably been given its name because the conditions specified in the hypotheses are easily verified in any particular instance and in recognition of Picard's contribution to the theory of differential equations.

EXAMPLE Solve the initial value problem $y' = x + y = f(x, y)$; $y(0) = 1$ using Picard's method of successive approximations.

First, we set $y_0(x) = c_0 = 1$ and then compute:

$$y_1(x) = c_0 + \int_{x_0}^{x} f(t, y_0(t))\,dt = 1 + \int_0^x (t + 1)\,dt = 1 + x^2/2 + x$$

$$y_2(x) = c_0 + \int_{x_0}^{x} f(t, y_1(t))\,dt = 1 + \int_0^x (t + 1 + t^2/2 + t)\,dt$$

$$= 1 + \frac{x^2}{2} + x + \frac{x^3}{3!} + \frac{x^2}{2}$$

$$y_3(x) = 1 + \int_{x_0}^{x} \left(t + 1 + \frac{t^2}{2} + t + \frac{t^3}{3!} + \frac{t^2}{2}\right) dt$$

$$= 1 + \frac{x^2}{2} + x + \frac{x^3}{3!} + \frac{x^2}{2} + \frac{x^4}{4!} + \frac{x^3}{3!}$$

$$= \left(1 + x + \frac{x^2}{2} + \frac{x^3}{3!}\right) + \left(1 + x + \frac{x^2}{2} + \frac{x^3}{3!} + \frac{x^4}{4!}\right) - 1 - x.$$

In general, we see that

$$y_n(x) = \left(1 + x + \cdots + \frac{x^n}{n!}\right) + \left[1 + x + \cdots + \frac{x^{n+1}}{(n+1)!}\right] - 1 - x.$$

Since

$$e^x = 1 + x + \frac{x^2}{2!} + \cdots = \sum_{n=0}^{\infty} \frac{x^n}{n!}.$$

We conclude that the sequence $\langle y_n \rangle_{n=0}^{\infty}$ converges to

$$\bar{y}(x) = 2e^x - 1 - x.$$

The reader should check to see that $\bar{y}$ indeed satisfies the given initial value problem.

A set S of the xy-plane is *convex* if the line segment joining any two points of S lies entirely in S. The set $A = \{(x, y) \mid x^2 + y^2 < 1\}$ which is the interior of the unit circle with center at the origin is convex. However, the set $B = \{(x, y) \mid 0 < x^2 + y^2 < 1\}$ which is the set obtained by removing the origin—the point $(0, 0)$—from the set A is not convex. Consider the line segment joining the points $(-\frac{1}{2}, 0) \in B$ and $(\frac{1}{2}, 0) \in B$. This line segment is a segment of the x-axis and contains the point $(0, 0)$. But $(0, 0) \notin B$, so the set B is not convex.

A point p is a *boundary point* of a domain D of the xy-plane if every circle about p contains both points in D and points not in D. The set of all boundary points of a domain D is called the *boundary of* D and often denoted by $\mathscr{B}(D)$. The *closure* of a domain D is denoted by $\bar{D}$ and is the union of the set D and its boundary. That is, $\bar{D} = D \cup \mathscr{B}(D)$. The boundary of the set A is

$$\mathscr{B}(A) = \{(x, y) \mid x^2 + y^2 = 1\}.$$

The boundary of the set B is

$$\mathscr{B}(B) = \{(x, y) \mid x^2 + y^2 = 1\} \cup \{(0, 0)\}.$$

The closure of both the domain A and the domain B is the set

$$\bar{A} = A \cup \mathscr{B}(A) = \bar{B} = B \cup \mathscr{B}(B) = \{(x, y) \mid x^2 + y^2 \leq 1\}.$$

A set of real numbers is *bounded* if there exists a positive constant M such that $|s| \leq M$ for all $s \in S$. The intervals (a, b), $[a, b)$, $(a, b]$, and $[a, b]$, where a and b are real numbers are bounded sets. The intervals $(-\infty, b)$, $(-\infty, b]$, (a, ∞), and $[a, \infty)$ are examples of sets that are not bounded. A real number u is an *upper bound* of a set of real numbers S iff $u \geq s$ for all $s \in S$. All bounded sets have an upper bound. However, not all sets that have an upper bound are bounded. For instance, the interval $(-\infty, 3)$ has an upper bound of 4 (and 3, and $3\frac{1}{2}$, and $\pi, \ldots$), but the interval is not bounded. A real number l is the *supremum* (*least upper bound*) of a set of real numbers S (written as $l = \sup S$) if (i) l is an upper bound of S and (ii) if u is an upper bound of S, then $l \leq u$. For $S = (-\infty, 3)$, $\sup S = 3$.

We shall be interested in determining values for expressions such as $\sup_D |x + y|$ where $D = \{(x, y) \mid x^2 + y^2 < 4\}$. The notation $l = \sup_D |x + y|$

means that $l = \sup\{|x + y| \mid (x, y) \in D\}$. In order to determine $l = \sup_D |x + y|$ graphically, we draw the domain D—a circle of radius 2 with center at the origin. Then we graph a few members of the one-parameter family of curves $|x + y| = C$. In this case the members are pairs of straight lines when $C \neq 0$ and a single straight line when $C = 0$. The largest value of C for which $(x, y) \in D$ or for which "(x, y) touches the boundary of D" and $|x + y| = C$ is the $\sup_D |x + y|$. From Figure 4.2 we see that $l = \sup_D = |x + y| = 2\sqrt{2}$.

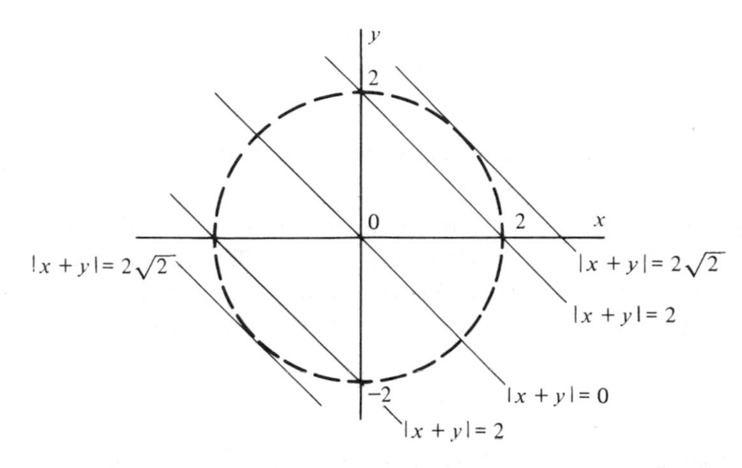

Figure 4.2 Graphical determination of $\sup_D |x + y|$.

EXAMPLE Let $f(x, y) = x^2 + xy^2$. Find $\sup_D |\partial f/\partial y|$, where $D = \{(x, y) \mid |x| < 2$ and $|y| < 2\}$.

Differentiating f with respect to y, we find that $\partial f/\partial y = 2xy$. The graph of D is a square with sides of length 4 and center at the origin. The graph of members of the one-parameter family of curves $|2xy| = C$ are the coordinate axes for $C = 0$ and are the four branches of two conjugate hyperbolas for $C \neq 0$. The graph of D and a few members of the family of curves $|2xy| = C$ are shown in Figure 4.3. From this figure we see that $\sup_D |\partial f/\partial y| = \sup_D |2xy| = 8$.

Theorem 4.3 Let D be a bounded convex domain and let $f(x, y)$ and $\partial f/\partial y$ be continuous on the closure of D. Then f satisfies a Lipschitz condition in y on D and a suitable Lipschitz constant is $M = \sup_D |\partial f/\partial y|$.

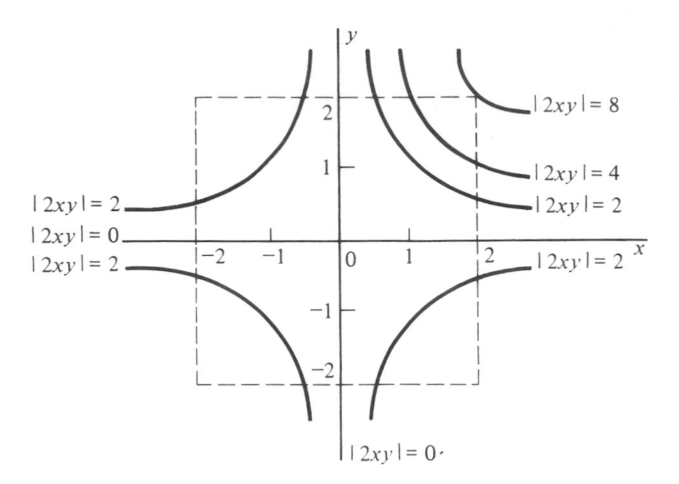

Figure 4.3 Graphical determination of $\sup_D |2xy|$.

Proof Let (x, y_1), $(x, y_2) \in D$. Since D is convex, the line segment joining these two points lies entirely in D. The Mean Value Theorem for Derivatives applies to $f(x, \eta)$ considered as a function of η on this line segment. Hence, there exists an $\bar{\eta}$ between y_1 and y_2 such that

$$|f(x, y_1) - f(x, y_2)| = \left|\frac{\partial f(x, \bar{\eta})}{\partial y}\right| |y_1 - y_2| \leq \sup_D \left|\frac{\partial f}{\partial y}\right| |y_1 - y_2|.$$

Theorem 4.4 If $f(x, y)$ and $\partial f/\partial y$ are continuous on the closure of D, where D is a bounded domain, then the initial value problem $y' = f(x, y)$; $y(x_0) = c_0$ with $(x_0, c_0) \in D$ has a unique solution on some interval I with center x_0. This theorem follows from Theorems 4.2 and 4.3.

It is generally easier to verify that a function f has a continuous first partial derivative with respect to y than to show that f satisfies a Lipschitz condition with respect to y. So in considering the IVP (1) one first determines the domain D on which f and $\partial f/\partial y$ are continuous. If (x_0, c_0) is in this domain, then one seeks, by various means, the solution that will be unique as long as it remains in D. If $f_y(x_0, c_0)$ does not exist, then one checks to see if f satisfies a Lipschitz condition with respect to y in a neighborhood of (x_0, c_0). If so, then one proceeds to find the solution to the initial value problem in this neighborhood. If not, then one is forewarned that any solution to the initial value problem that is produced by any means may not be unique.

Since Theorem 4.2 guarantees the existence and uniqueness of the solution $\bar{y}$ of the IVP (1) only on a "small" interval I with center x_0, Theorem 4.2 is referred to as a local existence theorem. Suppose the hypotheses of Theorem 4.2 hold. A solution $\bar{y}$ exists and is unique on some interval $I = [\alpha, \beta]$ with center x_0. This solution may be extended in the following manner. Since the points $(\alpha, y(\alpha))$ and $(\beta, y(\beta))$ are in D, there exist unique solutions y_α and y_β of the IVP (1) through these points, by Theorem 4.2. Since each of these solutions is unique on some closed interval I_α and I_β with centers α and β, respectively, and since neither $J_\alpha = I \cap I_\alpha$ nor $J_\beta = I \cap I_\beta$ is empty, y_α and y_β equal $\bar{y}$ on J_α and J_β, respectively. So y_α and y_β extend $\bar{y}$ to the left and right. The solution may be continued in this fashion until the solution approaches the boundary of D. See Figure 4.4.

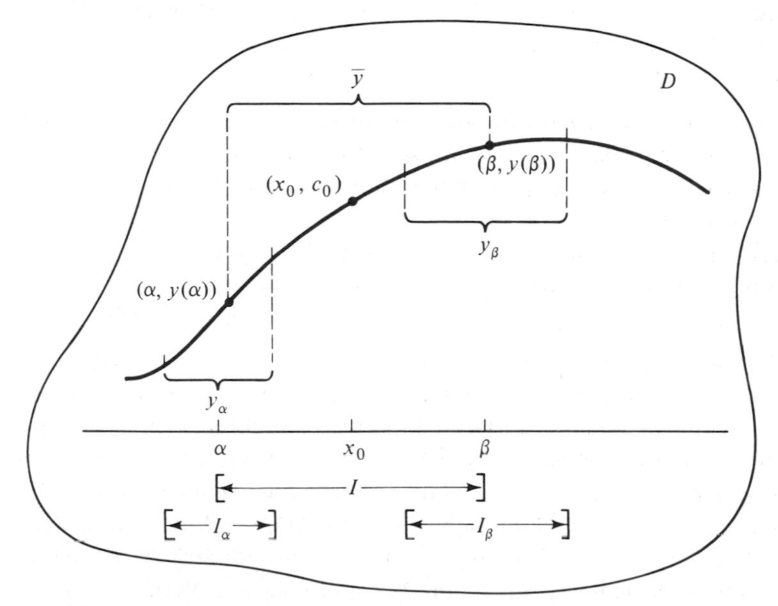

Figure 4.4 Continuation of a solution.

We state the following continuation theorem without proof.

Theorem 4.5 Under the hypotheses of Theorem 4.2, the solution $\bar{y}$ of the IVP (1) can be extended uniquely until the boundary of D is reached.

EXERCISES

1. Let $D_1 = \{(x, y) \mid |x| < 2 \text{ and } |y| < 2\}$,
$D_2 = \{(x, y) \mid x^2 + y^2 < 4\}$,
$D_3 = \{(x, y) \mid 1 < x^2 + y^2 < 4\}$,
$D_4 = \{(x, y) \mid y > \epsilon > 0\}$

Find:

(a) $\sup_{D_1} |x + y|$ (b) $\sup_{D_2} |2xy|$

(c) $\sup_{D_3} |1/(x^2 + y^2)|$ (d) $\sup_{D_4} |1/(x^2 + y^2)|$

(e) $\sup_{D_4} |1/y|$ (f) $\sup_{D_4} |1/\sqrt{y}|$

2. In what domain D does each of the following functions satisfy a Lipschitz condition with respect to y? And what is a suitable Lipschitz constant M for the domain?
(a) $f(x, y) = x$ (b) $f(x, y) = g(x)$
(c) $f(x, y) = x + y$ (d) $f(x, y) = g(x) + y$
(e) $f(x, y) = xy$ (f) $f(x, y) = y/x$
(g) $f(x, y) = yg(x)$ (h) $f(x, y) = \sin xy$
(i) $f(x, y) = y^2$ (j) $f(x, y) = y^{1/2}$
(k) $f(x, y) = 1/xy$ (l) $f(x, y) = (xy)^{1/2}$

3. For each of the following differential equations, determine the domain D of the xy-plane for which the fundamental theorem guarantees the existence and uniqueness of a solution passing through any specified point in D.
(a) $y' = 2xy + 1$ (b) $y' = y/x + \tan x$
(c) $y' = a(x)y + b(x)$ (d) $y' = y^2$
(e) $y' = y^{2/3}$ (f) $y' = xy/(x^2 + y^2)$

4. Consider the initial value problem $y' = y^2$; $y(x_0) = c_0$, where (x_0, c_0) is any point of the xy-plane. Since $f(x, y) = y^2$ is continuous on the entire xy-plane and since f satisfies a Lipschitz condition with respect to y on any bounded domain D of the xy-plane, Theorem 4.2 guarantees the existence of a unique solution of the initial value problem through (x_0, c_0) in some neighborhood of x_0. It is tempting to conclude that since the conditions of Theorem 4.2 are satisfied on the entire plane, the solution through (x_0, c_0) should be valid for all x. Verify that $y = -1/x$ is a solution of the IVP $y' = y^2$; $y(-1) = 1$. On what interval is the solution valid? Verify that $y = 0$ is a solution of the IVP $y' = y^2$; $y(-1) = 0$. Where is the solution valid? In general, verify that $y = -1/(x - x_0 - 1/c_0)$ is a solution to the IVP $y' = y^2$; $y(x_0) = c_0 \neq 0$. On what interval is the solution valid?

5. Show that $y = \tan x$ is the solution of the initial value problem $y' = 1 + y^2$; $y(0) = 0$. On what interval is the solution valid?

6. Use Picard's method of successive approximations to calculate $y_0(x)$, $y_1(x)$, $y_2(x)$, and $y_3(x)$ for the following initial value problems. See if you can determine the function to which the sequence $\langle y_n \rangle$ converges in each case.
(a) $y' = x$; $y(0) = 1$ (b) $y' = y$; $y(0) = 1$
(c) $y' = 2xy + 1$; $y(1) = -1$ (d) $y' = 1 + y^2$; $y(0) = 0$

4.2 COMPARISON THEOREMS AND CONTINUOUS DEPENDENCE OF SOLUTIONS ON x_0, c_0, f

In this section we wish to examine the effect of some "small" change in the initial value x_0 or c_0 or in the function $f(x)$ on the solution of the initial value problem

$$(5) \qquad y' = f(x); \qquad y(x_0) = c_0.$$

And we also want to examine the effect of some "small" change in the initial value x_0 or c_0 or in the function $f(x, y)$ on the solution of the initial value problem

$$(1) \qquad y' = f(x, y); \qquad y(x_0) = c_0.$$

The types of theorems that we shall prove and utilize in this section are known as *comparison theorems*. Sometimes these theorems are also called *continuity theorems* since an immediate consequence of such theorems is the continuous dependence of the solution on the initial values x_0 and c_0 and the function f.

4.2.1 $y' = f(x)$; $y(x_0) = c_0$

Suppose that we perform an experiment in which the independent variable x represents time and in which we observe some dependent variable y. Assume that during the course of the experiment we are able to determine that the rate of change of the variable y with respect to time is

$$(6) \qquad y' = 3x^{-.5}$$

and at 1.21 units of time after the start of the experiment, y has the value 5. We would like to be able to accurately predict y at any time $x > 0$ after the start of the experiment. To do this, we must solve the initial value problem

$$(7) \qquad y' = 3x^{-.5}; \qquad y(1.21) = 5,$$

which is of the form (5). In this instance $f(x) = 3x^{-.5}$, $x_0 = 1.21$, and $c_0 = 5$. Writing the IVP (7) as an integral equation, we easily find the solution to be

$$(8) \qquad y(x) = 5 + \int_{1.21}^{x} 3t^{-.5}\,dt = 5 + [6t^{.5}]|_{1.21}^{x}$$
$$= 5 + 6x^{.5} - 6(1.21)^{.5} = 6x^{.5} - 1.6.$$

So equation (8) is the equation for predicting y at any time $x > 0$.

Now suppose that at times $x = 4$ we measure y and find the measured value to be $y_m(4) = 9.8$. Substituting $x = 4$ into equation (8), we find the predicted value of y to be $y(4) = 10.4$. Notice that $y_m(4) = 9.8 \neq 10.4 =$

$y(4)$. What could be wrong? Either (i) we measured the value of y improperly at $x = 1.21$ or at $x = 4$, or (ii) the time 1.21 or 4 was read improperly, or (iii) the function $f(x) = 3x^{-.5}$ does not truly represent the rate of change of the variable y with respect to time, x, or (iv) most likely, a combination of all these factors.

If, for instance, only the initial time $x_0 = 1.21$ were improperly read and if the proper initial time were $x_1 = 1.44$, then we would need to solve the initial value problem

$$y' = 3x^{-.5}; \qquad y(1.44) = 5 \tag{9}$$

instead of the IVP (7). The solution of (9) is

$$\begin{aligned} y(x) &= 5 + \int_{1.44}^{x} 3t^{-.5}\,dt = 5 + [6t^{.5}]\Big|_{1.44}^{x} \\ &= 5 + 6x^{.5} - 6(1.44)^{.5} = 6x^{.5} - 2.2. \end{aligned} \tag{10}$$

And $y(4) = 9.8$, which is the value that was measured at $x = 4$. A graph of portions of equations (8) and (10) is shown in Figure 4.5. Observe that if z were allowed to vary from $x_0 = 1.21$ to $x_1 = 1.44$, then the solution of the initial value problem

$$y' = 3x^{-.5}; \qquad y(z) = 5, \tag{11}$$

which is

$$\begin{aligned} y(x) &= 5 + \int_{z}^{x} 3t^{-.5}\,dt = 5 + [6t^{.5}]\Big|_{z}^{x} \\ &= 5 + 6x^{.5} - 6z^{.5}, \end{aligned} \tag{12}$$

would vary from equation (8) to equation (10). Observe that the solution (12) of the IVP (11) varies continuously with respect to the initial value z, since the function $y(x)$ of equation (12) is a continuous function of z.

If, on the other hand, only the initial value $c_0 = 5$ were improperly measured and if the proper initial value were $c_1 = 4.4$, then we would need to solve the initial value problem

$$y' = 3x^{-.5}; \qquad y(1.21) = 4.4 \tag{13}$$

instead of the IVP (7). The solution of (13) is

$$\begin{aligned} y(x) &= 4.4 + \int_{1.21}^{x} 3t^{-.5}\,dt = 4.4 + [6t^{.5}]\Big|_{1.21}^{x} \\ &= 4.4 + 6x^{.5} - 6(1.21)^{.5} = 6x^{.5} - 2.2. \end{aligned} \tag{14}$$

Notice that equation (14) is identical to equation (10). So again $y(4) = 9.8$, as required. A graph of equations (8) and (14) for $x \in [1.21, 4]$ is shown in Figure 4.6. Observe that if z were allowed to vary from $c_0 = 5$ to $c_1 = 4.4$ in a continuous manner, then the solution of the initial value problem

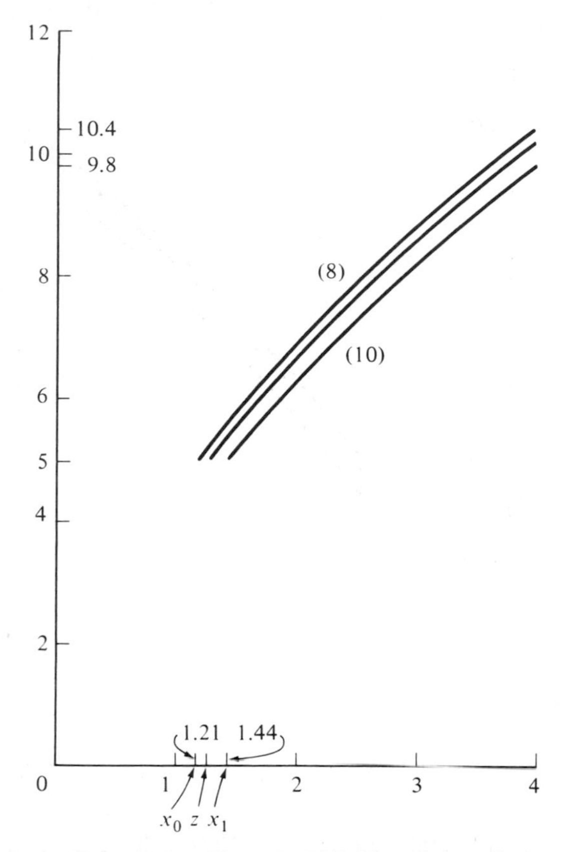

Figure 4.5 The solution (8) to the IVP (7) and the solution (10) to the IVP (9).

(15) $$y' = 3x^{-.5}; \qquad y(1.21) = z,$$

which is

(16) $$y(x) = z + \int_{1.21}^{x} 3t^{-.5}\, dt = 6x^{.5} + z - 6.6,$$

would vary from equation (8) to equation (14) in a continuous manner, since $y(x)$ of equation (16) is a continuous function of z.

Let us consider again the general initial value problem (5) $y' = f(x)$; $y(x_0) = c_0$. Recall that the solution of (5) written as an integral equation is

$$y(x) = c_0 + \int_{x_0}^{x} f(t)\, dt.$$

Since the function $y(x)$ is a continuous function of c_0 and a continuous function of x_0 provided that the integral exists, the IVP (5) depends continuously on the initial values c_0 and x_0.

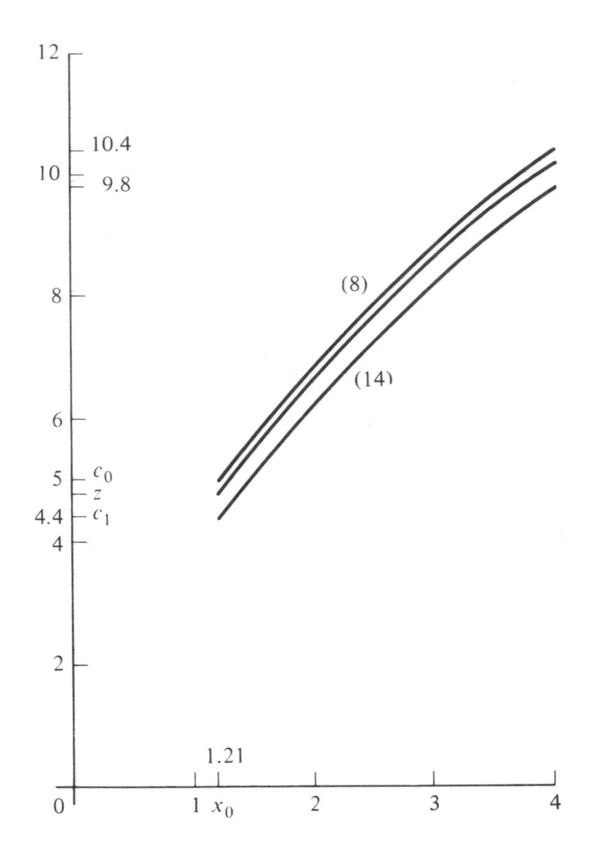

Figure 4.6 The solution (8) to the IVP (7) and the solution (14) to the IVP (13).

Returning now to the specific example under consideration, suppose that the initial values—$x_0 = 1.21$ and $c_0 = 5$—are correct and that $y = 9.8$ when $x = 4$. In this case the function $f(x) = 3x^{-.5}$ cannot truly represent the rate of change of the variable y with respect to time. Let us suppose that we do know from physical considerations that the function $f(x)$ has the form $3x^z$, where z is a constant. Our problem then becomes one of determining the constant z so that the solution to the initial value problem

(17) $$y' = 3x^z; \qquad y(1.21) = 5,$$

which is

(18) $$y(x) = 5 + \int_{1.21}^{x} 3t^z\,dt = 5 + \left[\frac{3t^{z+1}}{z+1}\right]\Bigg|_{1.21}^{x}$$

$$= 5 + \frac{3x^{z+1}}{z+1} - \frac{3(1.21)^{z+1}}{z+1},$$

satisfies $y(4) = 9.8$. That is, z must be chosen to satisfy

$$y(4) = 9.8 = 5 + \frac{3 \cdot 4^{z+1}}{z+1} - \frac{3(1.21)^{z+1}}{z+1}$$

or, equivalently,

$$4.8 = \frac{3}{z+1}(4^{z+1} - 1.21^{z+1}).$$

Solving this equation for z by trial and error, we find that $z \doteq -.640\ 312$, where "$\doteq$" denotes the phase "is approximately equal to." A graph of equation (8) and (18) with $z = -.640\ 312$ and $x \in [1.21, 4]$ is shown in Figure 4.7. Notice that since equation (18) is a continuous function of z, as z varies in a continuous manner from $-.5$ to $-.640\ 312$, equation (18) varies in a continuous manner from equation (8) to equation (18) with $z = -.640\ 312$.

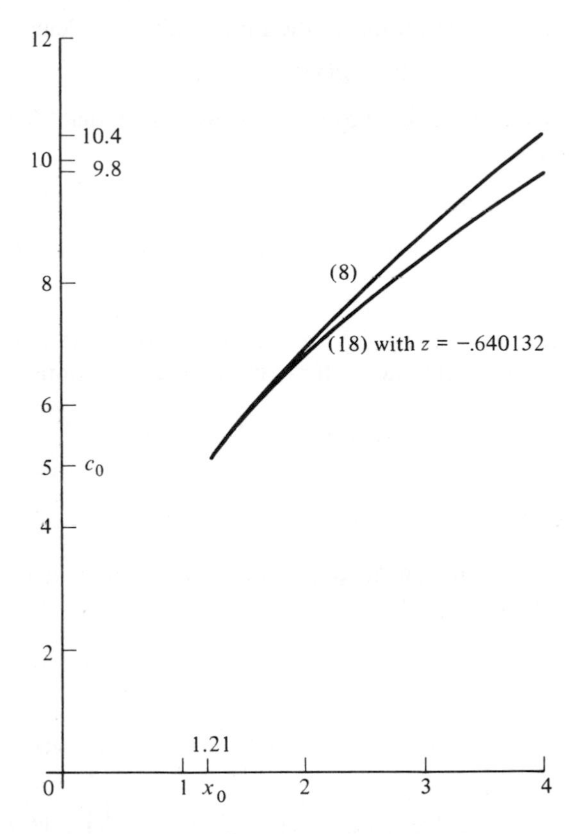

Figure 4.7 The solution (8) to the IVP (7) and the solution (18) to the IVP (17).

When we use an initial value problem of the form (5) $y' = f(x)$; $y(x_0) = c_0$ to model a physical situation, it is very likely that the initial values x_0 and c_0 will have an error associated with them since no measurement is exact, and that the function f will not be the true rate of change of y with respect to x. We have already stated that the solution of the IVP (5) depends continuously on x_0, c_0, and f—which means that a "small" change in any of these quantities results in a corresponding "small" change in the solution. The following theorem provides a relationship between a "small" change in any of the quantities c_0, x_0, and f and the corresponding "small" change that occurs at x in the solution of the IVP (5).

Theorem 4.6 Let $f(x)$ and $g(x)$ be continuous functions defined on some interval I which contains the points x_0 and x_1. Let $y_0(x)$ be the solution of the initial value problem

$$y' = f(x); \qquad y(x_0) = c_0 \tag{19}$$

and let $y_1(x)$ be the solution of the initial value problem

$$y' = g(x); \qquad y(x_1) = c_1. \tag{20}$$

Suppose that there exists a positive constant ϵ such that $|f(x) - g(x)| \leq \epsilon$ for $x \in I$. Then for $x \in I$,

$$|y_1(x) - y_0(x)| \leq |c_1 - c_0| + K|x_1 - x_0| + \epsilon|x - x_0|, \tag{21}$$

where $K = \max_{x \in J} |g(x)|$ and J is the closed interval with endpoints x_0 and x_1.

Proof Since $y_0(x)$ and $y_1(x)$ are the solutions of the IVP (19) and the IVP (20), respectively, they satisfy the integral equations

$$y_0(x) = c_0 + \int_{x_0}^{x} f(t)\,dt$$

$$y_1(x) = c_1 + \int_{x_1}^{x} g(t)\,dt.$$

Subtracting, taking the absolute value, and applying the triangle inequality, we find

$$\begin{aligned} |y_1(x) - y_0(x)| &\leq |c_1 - c_0| + \left| \int_{x_1}^{x} g(t)\,dt - \int_{x_0}^{x} f(t)\,dt \right| \\ &= |c_1 - c_0| + \left| -\int_{x_0}^{x_1} g(t)\,dt + \int_{x_0}^{x} [g(t) - f(t)]\,dt \right| \\ &\leq |c_1 - c_0| + \left| \int_{x_0}^{x_1} |g(t)|\,dt \right| + \left| \int_{x_0}^{x} |g(t) - f(t)|\,dt \right| \\ &\leq |c_1 - c_0| + K|x_1 - x_0| + \epsilon|x - x_0|. \end{aligned}$$

Theorem 4.6 provides an upper bound for the absolute value of the difference (or distance) between the solution y_0 of the IVP (19) and the solution y_1 of the IVP (20) at any point x in the interval I. If $x_0 = x_1$ and $f(x) = g(x)$, then the distance between the solutions remains a constant for all $x \in I$, namely, $|y_1(x) - y_0(x)| = |c_1 - c_0|$. If $c_0 = c_1$ and $f(x) = g(x)$, then the distance between the solutions is again a constant for all $x \in I$, namely,

$$|y_1(x) - y_0(x)| = \left| \int_{x_1}^{x_0} f(t)\, dt \right|.$$

If $x_0 = x_1$ and $c_0 = c_1$, then the distance between the solutions is

$$|y_1(x) - y_0(x)| = \left| \int_{x_0}^{x} [g(t) - f(t)]\, dt \right|,$$

which is not constant but a continuous function of x.

EXAMPLE Find an upper bound for the distance between the solution $y_0(x)$ of the initial value problem

$$y' = 2e^x; \qquad y(0) = 1 \tag{22}$$

and the solution $y_1(x)$ of the initial value problem

$$y' = 2e^{x^2}; \qquad y(.1) = 1.2 \tag{23}$$

on the interval [0, 1].

In this case $x_0 = 0$, $c_0 = 1$, $f(x) = 2e^x$, $x_1 = .1$, $c_1 = 1.2$, and $g(x) = 2e^{x^2}$. So

$$K = \max_{x \in [0,.1]} |2e^{x^2}| = 2e^{(.1)^2} \doteq 2.020\ 10.$$

and

$$|f(x) - g(x)| = |2e^x - 2e^{x^2}| = 2|e^x - e^{x^2}|.$$

Let $w(x) = e^x - e^{x^2}$. Differentiating and setting $w'(x) = 0$, we find that the maximum value of the function $w(x)$ on the interval [0, 1] occurs at the value of x which satisfies the equation

$$w'(x) = e^x - 2xe^{x^2} = 0. \tag{24}$$

Examining a table of values for e^x and e^{x^2}, we see that the maximum difference between these functions on the interval [0, 1] occurs between .6 and .7. Starting with $x = .6$ and solving equation (24) by trial and error, we find that the maximum of $w(x)$ occurs at $x \doteq .631\ 076$. So for $x \in [0, 1]$,

$$|f(x) - g(x)| = 2|e^x - e^{x^2}| \leq 2|e^{.631\ 076} - e^{(.631\ 076)^2}| \doteq .780\ 811 = \epsilon.$$

Applying Theorem 4.6, we find, for $x \in [0, 1]$, that the distance between the solutions y_0 and y_1 satisfies

$$(25) \qquad \begin{aligned} |y_1(x) - y_0(x)| &\leq |1.2 - 1| + 2.020\ 10|.1 - 0| + .780\ 811|x - 0| \\ &\leq .402\ 010 + .780\ 811x. \end{aligned}$$

The solution of the IVP (22) is

$$(26) \qquad y_0(x) = 1 + \int_0^x 2e^t\, dt = 2e^x - 1$$

and the solution of the IVP (23) is

$$y_1(x) = 1.2 + \int_{.1}^x 2e^{t^2}\, dt.$$

Since there is no elementary function $h(t)$ such that $h'(t) = e^{t^2}$, we cannot express the function $y_1(x)$ as we did the function $y_0(x)$. However, solving the inequality (25) for $y_1(x)$, we find, for $x \in [0, 1]$, that

$$(27) \qquad y_0(x) - .402\ 010 - .780\ 811x \leq y_1(x) \leq y_0(x) + .402\ 010 + .780\ 811x.$$

Substituting (26) into (27), we find that, for $x \in [0, 1]$, $y_1(x)$ satisfies

$$(28) \qquad \begin{aligned} y_2(x) &= 2e^x - 1.402\ 010 - .780\ 811x \\ &\leq y_1(x) \leq 2e^x - .597\ 99 + .780\ 811x = y_3(x). \end{aligned}$$

A graph of $y_0(x)$, $y_2(x)$, and $y_3(x)$ is shown in Figure 4.8. The solution $y_1(x)$ of the IVP (23) lies between the curves $y_2(x)$ and $y_3(x)$.

EXERCISES

1. Find the distance between the solution of the initial value problem

$$y' = 3x^2 + 2; \qquad y(0) = 1$$

and the solution of the initial value problem

$$y' = 3x^2 + 2; \qquad y(0) = 1.5.$$

2. Find the distance between the solution of the initial value problem

$$y' = 3x^2 + 2; \qquad y(0) = 1$$

and the solution of the initial value problem

$$y' = 3x^2 + 2; \qquad y(1) = 1.$$

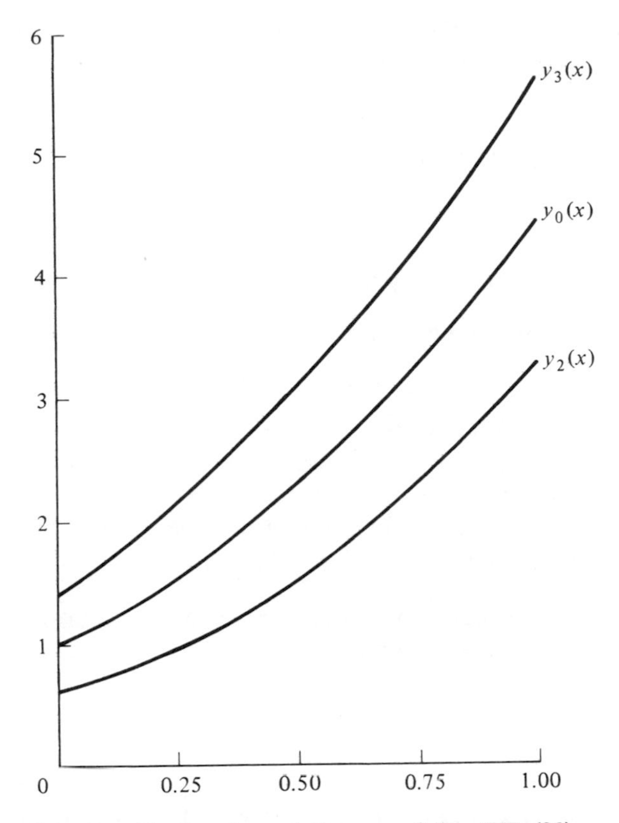

Figure 4.8 Bounds for the solution y_1 of the IVP (23) on the interval [0, 1].

3. Find an upper bound for the distance between the solution of the initial value problem

$$y' = 3x^2 + 2; \qquad y(0) = 1$$

and the solution of the initial value problem

$$y' = 3x^2 + 2; \qquad y(1) = 0.$$

4. Find an expression that is valid on the interval [0, 1] for an upper bound for the distance between the solution of the initial value problem

$$y' = 3x^2 + 2; \qquad y(0) = 1$$

and the solution of the initial value problem

$$y' = 4x^3 + 2; \qquad y(0) = 1.5.$$

What is an upper bound for the distance between solutions at $x = \frac{1}{4}$? $x = \frac{1}{2}$? $x = \frac{3}{4}$?

5. Find an expression that is valid on the interval $[0, \pi/2]$ for an upper bound for the distance between the solution of the initial value problem

$$y' = \sin x; \qquad y(0) = .5$$

and the solution of the initial value problem

$$y' = x \sin x; \qquad y(\pi/6) = .75.$$

4.2.2 $y' = f(x, y); y(x_0) = c_0$

In this section we prove a comparison theorem for the initial value problem

$$y' = f(x, y); \qquad y(x_0) = c_0. \tag{1}$$

First we prove a version of Gronwall's lemma, which is used within the proof of the comparison theorem. Some readers may wish to forego reading the proof of Gronwall's lemma, and others may wish to forego reading the proof of the comparison theorem and be content with employing the results of the theorem without the satisfaction of proof.

Lemma 4.1 Let $d(x)$, $p(x)$, and $q(x)$ be continuous nonnegative functions on an interval I. Let $x_0, x \in I$. If

$$d(x) \le p(x) + \left| \int_{x_0}^{x} q(t)d(t)\, dt \right|, \tag{29}$$

then

$$d(x) \le p(x) + \left| \int_{x_0}^{x} q(t)p(t) \exp\left\{ \left| \int_{t}^{x} q(s)\, ds \right| \right\} dt \right|. \tag{30}$$

Proof Let $V(x) = \int_{x_0}^{x} q(t)d(t)\, dt$. Then $V'(x) = q(x)d(x)$. For $x \ge x_0$, (29) is

$$d(x) \le p(x) + V(x). \tag{31}$$

Multiplying equation (31) by $q(x)$ and subtracting $q(x)V(x)$, we get

$$q(x)d(x) - q(x)V(x) = V'(x) - q(x)V(x) \le q(x)p(x). \tag{32}$$

Multiplying equation (32) by the integrating factor

$$\exp\left\{ -\int_{x_0}^{x} q(s)\, ds \right\},$$

we find

$$\frac{d}{dx}\left[V(x) \exp\left\{ -\int_{x_0}^{x} q(s)\, ds \right\} \right] \le q(x)p(x) \exp\left\{ -\int_{x_0}^{x} q(s)\, ds \right\}. \tag{33}$$

Integration from x_0 to x yields

$$V(x) \exp\left\{-\int_{x_0}^{x} q(s)\, ds\right\} \le \int_{x_0}^{x} q(t)p(t) \exp\left\{-\int_{x_0}^{t} q(s)\, ds\right\} dt. \tag{34}$$

Hence, for $x \ge x_0$,

$$d(x) \le p(x) + V(x) \le p(x) + \int_{x_0}^{x} q(t)p(t) \exp\left\{\int_{t}^{x} q(s)\, ds\right\} dt.$$

For $x < x_0$, (29) is

$$d(x) \le p(x) - V(x). \tag{35}$$

Multiplying equation (35) by $q(x)$ and adding $q(x)V(x)$, we get

$$q(x)d(x) + q(x)V(x) = V'(x) + q(x)V(x) \le q(x)p(x). \tag{36}$$

Multiplying equation (36) by the integrating factor

$$\exp\left\{\int_{x_0}^{x} q(s)\, ds\right\}$$

results in

$$\frac{d}{dx}\left[V(x) \exp\left\{\int_{x_0}^{x} q(s)\, ds\right\}\right] \le q(x)p(x) \exp\left\{\int_{x_0}^{x} q(s)\, ds\right\}. \tag{37}$$

Integration from x to x_0 preserves the order of the inequality and yields

$$-V(x) \exp\left\{\int_{x_0}^{x} q(s)\, ds\right\} \le \int_{x}^{x_0} q(t)p(t) \exp\left\{\int_{x_0}^{t} q(s)\, ds\right\} dt. \tag{38}$$

Thus, for $x < x_0$,

$$d(x) \le p(x) - V(x) \le p(x) + \int_{x}^{x_0} q(t)p(t) \exp\left\{\int_{x}^{t} q(s)\, ds\right\} dt.$$

and the result is proved.

Gronwall's lemma is useful because it allows one to replace an upper bound for the function $d(x)$ that involves $d(x)$ by an upper bound that does not involve $d(x)$. This lemma is often employed in proving uniqueness theorems, in proving continuous dependence of solutions on the initial conditions and the function, and in proving boundedness and stability theorems.

Theorem 4.7 Let $f(x, y)$ be continuous and satisfy a Lipschitz condition with respect to y on some bounded domain D. Let the Lipschitz constant be M. Let $(x_0, c_0) \in D$. By Theorem 4.2, there exists a closed interval I centered about x_0 and a unique function $y_0(x)$ defined on I which satisfies the initial value problem

$$y' = f(x, y); \qquad y(x_0) = c_0. \tag{39}$$

Suppose that $x_1 \in I$, $(x_1, c_1) \in D$, and $g(x, y)$ is continuous on D. Suppose that there exists a positive constant δ such that $|f(x, y) - g(x, y)| \le \delta$ for $(x, y) \in D$. And suppose that for $x \in I$, $y_1(x)$ satisfies the initial value problem

$$y' = g(x, y); \qquad y(x_1) = c_1 \tag{39'}$$

and $(x, y_1(x)) \in D$. Then for $x \in I$,

$$|y_1(x) - y_0(x)| \le (|c_1 - c_0| + K|x_1 - x_0|)e^{M|x - x_0|} + \frac{\delta}{M}(e^{M|x-x_0|} - 1),$$

where $K = \max_{x \in J} |g(x, y_1(x))|$ and J is the interval with endpoints x_0 and x_1.

Proof For $x \in I$, define $d(x) = |y_1(x) - y_0(x)|$. Since y_0 and y_1 satisfy the initial value problems (39) and (39′), respectively, they satisfy the following integral equations for $x \in I$:

$$y_0(x) = c_0 + \int_{x_0}^{x} f(t, y_0(t))\, dt \tag{40}$$

$$y_1(x) = c_1 + \int_{x_1}^{x} g(t, y_1(t))\, dt. \tag{40'}$$

Subtracting, taking the absolute value, and applying the triangle inequality, we find

$$\begin{aligned}
d(x) &\le |c_1 - c_0| + \left| \int_{x_1}^{x} g(t, y_1(t))\, dt - \int_{x_0}^{x} f(t, y_0(t))\, dt \right| \\
&= |c_1 - c_0| + \Bigg| -\int_{x_0}^{x_1} g(t, y_1(t))\, dt + \int_{x_0}^{x} g(t, y_1(t))\, dt \\
&\qquad - \int_{x_0}^{x} f(t, y_1(t))\, dt + \int_{x_0}^{x} f(t, y_1(t))\, dt \\
&\qquad - \int_{x_0}^{x} f(t, y_0(t))\, dt \Bigg| \\
&\le |c_1 - c_0| + \left| \int_{x_0}^{x_1} |g(t, y_1(t))|\, dt \right| + \left| \int_{x_0}^{x} |g(t, y_1(t)) - f(t, y_1(t))|\, dt \right| \\
&\qquad + \left| \int_{x_0}^{x} |f(t, y_1(t)) - f(t, y_0(t))|\, dt \right| \\
&\le |c_1 - c_0| + K|x_1 - x_0| + \delta|x - x_0| + \left| \int_{x_0}^{x} Md(t)\, dt \right|
\end{aligned}$$

since f satisfies a Lipschitz condition with respect to y on D and $(x, y_0(x))$, $(x, y_1(x)) \in D$ for $x \in I$. Letting $p(x) = |c_1 - c_0| + K|x_1 - x_0| + \delta|x - x_0|$ and $q(x) = M$ and invoking Lemma 4.1, we find

$$d(x) \leq |c_1 - c_0| + K|x_1 - x_0| + \delta|x - x_0| + \left|\int_{x_0}^{x} M(|c_1 - c_0| + K|x_1 - x_0| + \delta|x - x_0|)e^{M|x-t|}\,dt\right|.$$

Integration produces the desired result. The actual integration is most easily performed by considering the cases $x \geq x_0$ and $x < x_0$ separately and then combining the results by using absolute values.

Theorem 4.7 provides a proof of the uniqueness portion of Theorem 4.2, since assuming $y_0(x)$ and $y_1(x)$ are distinct solutions of the IVP (1) on I and taking $g(x, y) = f(x, y)$, $x_1 = x_0$, and $c_1 = c_0$ implies that $0 \leq |y_1(x) - y_0(x)| \leq 0$ for all $x \in I$.

EXAMPLE Find an expression that is valid on the interval $[0, 1]$ for an upper bound for the distance between the solution of the initial value problem

$$y' = x \sin y^2; \qquad y(0) = .1 \tag{41}$$

and the solution of the initial value problem

$$y' = xy^2; \qquad y(0) = .1. \tag{42}$$

Calculate a specific upper bound for the distance between the solutions for $x = \frac{1}{2}$ and $x = 1$.

In this case we choose $f(x, y) = xy^2$. Hence, we will let $y_0(x)$ be the solution of the IVP (42) and $y_1(x)$ be the solution of the IVP (41). Making this choice and using the notation convention of Theorem 4.7, we have $x_0 = x_1 = 0$, $c_0 = c_1 = .1$, and $g(x, y) = x \sin y^2$. In order to calculate values for the constants M and δ, which we will need when applying Theorem 4.7 (a value for K will not be needed in this case since $x_0 = x_1$), we must specify a domain D and verify that $(x, y_0(x))$ and $(x, y_1(x)) \in D$ for all $x \in (0, 1)$. Let $D = \{(x, y) \mid 0 < x < 1 \text{ and } 0 < y < .5\}$. For $(x, y) \in D$ we have $0 < g(x, y) = x \sin y^2 < xy^2 = f(x, y) \leq .25$. Thus, $0 < y_1' < y_0' \leq .25$ for $(x, y) \in D$. Hence both y_1 and y_0 are monotone-increasing functions on the interval $(0, 1)$. Integrating, we see that $y_1(x) \leq y_0(x) \leq .25x + c$ for $x \in (0, 1)$. Since $y_1(0) = y_0(0) = .1$, we have $.1 \leq y_1(x) \leq y_0(x) \leq .25x + .1 < .5$ for $x \in (0, 1)$. So $(x, y_0(x)) \in D$ and $(x, y_1(x)) \in D$ for $x \in (0, 1)$. The function $f(x, y) = xy^2$ satisfies a Lipschitz condition on D with Lipschitz constant

$$M = \sup_D |\partial f/\partial y| = \sup_D |2xy| = 1.$$

And for $(x, y) \in D$, we have

$$|f(x, y) - g(x, y)| = |xy^2 - x \sin y^2| \leq |(.5)^2 - \sin (.5)^2| < .0026 = \delta.$$

Applying Theorem 4.7, we find for $x \in [0, 1]$,

$$|y_1(x) - y_0(x)| \leq .0026(e^x - 1).$$

Thus,

$$|y_1(\tfrac{1}{2}) - y_0(\tfrac{1}{2})| \leq .0026(e^{.5} - 1) \doteq .001\ 69$$

and

$$|y_1(1) - y_0(1)| \leq .0026(e - 1) \doteq .004\ 64.$$

From these results we see that on the interval [0, 1] the solution of the IVP (42), which we are able to solve explicitly using techniques presented earlier, is a very good approximation of the solution of the IVP (41), which can only be solved using numerical techniques.

EXAMPLE Find an expression that is valid on the interval [0, 1] for an upper bound for the distance between the solution of the initial value problem

$$(43) \qquad y' = x^3 - y^3; \qquad y(.1) = .15$$

and the solution of the initial value problem

$$(44) \qquad y' = x - y; \qquad y(0) = .1.$$

Let $y_0(x)$ be the solution of the IVP (43) and let $y_1(x)$ be the solution of the IVP (44). According to the notation we have adopted, $f(x, y) = x^3 - y^3$, $x_0 = .1$, $c_0 = .15$, $g(x, y) = x - y$, $x_1 = 0$, and $c_1 = .1$. The functions f and g are both continuous and have a continuous first partial derivative with respect to y on any domain of the xy-plane. So $y_0(x)$ and $y_1(x)$ exist and, as a result of Theorem 4.5, may be continued until they become infinitely large in absolute value. We choose $D = \{(x, y) \mid 0 < x < 1 \text{ and } 0 < y < 1\}$ and let $N = \{(x, y) \mid 0 < x < y < 1\}$ and $P = \{(x, y) \mid 0 < y < x < 1\}$. So N is the subset of the square D above the line $y = x$ and P is the subset of the square D below the line $y = x$. See Figure 4.9. On the set N, $f(x, y) < 0$ and $g(x, y) < 0$. So the solutions $y_0(x)$ and $y_1(x)$ are both monotone-decreasing functions on the set N, since $y_0' < 0$ and $y_1' < 0$ for $(x, y) \in N$. When the solution $y_0(x)$ or $y_1(x)$ reaches the line $y = x$, the slope $y_0'(x) = f(x, y_0(x)) = f(x, x) = 0$ and the slope $y_1'(x) = g(x, y_1(x)) = g(x, x) = 0$. Therefore, the point at which the graph of the solution $y_0(x)$ touches the line $y = x$ is a critical point of the function $y_0(x)$. Likewise, the point at which the graph of the solution $y_1(x)$ touches the line $y = x$ is a critical point of the function $y_1(x)$. Since $f(x, y) > 0$ and $g(x, y) > 0$ for $(x, y) \in P$, both $y_0(x)$ and $y_1(x)$ are monotone increasing on P. So the critical point of $y_0(x)$ and the critical point of $y_1(x)$ which lies in D is a minimum. Thus, both $y_0(x)$ and $y_1(x)$ exist and remain in D for $x \in (0, 1)$. A graph of a function y^* is sketched in Figure

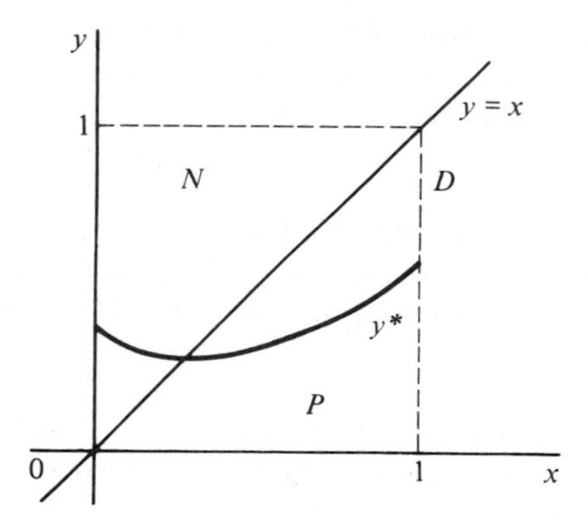

Figure 4.9 Sketch of the solution of the IVP (43) and (44) on D.

4.9 to represent the general appearance of the graph of both $y_0(x)$ and $y_1(x)$ on the domain D.

The constant K required by Theorem 4.7 satisfies

$$K = \max_{x \in [0,.1]} |g(x, y_1(x))| = \max_{x \in [0,.1]} |x - y_1(x)|.$$

We easily find the solution of the linear initial value problem (44) to be

$$y_1(x) = 1.1e^{-x} + x - 1.$$

So

$$K = \max_{x \in [0,.1]} |x - y_1(x)| = \max_{x \in [0,.1]} |1.1e^{-x} - 1| = .1.$$

Taking the partial derivative of f with respect to y, we find

$$M = \sup_D \left|\frac{\partial f}{\partial y}\right| = \sup_D |-3y^2| = 3.$$

And for $(x, y) \in D$, we have

$$\begin{aligned} |f(x, y) - g(x, y)| &= |(x^3 - y^3) - (x - y)| \\ &= |x - y||x^2 + xy + y^2 - 1| \leq 2 = \delta. \end{aligned}$$

Applying Theorem 4.7, we find, for $x \in [0, 1]$,

$$\begin{aligned} |y_1(x) - y_0(x)| &\leq [.05 + .1(.1)]e^{3|x-.1|} + \tfrac{2}{3}(e^{3|x-.1|} - 1) \\ &\doteq .726\ 667e^{3|x-.1|} - .666\ 667. \end{aligned}$$

EXERCISES

1. Suppose that $y_0(x)$ satisfies the initial value problem $y' = 2xy + x^2$; $y(0) = 1$ and $y_1(x)$ satisfies the initial value problem $y' = 2xy + x^2$; $y(0) = 2$. Use Theorem 4.7 to produce an upper bound for the distance between the solutions $y_0(x)$ and $y_1(x)$ when $x = \frac{1}{2}$; when $x = 1$.

2. Suppose that $y_0(x)$ satisfies the initial value problem $y' = 2xy + x^2$; $y(0) = 1$ and $y_1(x)$ satisfies the initial value problem $y' = 2xy$; $y(0) = 1$. Use Theorem 4.7 to produce an upper bound for the distance between the solutions $y_0(x)$ and $y_1(x)$ when $x = \frac{1}{2}$; when $x = 1$.

3. Suppose that $y_0(x)$ satisfies the initial value problem $y' = 2xy$; $y(.1) = 1$. Verify that $y_1(x) = e^{x^2}$ satisfies the initial value problem $y' = 2xy$; $y(0) = 1$. Find an upper bound for the distance between y_0 and y_1 when $x = \frac{1}{2}$; when $x = 1$.

4. Suppose that $y_0(x)$ satisfies the initial value problem $y' = 2xy + x^2$; $y(.1) = 1.1$. Find an expression that is valid on the interval $[0, 1]$ for an upper bound for the distance between $y_0(x)$ and $y_1(x) = e^{x^2}$ which is the solution of the initial value problem $y' = 2xy$; $y(0) = 1$.

4.3 NUMERICAL SOLUTIONS

In this section we will present some of the simpler methods for computing numerical solutions to the initial value problem

$$y' = f(x, y); \qquad y(x_0) = c_0 \tag{1}$$

and for estimating the error of the computed solution. As before, we will denote the exact solution $y(x)$ of the initial value problem at x_n by $y(x_n)$ and the numerical solution at x_n by y_n. All the computational algorithms presented in this section may be performed manually, on a desk calculator, or on an electronic computer. The majority of the computational exercises are designed to be solved using a desk calculator. However, some of the examples presented in the text are solved on the computer using the FORTRAN language. In each such instance the computer program is explained in detail. Students who are familiar with a scientific-oriented programming language such as FORTRAN are encouraged to solve the computational exercises using the computer.

Applying Theorem 4.1, we can symbolically write the solution of the IVP (1) on the interval $[x_n, x_{n+1})$ as

$$y(x_{n+1}) = y(x_n) + \int_{x_n}^{x_{n+1}} f(t, y(t))\, dt. \tag{45}$$

When f was a function of the dependent variable, x, alone—that is, when we were considering the initial value problem $y' = f(x)$; $y(x_0) = c_0$, we approximated f by a step function or polynomial in the dependent variable and integrated in order to obtain an approximate solution y_{n+1}. In this case the value of the approximate solution y_{n+1} depends on the solution, $y(x)$, on the interval $[x_n, x_{n+1})$ and the function f on the set

$$S = \{(x, y) \mid x_n \leq x \leq x_{n+1}, y \in \{y(x) \mid x_n \leq x \leq x_{n+1}\}\}.$$

Thus, we must approximate $y(x)$ on $[x_n, x_{n+1})$ and f on S in order to obtain an approximate solution y_{n+1} at x_{n+1}. If our approximation y_{n+1} depends only on x_n, y_n, and our approximation to f, then the procedure for obtaining a numerical solution is called a *single-step*, *one-step*, *stepwise*, or *starting method.* If the approximation y_{n+1} is based on $x_n, y_n, x_{n-1}, y_{n-1}, \ldots, x_{n-m}, y_{n-m}$, where $m \geq 1$ and an approximation to f, then the procedure for generating a numerical solution is known as a *multistep* or *continuing method.*

4.3.1 Single-Step Methods

For single-step methods it is convenient to symbolize the approximate solution which corresponds to the exact solution (45) by the recursive formula

$$y_{n+1} = y_n + \phi(x_n, y_n, h_n), \tag{46}$$

where $h_n = x_{n+1} - x_n$.

4.3.1.1 Taylor series expansion

If $y(x)$ has $m + 1$ continuous derivatives on an interval I containing x_n, then by *Taylor's Formula with a Remainder*,

$$\begin{aligned} y(x) &= y(x_n) + y'(x_n)(x - x_n) + \frac{y''(x_n)}{2}(x - x_n)^2 + \cdots \\ &\quad + \frac{y^{(m)}(x_n)}{m!}(x - x_n)^m + \frac{y^{(m+1)}(\xi)}{(m+1)!}(x - x_n)^{m+1}, \end{aligned} \tag{47}$$

where ξ is between x and x_n. In particular, if $y(x)$ is a solution to the IVP (1) and $y(x)$ has $m + 1$ continuous derivatives, then

$$\begin{aligned} y'(x_n) &= f(x_n, y(x_n)) \\ y''(x_n) &= f'(x_n, y(x_n)) = f_x + f_y y' = f_x + f_y f \\ y'''(x_n) &= f''(x_n, y(x_n)) = f_{xx} + f_{xy}y' + (f_{yx} + f_{yy}y')y' + f_y y'' \\ &= f_{xx} + 2f_{xy}f + f_{yy}f^2 + f_x f_y + f_y^2 f, \end{aligned}$$

where f and its partial derivatives are all evaluated at x_n, $y(x_n)$. We could continue in this manner and eventually write any derivative of y evaluated at x_n, $y^{(k)}(x_n)$, in terms of f and its partial derivatives evaluated at x_n, $y(x_n)$. However, it is apparent that the evaluation of each successive higher order derivative by this technique usually becomes increasingly difficult unless the function f is very simple. Hence, one chooses m in equation (47) to be reasonably small and approximates $y(x_{n+1})$ by

$$y_{n+1} = y_n + f(x_n, y_n)h_n + \tfrac{1}{2}f'(x_n, y_n)h_n^2 + \cdots + \frac{1}{m!}f^{(m-1)}(x_n, y_n)h_n^m. \tag{48}$$

This single-step method of numerical approximation to the solution is known as the *Taylor series expansion method of order m* and the truncation, or formula, error is given by

$$E_n = \frac{1}{(m+1)!}f^{(m)}(\xi, y(\xi))h_n^{m+1}, \tag{49}$$

where $\xi \in (x_n, x_{n+1})$.

A derivation of the series that bears his name was published by the English mathematician Brook Taylor (1685–1731) in 1715. However, both Gregory and Leibniz knew the series before Taylor, and John Bernoulli had published a similar result in 1694. The series was published without any discussion of convergence and without the truncation error term—equation (49)—being given.

EXAMPLE Find an approximate solution to the initial value problem

$$y' = y + x = f(x, y); \qquad y(0) = 1 \tag{50}$$

on the interval [0, 1] by using a Taylor series expansion of order 3 and a constant stepsize $h_n = .1$ and estimate the maximum truncation error per step on this interval.

We find:

$$\begin{aligned}
y' = f(x, y) &= y + x, & \text{so} \quad f(x_n, y_n) &= y_n + x_n;\\
f'(x, y) &= y' + 1 = y + x + 1, & \text{so} \quad f'(x_n, y_n) &= y_n + x_n + 1;\\
f^{(2)}(x, y) &= y' + 1 = y + x + 1, & \text{so} \quad f^{(2)}(x_n, y_n) &= y_n + x_n + 1;\\
f^{(3)}(x, y) &= y' + 1 = y + x + 1, & \text{so} \quad f^{(3)}(x_n, y_n) &= y_n + x_n + 1.
\end{aligned}$$

So equation (48) becomes

$$\begin{aligned}
y_{n+1} &= y_n + (y_n + x_n)(.1) + \tfrac{1}{2}(y_n + x_n + 1)(.01) + \tfrac{1}{6}(y_n + x_n + 1)(.001)\\
&= .005\ 166\ 67 + .105\ 167x_n + 1.105\ 17y_n.
\end{aligned}$$

In this case each constant was rounded to six significant digits. The third order Taylor series solution to the IVP (50) on the interval [0, 1] which was obtained by performing calculations to six significant digits is shown in Table 4.1.

Table 4.1 Third order Taylor series solution to the IVP (50) with $h = .1$

x_n	y_n	x_n	y_n
.0	1.000 00	.6	2.044 17
.1	1.110 34	.7	2.327 42
.2	1.242 79	.8	2.650 98
.3	1.399 69	.9	3.019 08
.4	1.583 61	1.0	3.436 41
.5	1.797 39		

The differential equation of the IVP (50) is linear, and in this instance we can find the true solution. Therefore, we could use the true solution when estimating the truncation error. However, since we will not normally be able to obtain the true solution, we will do what one must usually do in practice. We will use information obtained from the approximate solution values of Table 4.1 to estimate the truncation error. Examining the values in Table 4.1, we see that the Taylor series solution of order 3 to the IVP (50) satisfies $|y| < 3.45$ for $x \in [0, 1]$. Assuming that for $x \in [0, 1]$, $|y| < 7$ (which is slightly more than twice the largest value appearing in Table 4.1), we find

$$|f^{(3)}(x, y)| \leq 1 + |x| + |y| < 9 \quad \text{for } x \in [0, 1].$$

From equation (49) with $m = 3$ and the upper bound shown above, we obtain the following estimate of the local truncation error on the interval [0, 1]:

$$|E| \leq \frac{1}{4!} \max_{0 \leq x \leq 1} |f^{(3)}(x, y)|(.1)^4 < \frac{1}{4!} 9(.1)^4 < .000\ 0375.$$

The computational disadvantages of using a Taylor series expansion to approximate the solution to an initial value problem are fairly obvious. For any given f one must calculate the derivatives $f', f'', \ldots, f^{(m-1)}$, and these functions must all be evaluated at (x_n, y_n). However, the Taylor series expansion is of theoretical value, since other numerical schemes were generally derived by attempts to achieve a given order of accuracy without having to calculate higher derivatives. As a matter of fact, an approximation technique is said to be of order m if the truncation error per step is proportional to the truncation error of a Taylor series expansion of order m. Hence, the object is to produce an approximation scheme that has an error proportional to h^{m+1} without having to compute any derivatives of f.

4.3.1.2 Euler's method

Consider again the IVP (1) $y' = f(x, y)$; $y(x_0) = c_0$. Given the initial condition $y(x_0) = c_0 = y_0$, we can calculate $y'(x_0) = f(x_0, y(x_0)) = f(x_0, y_0)$, which is the slope of the tangent line of the exact solution $y(x)$ at $x = x_0$. If in equation

$$(45) \qquad y(x_{n+1}) = y(x_n) + \int_{x_n}^{x_{n+1}} f(t, y(t))\, dt,$$

we approximate $y(t)$ by y_0 and $f(t, y(t))$ by $f(x_0, y_0)$ for $t \in (x_0, x_1)$, then we obtain the following approximation to $y(x_1)$:

$$y_1 = y_0 + \int_{x_0}^{x_1} f(x_0, y_0)\, dt = y_0 + f(x_0, y_0)(x_1 - x_0).$$

Knowing y_1 we can compute $f(x_1, y_1)$, which is an approximation to the slope of the tangent line of the exact solution $y(x)$ at $x = x_1$. Notice that $f(x_1, y_1)$ is only an approximation to $y'(x_1) = f(x_1, y(x_1))$, the slope of the tangent line of the exact solution $y(x)$ at $x = x_1$, since in general $y_1 \neq y(x_1)$. See Figure 4.10. Using y_1 to approximate $y(t)$ and $f(x_1, y_1)$ to approximate $f(t, y(t))$ for $t \in (x_1, x_2)$ in equation (45), we obtain

$$y_2 = y_1 + \int_{x_1}^{x_2} f(x_1, y_1)\, dt = y_1 + f(x_1, y_1)(x_2 - x_1).$$

The general recursive formula that we derive by this method is

$$(51) \qquad y_{n+1} = y_n + f(x_n, y_n)(x_{n+1} - x_n).$$

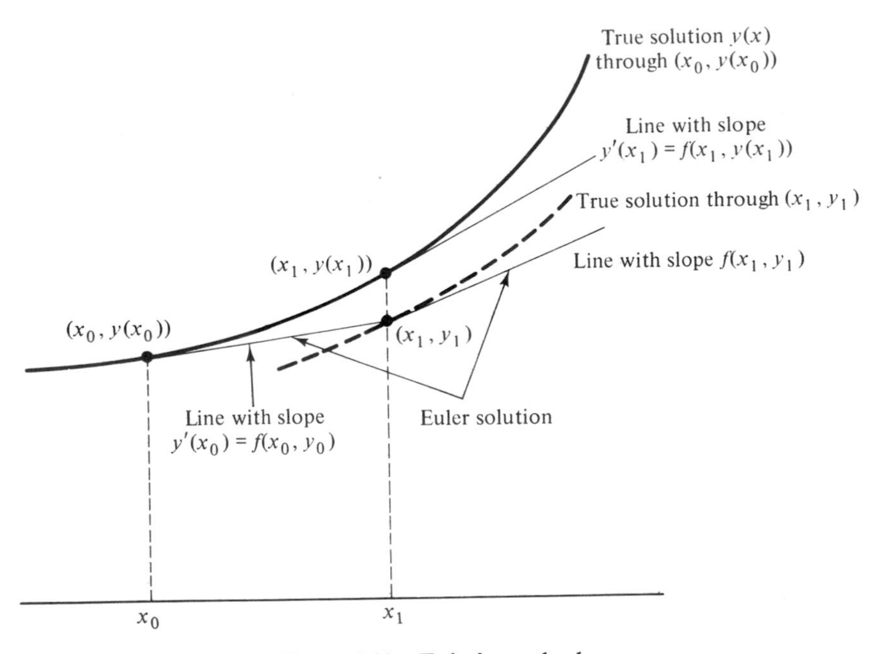

Figure 4.10 Euler's method.

Equation (51) is known as *Euler's method* or the *tangent line method.* We also see that this is the Taylor series expansion method of order 1, so the local truncation error is

$$E_n^E = \frac{1}{2!} f'(\xi, y(\xi))h_n^2 = \frac{h_n^2}{2} y''(\xi), \tag{52}$$

where $\xi \in (x_n, x_{n+1})$ and $h_n = x_{n+1} - x_n$. Equation (51) is a simple formula to use—especially for hand calculations—but, in general, it is not very accurate.

The Swiss mathematician Leonhard Euler (1707–1783) was one of the most, if not the most, prolific authors in the field of mathematics. He introduced the now common notations of e for the base of natural logarithms, π for the ratio of the circumference of a circle to its diameter, $\sum$ for the sign for summation, i for the imaginary unit, and $\sin x$ and $\cos x$ for the trigonometric functions. Euler's famous formula $e^{i\theta} = \cos\theta + i\sin\theta$ expresses a relation between the exponential function and trigonometric functions. The special case $\theta = \pi$ provides the following relation between the numbers e, π, and i: $e^{\pi i} = -1$. Euler's contribution to differential equations includes the concept of an integrating factor, the method for solving linear differential equations with constant coefficients, the method of reduction of order, and power series solutions, to name a few.

The error in equation (52) is a local error: that is, the error per step. The total error usually builds up as one moves away from the initial point x_0. The total error at any point is composed of the accumulated local truncation error and roundoff error. Conte and de Boor [6] show on pages 333–334 that if the exact solution to the IVP (1), $y(x)$, has a continuous second derivative on $[x_0, x_n]$ and if $|f_y(x, y)| \le L$ and $|y''(x)| \le Y$ for $x \in [x_0, x_n]$, then the total truncation error at x_n, $y_n - y(x_n)$, is bounded as follows:

$$|y_n - y(x_n)| \le \frac{hY}{2L}(e^{(x_n - x_0)L} - 1) \tag{53}$$

for a fixed stepsize h. It should be remembered that this error bound is an upper bound and not a realistic error estimate. Equation (53) does, however, show that as $h \to 0$, the accumulated truncation error also goes to zero. Thus, as $h \to 0$, the Euler approximation converges to the true solution—neglecting roundoff errors.

EXAMPLE Use Euler's method with $h = .2$ and $h = .1$ to generate an approximate solution to the initial value problem

$$y' = y + x; \qquad y(0) = 1 \tag{50}$$

on the interval [0, 1] and find an upper bound for the truncation error at $x = 1$ in each case.

Euler's solution to the IVP (50) for $h = .2$ and $h = .1$ is displayed in

Table 4.2 Euler's solution to the IVP (50)

$h = .2$

n	x_n	y_n	$f(x_n, y_n) = y_n + x_n$	$hf(x_n, y_n)$	$y_{n+1} = y_n + hf(x_n, y_n)$
0	.0	1.0 →	→ 1.0 →	→ .2 →	→ 1.2
1	.2	1.2 ← →	→ 1.4	.28	1.48
2	.4	1.48	1.88	.376	1.856
3	.6	1.856	2.456	.491 2	2.347 2
4	.8	2.347 2	3.147 2	.629 44	2.976 64
5	1.0	2.976 64			

$h = .1$

n	x_n	y_n	$f(x_n, y_n) = y_n + x_n$	$hf(x_n, y_n)$	$y_{n+1} = y_n + hf(x_n, y_n)$
0	.0	1.0	1.0	.1	1.1
1	.1	1.1	1.2	.12	1.22
2	.2	1.22	1.42	.142	1.362
3	.3	1.362	1.662	.166 2	1.528 2
4	.4	1.528 2	1.928 2	.192 82	1.721 02
5	.5	1.721 02	2.221 02	.221 02	1.943 12
6	.6	1.943 12	2.543 12	.254 312	2.197 43
7	.7	2.197 43	2.897 43	.289 743	2.487 18
8	.8	2.487 18	3.287 18	.328 718	2.815 90
9	.9	2.815 90	3.715 90	.371 590	3.187 48
10	1.0	3.187 48			

Table 4.2. Taking the partial derivative of $f(x, y) = x + y$ with respect to y, we find $f_y = 1$, so $|f_y| \leq 1 = L$ on $[0, 1]$ and

$$y''(x) = f_x + f_y f = 1 + x + y.$$

Assuming, as we did before in the Taylor series expansion example, that $|y| < 7$ for $x \in [0, 1]$, we have $|y''| < 9 = Y$ on $[0, 1]$. Therefore, for $h = .2$,

$$|y_5 - y(1)| \leq \frac{hY}{2L}(e^L - 1) = \frac{(.2)(9)}{2}(e - 1) \doteq 1.546\ 45,$$

and for $h = .1$, $|y_{10} - y(1)| \leq .773\ 227$. Since the given differential equation is linear, we easily calculate the exact solution of the IVP (50) to be $y = 2e^x - x - 1$. So the value of the exact solution to the IVP (50) at $x = 1$ is $y(1) = 2e - 2 = 3.436\ 56$. Hence, the actual total error due to truncation and roundoff for $h = .2$ is $3.436\ 56 - 2.976\ 64 = .459\ 92$, which is about one-third of the estimated error, 1.546 45.

4.3.1.3 Improved Euler's method or second order Runge-Kutta method

In deriving Euler's method we approximated the integrand of equation (45), $f(t, y(t))$, over the entire interval $[x_n, x_{n+1})$ by its approximate value at the left endpoint, $f(x_n, y_n)$. Upon integrating, we obtained Euler's recursive formula for solving the IVP (1) $y' = f(x, y)$; $y(x_0) = c_0$. A more accurate approximation may be obtained, if, instead of approximating the integrand, $f(t, y(t))$, by its approximate value at the left endpoint of the interval of integration, we approximate it by the average of its approximate values at the left endpoint and the right endpoint. That is, in equation (45) we replace the integrand $f(t, y(t))$ by $\frac{1}{2}(f(x_n, y_n) + f(x_{n+1}, y_{n+1}))$. After making this substitution in equation (45) and integrating, we obtain the following equation for y_{n+1}:

$$y_{n+1} = y_n + \tfrac{1}{2}(f(x_n, y_n) + f(x_{n+1}, y_{n+1}))(x_{n+1} - x_n).$$

This equation involves the unknown y_{n+1} as an argument of f on the right-hand side of the equation and, therefore, will generally be difficult to solve explicitly for y_{n+1}. Instead of trying to solve this equation for y_{n+1}, we simply replace y_{n+1} on the right-hand side by the approximation which we obtain using Euler's method. The recursive formula that results is known as the *improved Euler's method*, the *second order Runge-Kutta method*, or the *Heun method*:

$$(54) \qquad y_{n+1} = y_n + \tfrac{1}{2}(f(x_n, y_n) + f(x_{n+1}, y_n + f(x_n, y_n)h_n))h_n,$$

where $h_n = x_{n+1} - x_n$. The local truncation error for this method is

$$(55) \qquad E_n^I = -\tfrac{1}{12}f''(\xi, y(\xi))h_n^3 = -\tfrac{1}{12}y'''(\xi)h_n^3$$

where $\xi \in (x_n, x_{n+1})$. Thus, the local truncation error for the improved Euler's method is proportional to the cube of the stepsize; whereas the local truncation error for Euler's method is proportional to the square of the stepsize. The greater accuracy of the improved Euler's method must be paid for by an increase in the number of computations and function evaluations which must be performed at each step. Notice that f must be evaluated once for each step when using Euler's method and that f must be evaluated twice for each step when using the improved Euler's method.

EXAMPLE Calculate an approximate solution to the initial value problem

$$(50) \qquad y' = y + x = f(x, y); \qquad y(0) = 1$$

on the interval [0, 1] using the improved Euler's method with a constant stepsize $h = .1$ and estimate the local truncation error per step.

A flowchart of the computer program to solve an initial value problem by the improved Euler's method is displayed in Figure 4.11. The number of the corresponding computer program statement which accomplishes each of the operations indicated by the flowchart is also shown in Figure 4.11. The computer program used to produce the improved Euler solution to the IVP (50) is displayed in Figure 4.12 and the results of the calculations are given in Table 4.3. The first statement of the computer program specifies the differential equation to be integrated. This statement must be changed by the person using the program each time the differential equation that is to be integrated changes. The fourth statement of the program instructs the computer to read four values—H, X, Y, XF—from a punched data card according to format 900—the third statement in the program. The variable H denotes the step size that is to be used to generate the solution and must be punched in columns 2–18 inclusive of the data card; X denotes the initial value of the independent variable and must be punched in columns 19–35; Y denotes the initial value of the dependent variable and must be punched in columns 36–52; and XF denotes the final value of the independent variable and must be punched in columns 53–69. All the values—H, X, Y, XF—to be punched in the data card must be punched using E-format (FORTRAN scientific notation) or as real numbers—that is, as numbers with decimal points. The fifth statement of the program checks to see if H is identically zero. If H is zero, then program execution is terminated. So a blank card at the end of a data deck will indicate to the computer that processing is to be terminated, since H will be read as zero. If H is not zero, the initial values of the independent variable, X, and dependent variable, Y, are written by the sixth statement of the program using format 900—the second statement—and the number of times the recursion (54) is to be performed, N, is calculated by the seventh statement. If the difference between X and XF is divisible by H, then the last value of the independent variable for which an approximate solution value is calculated is XF. Otherwise, the last value of the independent variable for which an approximate solution value is calculated is within H of XF. Notice that if XF is greater than X, H should be positive; whereas, if XF is less than X, H should be negative. Otherwise, N will be a nonpositive integer and program execution will be terminated at the eighth statement of the program, since the DO variable—I in this case—can only assume positive values. Program statements 8–14 recursively calculate equation (54) N times. An approximation to the slope of the true solution at the point x_n, $f(x_n, y_n)$, is calculated and denoted by XK1; an approximation to the slope of the true solution at the point x_{n+1}, $f(x_{n+1}, y_n + f(x_n, y_n)h)$, is calculated and denoted by XK2; and the approximate change in y as the independent variable changes from x_n to x_{n+1} is calculated and denoted by DY. The fourteenth statement of the program requests the computer to write XK1, XK2, DY, X—which corresponds to x_{n+1}, of equation (54)—and Y—which corresponds to y_{n+1}.

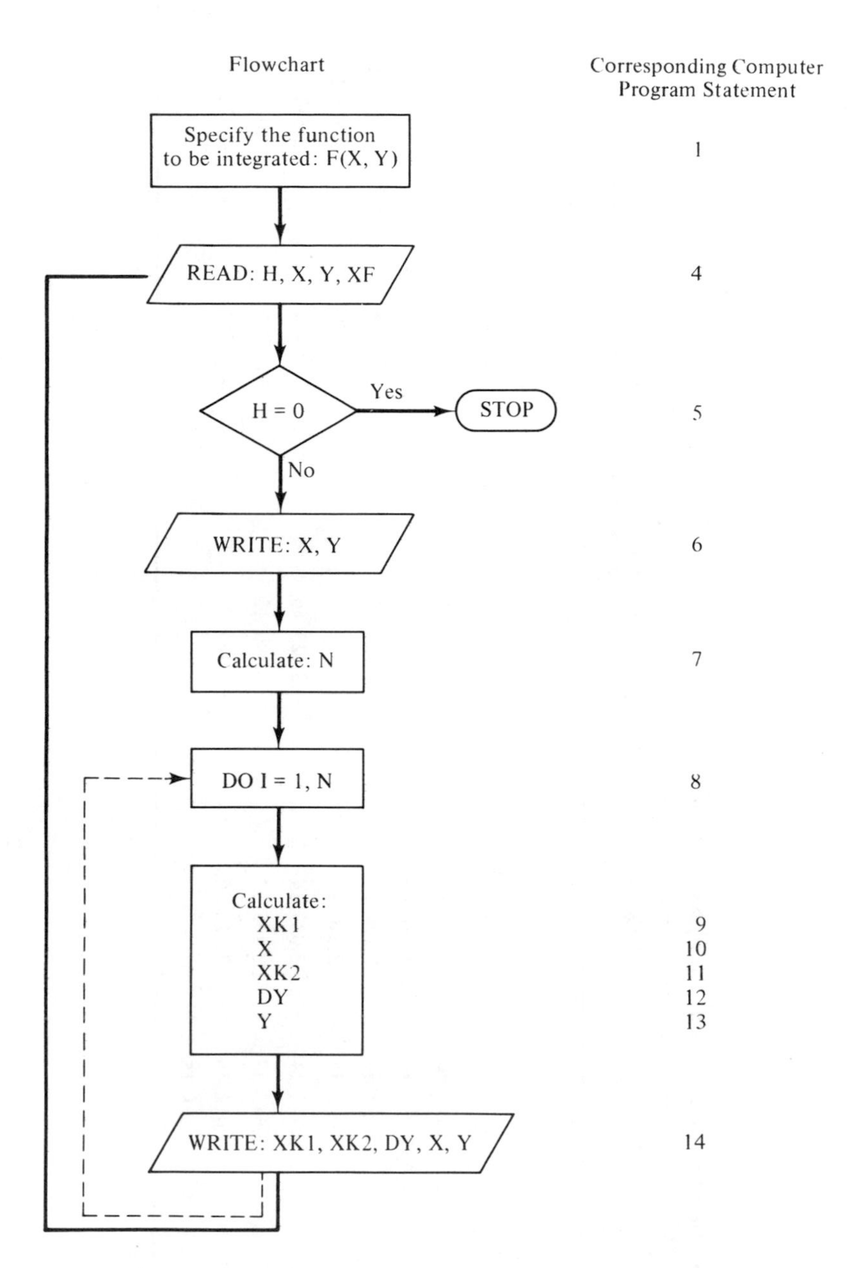

Figure 4.11 Flowchart of the improved Euler's method.

Table 4.3 Improved Euler's solution of the IVP (50) with $h = .1$

n	x_n	y_n	$K_1 = f(x_n, y_n)$ $= x_n + y_n$	$K_2 = f(x_{n+1}, y_n + f(x_n, y_n)h)$ $= x_{n+1} + y_n + f(x_n, y_n)h$	$\frac{1}{2}(K_1 + K_2)h$	$y_{n+1} = y_n + \frac{1}{2}(K_1 + K_2)h$
0	.0	1.0	1.0	1.2	.11	1.11
1	.1	1.11	1.21	1.431	.132 05	1.242 05
2	.2	1.242 05	1.442 05	1.686 25	.156 415	1.398 46
3	.3	1.398 46	1.698 46	1.968 31	.183 339	1.581 80
4	.4	1.581 80	1.981 80	2.279 98	.213 089	1.794 89
5	.5	1.794 89	2.294 89	2.624 38	.245 963	2.040 85
6	.6	2.040 85	2.640 85	3.004 94	.282 289	2.323 14
7	.7	2.323 14	3.023 14	3.425 46	.322 430	2.645 57
8	.8	2.645 57	3.445 57	3.890 13	.366 785	3.012 36
9	.9	3.012 36	3.912 36	4.403 59	.415 797	3.428 15
10	1.0	3.428 15				

```
              C---IMPROVED EULER METHOD
0001            1 F(X,Y) = Y + X
0002          800 FORMAT('+',34X,3E17.10/1X,2E17.10)
0003          900 FORMAT(1X,4E17.10)
0004            2 READ(1,900) H, X, Y, XF
0005              IF(H.EQ.0.) GO TO 4
0006              WRITE(3,900) X, Y
0007              N = (XF - X)/H + .5
0008              DO 3 I = 1,N
0009              XK1 = F(X,Y)
0010              X = X + H
0011              XK2 = F(X,Y+XK1*H)
0012              DY = .5*H*(XK1+XK2)
0013              Y = Y + DY
0014            3 WRITE(3,800) XK1,XK2,DY,X,Y
0015              GO TO 2
0016            4 CALL EXIT
0017              END
```

Figure 4.12 Computer program to produce the improved Euler's solution.

After equation (54) has been calculated recursively N times, the fifteenth statement of the program instructs the computer to return to the fourth statement of the program—which has been assigned statement number 2. This statement instructs the computer to read the next data card. If there is no data card present, program execution is terminated. If there is a data card present and if the value of H on this data card is not zero, then the solution of the differential equation specified by the first statement which has initial value Y at X is produced on the interval from X to XF using a step size H. If there is a data card present and if the value of H on the data card is zero, then the computer is instructed to terminate execution.

We shall assume, as we did in the Taylor series expansion example, that $|y(x)| < 7$ on $[0, 1]$. Since $y''' = y + x + 1$ and since $h = .1$, we see from equation (55) that the local truncation error per each step on $[0, 1]$ satisfies

$$|E_n^I| = |y(x_n) - y_n| \leq \tfrac{1}{12} \max_{x \in [0,1]} |y'''(x)| h^3 \leq \tfrac{1}{12} \cdot 9(.1)^3 = .000\ 75.$$

Analyzing the form of the recursive formula (54) might lead one to try to devise a general recursion of the form

$$(56) \qquad y_{n+1} = y_n + (af(x_n, y_n) + bf(x_n + ch_n, y_n + df(x_n, y_n)h_n))h_n,$$

in which the constants a, b, c, and d are to be determined in such a manner that (56) will agree with a Taylor series expansion of as high an order as possible. As we have seen the Taylor series expansion for $y(x_{n+1})$ about x_n is

$$(57) \qquad y(x_{n+1}) = y(x_n) + fh_n + \tfrac{1}{2}(f_x + f_y f)h_n^2 + \tfrac{1}{6}(f_{xx} + 2f_{xy}f + f_{yy}f^2 + f_x f_y + f_y^2 f)h_n^3 + O(h_n^4),$$

where f and its partial derivatives are all evaluated at (x_n, y_n) and $O(h_n^4)$

indicates that the error made by omitting the remainder of the terms in the expansion is proportional to the fourth power of the stepsize.

Let $k_1 = f(x_n, y_n)$ and $k_2 = f(x + ch_n, y_n + dk_1h_n)$. Using the Taylor series expansion for a function of two variables to expand k_2 about (x_n, y_n), we obtain

$$(58) \qquad \begin{aligned} k_2 &= f(x_n + ch_n, y_n + dk_1h_n) \\ &= f(x_n, y_n) + ch_nf_x + dk_1h_nf_y + \frac{c^2h_n^2}{2} f_{xx} + cdh_n^2k_1f_{xy} \\ &\quad + \frac{d^2h_n^2k_1^2}{2} f_{yy} + O(h^3). \end{aligned}$$

Substituting $k_1 = f$ into (58), substituting (58) into (56), and rearranging in ascending powers of h_n, we obtain

$$(59) \qquad \begin{aligned} y_{n+1} &= y_n + (a + b)fh_n + b(cf_x + df_yf)h_n^2 \\ &\quad + b\left(\frac{c^2}{2} f_{xx} + cdff_{xy} + \frac{d^2}{2} f^2f_{yy}\right)h_n^3 + O(h_n^4). \end{aligned}$$

Comparing (57) with (59) we see that for the corresponding coefficients of h_n and h_n^2 to agree we must have

$$(60) \qquad a + b = 1, \quad bc = \tfrac{1}{2} \quad \text{and} \quad bd = \tfrac{1}{2}.$$

Thus, we have three equations in four unknowns. Hence, we might hope to be able to choose the constants in such a manner that the coefficients of h_n^3 in (57) and (59) agree. However, for these coefficients to agree we must have $bc^2/2 = \frac{1}{6}$, $bcd = \frac{1}{3}$, $bd^2/2 = \frac{1}{6}$, and $f_xf_y + f_y^2f = 0$. Obviously, the last equality is not satisfied by all functions f.

There are an infinite number of solutions to (60). The choice $a = \frac{1}{2}$, $b = \frac{1}{2}$, $c = 1$, and $d = 1$ yields the improved Euler's method, the second order Runge-Kutta method, or the Heun method recursive formulas (54). Choosing $a = 0$, $b = 1$, $c = \frac{1}{2}$, and $d = \frac{1}{2}$ results in the following recursion, known as the *modified Euler's method*:

$$(61) \qquad y_{n+1} = y_n + f(x_n + \tfrac{1}{2}h_n, y_n + \tfrac{1}{2}f(x_n, y_n)h_n)h_n.$$

4.3.1.4 Fourth order Runge-Kutta method

If one tries to develop a general recursion of the form

(62)

$$\begin{aligned} y_{n+1} = y_n + h_n[a_1f(x_n, y_n) &+ a_2f(x_n + b_1h_n, y_n + b_1h_nk_1) \\ &+ a_3f(x_n + b_2h_n, y_n + b_2h_nk_2) + a_4f(x_n + b_3h_n, y_n + b_3h_nk_3)], \end{aligned}$$

where $k_1 = f(x_n, y_n)$ and $k_i = f(x_n + b_{i-1}h_n, y_n + b_{i-1}h_nk_{i-1})$ for $i = 2, 3, 4$, by determining the constants $a_1, a_2, a_3, a_4, b_1, b_2$, and b_3 in such a manner that (62) will agree with a Taylor series expansion of as high an order as possible, one obtains a system of algebraic equations in the constants. In this case, as before, there are an infinite number of solutions to the system of equations. The choice for the constants which leads to the classical fourth order Runge-Kutta recursions is $a_1 = a_4 = \frac{1}{6}$, $a_2 = a_3 = \frac{1}{3}$, $b_1 = b_2 = \frac{1}{2}$, and $b_3 = 1$. One usually finds the recursion for y_{n+1} written as

$$y_{n+1} = y_n + \frac{h_n(k_1 + 2k_2 + 2k_3 + k_4)}{6}, \tag{63}$$

where

$$k_1 = f(x_n, y_n)$$
$$k_2 = f\left(x_n + \frac{h_n}{2}, y_n + \frac{h_n k_1}{2}\right)$$
$$k_3 = f\left(x_n + \frac{h_n}{2}, y_n + \frac{h_n k_2}{2}\right)$$
$$k_4 = f(x_n + h_n, y_n + h_n k_3).$$

Hence, the fourth order Runge-Kutta method may be viewed as a weighted average of approximate values of $f(t, y(t))$ at different points within the interval of integration $[x_n, x_{n+1}]$. The value k_1 is an approximation of the slope of the true solution at the left endpoint of the interval of integration; k_2 is an approximation to the slope at the midpoint of the interval of integration which is obtained by using Euler's method to approximate $y(x_n + h_n/2)$; k_3 is another approximation to the slope at the midpoint of the interval of integration; and k_4 is an approximation to the slope at the right endpoint of the interval of integration. The local truncation error of this method is proportional to h_n^5 and if f is a function of x alone, then the fourth order Runge-Kutta recursion (63) reduces to Simpson's rule. Because of its relative high order of accuracy, the fourth order Runge-Kutta method is one of the most commonly used single-step methods.

This general type of method for numerical integration was initially developed by the German mathematician Carl Runge (1856–1927) and published in 1895. In 1900 Heun published a paper concerning the improvement of Runge's method and in 1901 Wilhelm Kutta (1867–1944) published a more comprehensive paper. Several methods of different order for numerical integration resulted. These methods are generally known as Runge-Kutta methods. The method most commonly referred to and utilized is the fourth order Runge-Kutta method just presented.

EXAMPLE Use the fourth order Runge-Kutta method with a constant step-size $h = .1$ to calculate an approximate solution to the initial value problem

$$y' = y + x = f(x, y); \qquad y(0) = 1 \tag{50}$$

on the interval [0, 1].

Table 4.4 Fourth order Runge-Kutta solution to the IVP (50) with $h = .1$

n	x_n	y_n	k_1	k_2	k_3	k_4	$h_n(k_1 + 2k_2 + 2k_3 + k_4)/6$
0	.0	1.0	1.0	1.1	1.105	1.210 5	.110 341
1	.1	1.110 34	1.210 34	1.320 86	1.326 38	1.442 98	.132 463
2	.2	1.242 80	1.442 80	1.564 94	1.571 05	1.699 91	.156 911
3	.3	1.399 71	1.699 71	1.834 70	1.841 45	1.983 86	.183 931
4	.4	1.583 64	1.983 64	2.132 83	2.140 28	2.297 67	.213 792
5	.5	1.797 44	2.297 44	2.462 31	2.470 55	2.644 49	.246 794
6	.6	2.044 23	2.644 23	2.826 44	2.835 55	3.027 78	.283 266
7	.7	2.327 50	3.027 50	3.228 87	3.238 94	3.451 39	.323 574
8	.8	2.651 07	3.451 07	3.673 62	3.684 75	3.919 54	.368 122
9	.9	3.019 19	3.919 19	4.165 15	4.177 45	4.436 94	.417 355
10	1.0	3.436 55					

The computer program shown in Figure 4.13 was used to generate the fourth order Runge-Kutta solution to the IVP (50) which is displayed in Table 4.4. The program is very similar to the program used to produce the improved Euler solution, the only difference being that the improved Euler program statements 10–14, which are used to recursively calculate equation (54), are replaced by the fourth order Runge-Kutta program statements 10–17, which are used to recursively calculate equation (63).

```
              C---FOURTH ORDER RUNGE-KUTTA METHOD
0001            1 F(X,Y) = Y + X
0002          800 FORMAT('+',34X,5E17.10/1X,2E17.10)
0003          900 FORMAT(1X,4E17.10)
0004            2 READ(1,900) H, X, Y, XF
0005              IF(H.EQ.0.) GO TO 4
0006              WRITE(3,900) X, Y
0007              N = (XF - X)/H + .5
0008              DO 3 I = 1,N
0009              XK1 = F(X,Y)
0010              X = X + .5*H
0011              XK2 = F(X,Y+.5*H*XK1)
0012              XK3 = F(X,Y+.5*H*XK2)
0013              X = X + .5*H
0014              XK4 = F(X,Y+H*XK3)
0015              DY = H*(XK1 + 2.*XK2 + 2.*XK3 + XK4)/6.
0016              Y = Y + DY
0017            3 WRITE(3,800) XK1,XK2,XK3,XK4,DY,X,Y
0018              GO TO 2
0019            4 CALL EXIT
0020              END
```

Figure 4.13 Computer program to produce the fourth order Runge-Kutta solution.

A tabular comparison of the methods which we have used in this section to solve the IVP (50) $y' = y + x$; $y(0) = 1$ on the interval [0, 1] with a constant stepsize of $h = .1$ is made in Table 4.5. From this table it is obvious

Table 4.5 Solutions of the IVP (50) with $h = .1$

			y_n		
x_n	*Taylor series order 3*	*Euler's method*	*Improved Euler's method*	*Fourth order Runge-Kutta*	*Actual solution*
.0	1.0	1.0	1.0	1.0	1.0
.1	1.110 34	1.1	1.11	1.110 34	1.110 34
.2	1.242 79	1.22	1.242 05	1.242 80	1.242 81
.3	1.399 70	1.362	1.398 46	1.399 71	1.399 72
.4	1.583 62	1.528 2	1.581 80	1.583 64	1.583 65
.5	1.797 40	1.721 02	1.794 89	1.797 44	1.797 44
.6	2.044 18	1.943 12	2.040 85	2.044 23	2.044 24
.7	2.327 44	2.197 43	2.323 14	2.327 50	2.327 51
.8	2.651 00	2.487 18	2.645 57	2.651 07	2.651 08
.9	3.019 10	2.815 90	3.012 36	3.019 19	3.019 21
1.0	3.436 44	3.187 48	3.428 15	3.436 55	3.436 56

that the fourth order Runge-Kutta method is the most accurate method for this particular initial value problem. However, this method required approximately twice the computing time of the improved Euler's method and four times the computing time of Euler's method, since Euler's method requires only one evaluation of the function f per step while the improved Euler's method requires two evaluations and the fourth order Runge-Kutta method requires four evaluations. Consequently, one might anticipate that the solutions to the IVP (50) generated using Euler's method with $h = .025$, using the improved Euler's method with $h = .05$, and using the fourth order Runge-Kutta method with $h = .1$ would require approximately the same amount of computing time and have the same accuracy, since each method would then require 40 evaluations of the function f. Performing the necessary calculations, we obtain the results shown in Table 4.6.

Table 4.6 Solutions of the IVP (50) $y' = y + x$; $y(0) = 1$ on [0, 1]

	y_n			
x_n	*Euler* $h = .025$	*Improved Euler* $h = .05$	*Runge-Kutta* $h = .1$	*Actual solution*
.0	1.0	1.0	1.0	1.0
.1	1.107 62	1.110 25	1.110 34	1.110 34
.2	1.236 80	1.242 61	1.242 80	1.242 81
.3	1.389 77	1.399 39	1.399 71	1.399 72
.4	1.569 00	1.583 17	1.583 64	1.583 65
.5	1.777 22	1.796 78	1.797 44	1.797 44
.6	2.017 43	2.043 35	2.044 23	2.044 24
.7	2.292 97	2.326 37	2.327 50	2.327 51
.8	2.607 49	2.649 64	2.651 07	2.651 08
.9	2.965 04	3.017 42	3.019 19	3.019 21
1.0	3.370 09	3.434 37	3.436 55	3.436 56

These results emphasize that it is not only the number of function evaluations which are used in producing an approximate solution but also the manner in which these function evaluations are combined which ultimately determines the accuracy of the solution.

EXERCISES

1. Consider the initial value problem $y' = x^2 - y$; $y(0) = 1$.

(a) Derive the Taylor series expansion formula of order 3 for this initial value problem.

(b) Use the formula derived in (a) and a constant stepsize $h = .1$ to calculate an approximate solution on the interval [0, 1].

(c) Estimate the maximum truncation error per step on the [0, 1] for the stepsize $h = .1$.
(d) How small must the stepsize be in order to ensure six-decimal-place accuracy per step?

2. (a) Compute an approximate solution to the initial value problem $y' = x^2 - y$; $y(0) = 1$ on the interval [0, 1] using Euler's method and a constant stepsize $h = .1$.
(b) Find an upper bound for the total truncation error at $x = 1$.
(c) How small must the stepsize be to ensure six-decimal-place accuracy per step?
(d) How small must the stepsize be to ensure six-decimal-place accuracy over the interval [0, 1]?

3.* Use the improved Euler's formula, the modified Euler's formulas, and the fourth order Runge-Kutta formula with a stepsize $h = .1$ to generate approximate solutions to the initial value problem $y' = x^2 - y$; $y(0) = 1$ on the interval [0, 1].

4. (a) Find the exact solution of $y' = x^2 - y$; $y(0) = 1$.
(b) Compare the various approximate solutions generated in Exercises 1–3 with each other and the exact solution by producing a table of values.

5. Consider the general recursive formula (56). Suppose that in addition to satisfying equations (60), we require that the coefficients of $f^2 f_{yy}$ in equations (57) and (59) be equal. What is the solution of the resulting system of four equations in the four unknowns, a, b, c, and d?

4.3.2 Multistep Methods

Let $z(x)$ be a function which is defined on some interval containing the points $x_0, x_1, \ldots, x_n$. It is well known that there exists only one polynomial, $p(x)$, of degree less than or equal to n for which

$$p(x_i) = z(x_i), \qquad i = 0, 1, \ldots, n.$$

The polynomial $p(x)$ is called the *interpolating polynomial.* There are many ways to write an expression for the interpolating polynomial; however, we shall not present any of these expressions here. We are only interested in the fact that an interpolating polynomial exists and that it is unique.

Consider again the IVP (1) $y' = f(x, y)$; $y(x_0) = c_0$. Suppose that we have used one of the single-step methods presented in the preceding section to produce the approximations y_i to $y(x_i)$ for $i = 0, 1, \ldots, n$, where the x_i's are

---

* Most easily accomplished using programmable machines.

equally spaced. For each single-step method presented, we also found it necessary to compute $y_i' = f(x_i, y_i)$ in order to produce y_{i+1}. Henceforth let $y_i' = f(x_i, y_i) = f_i$. In deriving each of the single-step methods we integrated the differential equation $y' = f(x, y)$ of the IVP (1) over the interval $[x_n, x_{n+1})$ to obtain the integral equation

$$y(x_{n+1}) = y(x_n) + \int_{x_n}^{x_{n+1}} f(x, y(x))\, dx$$

and then approximated f on the interval $[x_n, x_{n+1})$. Multistep methods are generally derived by integrating the differential equation $y' = f(x, y)$ from x_{n-p} to x_{n+q} where $p, q \geq 0$ and by approximating the integrand $f(x, y(x))$ on the interval $[x_{n-p}, x_{n+q})$ by the interpolating polynomial $p(x)$ which interpolates f at the $m + 1$ points $x_{r-m}, x_{r-m+1}, \ldots, x_{r-1}, x_r$, where $r = n$ or $r = n + 1$. See Figure 4.14. If $r = n$, the resulting formula is said to be *open*; whereas, if $r = n + 1$, the resulting formula is called *closed*. Open formulas are explicit formulas for y_{n+1}. Closed formulas, on the other hand, are implicit formulas for y_{n+1}.

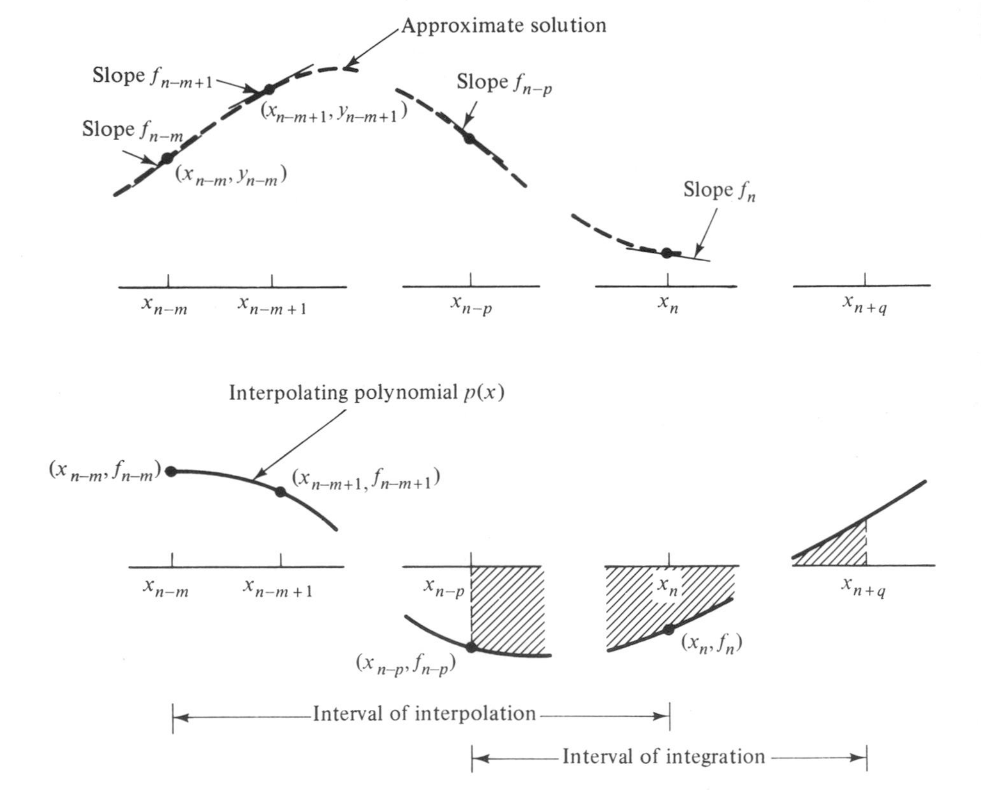

Figure 4.14 Multistep integration diagram for $r = n$.

4.3.2.1 Adams-Bashforth methods

Choosing $r = n$, $p = 0$, and $q = 1$ results in a set of formulas which are known as Adams-Bashforth (AB) formulas. Determining the interpolating polynomial for the first few values of m and integrating results in the following Adams-Bashforth formulas and their associated local errors:

m	*AB formula*	*Error*
0	$y_{n+1} = y_n + hf_n$	$\left(\frac{h^2}{2}\right)y^{(2)}(\xi)$
1	$y_{n+1} = y_n + \frac{h(3f_n - f_{n-1})}{2}$	$\left(\frac{5h^3}{12}\right)y^{(3)}(\xi)$
2	$y_{n+1} = y_n + \frac{h(23f_n - 16f_{n-1} + 5f_{n-2})}{12}$	$\left(\frac{9h^4}{24}\right)y^{(4)}(\xi)$
3	$y_{n+1} = y_n + \frac{h(55f_n - 59f_{n-1} + 37f_{n-2} - 9f_{n-3})}{24}$	$\left(\frac{251h^5}{720}\right)y^{(5)}(\xi)$

where in each case $\xi \in (x_{n-m}, x_n)$. The formula for $m = 0$ is a single-step method—Euler's method. Notice that each method utilizes f evaluated at $m + 1$ points and has a truncation error of order h^{m+2}. Hence each formula requires $m + 1$ starting values. The Adams-Bashforth method for numerical integration initially appeared in 1882 in an article on the theory of capillary action!

EXAMPLE Calculate an approximate solution to the initial value problem

$$y' = y + x; \qquad y(0) = 1 \tag{50}$$

on the interval [0, 1] using a stepsize $h = .1$ and Adams-Bashforth formulas for $m = 1, 2, 3$. Use starting values obtained by the fourth order Runge-Kutta method. Compare the results with the actual solution.

The results of the required calculations are shown in Table 4.7. Some of the error in the values for y_n obtained by the Adams-Bashforth formula with $m = 2$ and $m = 3$ is due to the error of the starting values, and some of the error is due to the formula itself. In order to determine how much of the error was due to the starting values, we generated Adams-Bashforth solutions to the IVP (50) and used the actual solution as starting values. The results differed from those shown by at most .000 01. Hence, the majority of the error is due to the formula and not the starting values.

Table 4.7 Adams-Bashforth solutions of the IVP (50) with $h = .1$

		y_n			
		Adams-Bashforth formulas			
x_n	*Runge-Kutta values*	$m = 1$	$m = 2$	$m = 3$	*Actual solution*
.0	1.0				1.0
.1	1.110 34				1.110 34
.2	1.242 80	1.241 89			1.242 81
.3	1.399 71	1.397 66	1.399 63		1.399 72
.4		1.580 21	1.583 45	1.583 64	1.583 65
.5		1.792 36	1.797 11	1.797 42	1.797 44
.6		2.037 20	2.043 74	2.044 20	2.044 24
.7		2.318 16	2.326 82	2.327 45	2.327 51
.8		2.639 03	2.650 18	2.651 00	2.651 08
.9		3.003 97	3.018 04	3.019 11	3.019 21
1.0		3.417 62	3.435 09	3.436 44	3.436 56

4.3.2.2 Nystrom methods

A set of formulas called Nystrom formulas is obtained by selecting $r = n$, $p = 1$, and $q = 1$. After calculating the interpolating formula and integrating, one obtains the following formulas:

m	*Formula*
0	$y_{n+1} = y_{n-1} + 2hf_n$
1	$y_{n+1} = y_{n-1} + 2hf_n$
2	$y_{n+1} = y_{n-1} + \dfrac{h(7f_n - 2f_{n-1} + f_{n-2})}{3}$
3	$y_{n+1} = y_{n-1} + \dfrac{h(8f_n - 5f_{n-1} + 4f_{n-2} - f_{n-3})}{3}$

These formulas were derived by the Finnish mathematician E. J. Nystrom and published in 1925.

The formula for $m = 0$—which is the same as the formula for $m = 1$—is known as the *midpoint rule* and has a local error of $h^3 y^{(3)}(\xi)/6$, where $\xi \in (x_{n-1}, x_{n+1})$. The midpoint rule has the simplicity of Euler's method and has a smaller local error; however, it requires one starting value and is generally less stable than Euler's method. (The term "stable" is discussed in Section 4.3.3.)

4.3.2.3 Milne's method

Milne's formula is derived by finding the polynomial which interpolates f at x_{n-3}, x_{n-2}, x_{n-1}, and x_n and integrating over the interval (x_{n-3}, x_{n+1}). That is, $r = n$, $p = 3$, $q = 1$, and $m = 3$. The resulting formula is

$$y_{n+1} = y_{n-3} + \frac{4h(2f_n - f_{n-1} + 2f_{n-2})}{3} \tag{64}$$

and the associated local error is $14h^5y^{(5)}(\xi)/45$. This method for numerical integration was published by the American mathematician William E. Milne (1890–) in 1926.

EXAMPLE Use a constant stepsize $h = .1$ and Nystrom formulas with $m = 1, 2, 3$ and Milne's formula to calculate approximate solutions to the initial value problem

$$y' = y + x; \qquad y(0) = 1 \tag{50}$$

on the interval [0, 1]. Use starting values obtained by the fourth order Runge-Kutta method. Compare the results to the actual solution.

The results of the desired calculations are displayed in Table 4.8.

Table 4.8 Nystrom and Milne solutions to the IVP (50)

		y_n				
		Nystrom formulas				
x_n	*Runge-Kutta values*	$m = 1$	$m = 2$	$m = 3$	*Milne's formula*	*Actual solution*
.0	1.0					1.0
.1	1.110 34					1.110 34
.2	1.242 80	1.242 07				1.242 81
.3	1.399 71	1.398 75	1.399 64			1.399 72
.4		1.581 82	1.583 54	1.583 64	1.583 64	1.583 65
.5		1.795 12	1.797 25	1.797 43	1.797 43	1.797 44
.6		2.040 84	2.043 98	2.044 21	2.044 22	2.044 24
.7		2.323 28	2.327 14	2.327 47	2.327 49	2.327 51
.8		2.645 50	2.650 63	2.651 04	2.651 05	2.651 08
.9		3.012 38	3.018 61	3.019 15	3.019 17	3.019 21
1.0		3.427 97	3.435 83	3.436 49	3.436 52	3.436 56

4.3.2.4 Adams-Moulton methods

All the foregoing multistep methods employed polynomials which interpolated f at x_n and preceding points $x_{n-1}, \ldots, x_{n-m}$. If we find an interpolating polynomial which interpolates f at $x_{n+1}, x_n, \ldots, x_{n-m}$ and integrate from x_n to x_{n+1}, we obtain a set of formulas known as Adams–Moulton formulas. Hence, in our notation $r = n + 1$, $p = 0$, and $q = 1$. The first few Adams-Moulton (AM) formulas and their associated local errors follow.

m	*AM formula*	*Error*
0	$y_{n+1} = y_n + \dfrac{h(f_{n+1} + f_n)}{2}$	$\dfrac{h^3 y^{(3)}(\xi)}{12}$
1	$y_{n+1} = y_n + \dfrac{h(5f_{n+1} + 8f_n - f_{n-1})}{12}$	$\dfrac{h^4 y^{(4)}(\xi)}{24}$
2	$y_{n+1} = y_n + \dfrac{h(9f_{n+1} + 19f_n - 5f_{n-1} + f_{n-2})}{24}$	$\dfrac{19h^5 y^{(5)}(\xi)}{720}$

In each case, $\xi \in (x_{n-m}, x_{n+1})$. All these formulas are implicit formulas for y_{n+1}, since y_{n+1} appears on the right hand of each equation in $f_{n+1} = f(x_{n+1}, y_{n+1})$. The formula obtained for $m = 0$ is the trapezoidal rule.

4.3.2.5 Milne-Simpson method

By selection $r = n + 1$, $p = 2$, and $q = 1$ we obtain a set of formulas which are called Milne-Simpson formulas. The same formula is obtained for $m = 2$ and $m = 3$, namely

$$y_{n+1} = y_{n-1} + \frac{h(f_{n+1} + 4f_n + f_{n-1})}{3}. \tag{65}$$

The local error of this formula is $h^5 y^{(5)}(\xi)/90$, and if f is a function of x alone, the formula reduces to Simpson's rule.

EXERCISES

1. (a) Compute an approximate solution to the initial value problem $y' = x^2 - y$; $y(0) = 1$ on the interval [0, 1] using the Adams-Bashforth formula for $m = 1$. Use a constant stepsize $h = .1$ and use actual solution values for starting values.

(b) Estimate the maximum truncation error per step on the interval [0, 1].

(c) How small must the stepsize be to ensure six-decimal-place accuracy per step?

2. (a) Use the midpoint rule and a stepsize $h = .1$ to produce an approximate solution to the initial value problem $y' = x^2 - y$; $y(0) = 1$ on the interval [0, 1].

(b) Estimate the maximum truncation error per step on the interval [0, 1].

(c) Compare this solution to Euler's solution—Exercise 2(a) of Section 4.3.1.

3. Estimate the maximum truncation per step when using Milne's method with a stepsize $h = .1$ to solve the initial value problem $y' = x^2 - y$; $y(0) = 1$ on the interval $[0, 1]$.
4. Consider the initial value problem $y' = x^2 - y$; $y(0) = 1$.
   (a) For the given initial value problem solve the Adams-Moulton formula for $m = 0$ explicitly for y_{n+1}.
   (b) Generate an approximate solution on the interval $[0, 1]$ using a stepsize $h = .1$.
5. Solve the Milne-Simpson equation (65) explicitly for y_{n+1} for the differential equation $y' = x^2 - y$.

4.3.3 Predictor-Corrector Methods

The Adams-Moulton formulas and the Milne-Simpson formula of the previous two sections are implicit formulas. In general, $f(x, y)$ will be nonlinear and it will be impossible to solve the equation explicitly for y_{n+1}. However, we can try to determine y_{n+1} by iteration. That is, we obtain in some manner a first approximation to y_{n+1}, call it y_{n+1}^0, and then we successively calculate $f_{n+1}^k = f(x_{n+1}, y_{n+1}^k)$ and use this approximation of f_{n+1} in the implicit formula to successively calculate y_{n+1}^{k+1} for $k = 0, 1, \ldots$. Under fairly general conditions on the function f it can be shown that for sufficiently small values of the stepsize h, the sequence $\langle y_{n+1}^k \rangle_{k=0}^{\infty}$ converges to a solution y_{n+1} of the implicit formula and that the solution y_{n+1} is unique. [Note that y_{n+1} will be the solution of the implicit formula but will not generally be the solution to the differential equation at x_{n+1}, $y(x_{n+1})$.] A formula that is used to obtain the first approximation, y_{n+1}^0, is called a *predictor formula* and an implicit formula used in the iteration procedure to calculate y_{n+1}^k for $k = 1, 2, \ldots$ is called a *corrector formula*.

One usually chooses the predictor and corrector formulas of the iteration procedure so that the order of the local error of each formula is nearly the same. Generally, the corrector formula is chosen so that the local error is smaller than the local of the predictor formula. Both single-step and multistep predictor-corrector methods may be devised.

A simple single-step predictor-corrector method, for instance, might employ Euler's formula for the predictor and the trapezoidal formula for the corrector. Hence, one would have the following iteration procedure:

$$\text{(66p)} \qquad y_{n+1}^0 = y_n + hf(x_n, y_n)$$

$$\text{(66c)} \qquad y_{n+1}^k = y_n + \frac{h[f(x_n, y_n) + f(x_{n+1}, y_{n+1}^{k-1})]}{2}, \qquad k = 1, 2, \ldots.$$

Notice that the local error of (66p) is $h^2 y^{(2)}(\xi)/2$ and the local error of (66c) is $h^3 y^{(3)}(\xi)/12$.

The following multistep predictor-corrector iteration procedure was introduced by Milne and uses equation (64) for the predictor and equation (65) for the corrector. Hence, the Milne iteration procedure is

$$(67\text{p}) \qquad y_{n+1}^{0} = y_{n-3} + \frac{4h(2f_n - f_{n-1} + 2f_{n-2})}{3}$$

$$(67\text{c}) \qquad y_{n+1}^{k} = y_{n-1} + \frac{h(f(x_{n+1}, y_{n+1}^{k-1}) + 4f_n + f_{n-1})}{3}, \qquad k = 1, 2, 3, \ldots.$$

The local error of each of these formulas is of order h^5.

Another commonly used predictor-corrector method for which each formula has a local error of order h^5 but a slightly smaller error coefficient than Milne's method employs the Adams-Bashforth formula with $m = 3$ as the predictor and the Adams-Moulton formula with $m = 2$ as the corrector. The iteration procedure is

$$(68\text{p}) \qquad y_{n+1}^{0} = y_n + \frac{h(55f_n - 59f_{n-1} + 37f_{n-2} - 9f_{n-3})}{24}$$

(68c)

$$y_{n+1}^{k} = y_n + \frac{h(9f(x_{n+1}, y_{n+1}^{k-1}) + 19f_n - 5f_{n-1} + f_{n-2})}{24}, \qquad k = 1, 2, 3, \ldots.$$

In order to solve an initial value problem on an interval $[a, b]$ using a predictor-corrector algorithm one needs to specify (i) the stepsize, h, to be taken; (ii) the maximum absolute iteration error, $E = |y_{n+1}^{k} - y_{n+1}^{k-1}|$, or the maximum relative iteration error, $\epsilon = |y_{n+1}^{k} - y_{n+1}^{k-1}|/|y_{n+1}^{k}|$, to be allowed per step; (iii) the maximum number of iterations, K, to be taken per step; and (iv) what to do if K is reached before the error requirement E or ϵ is satisfied. And, of course, if the predictor-corrector algorithm being utilized involves a multistep formula, one must obtain starting values by using some single-step method.

In selecting the stepsize, the maximum iteration error per step, and the maximum number of iterations per step one must keep in mind that these are not independent but are related through the algorithm local error formula, which in turn depends upon the differential equation. In general, the maximum iteration error desired per step is chosen, the maximum number of iterations per step is set at a small number—usually two or three, and the stepsize is then determined so that it is consistent with the maximum iteration error, the number of iterations per step, the algorithm, and the differential equation.

EXAMPLE Solve the initial value problem

$$(50) \qquad y' = y + x; \qquad y(0) = 1$$

on the interval $[0, 1]$ using the predictor-corrector algorithms of equations (66), (67), and (68). Use a fixed stepsize $h = .1$, use actual solution values as starting values, and set the maximum absolute iteration error, E, equal to 5×10^{-6}. Also record the number of iterations per step, k, required to achieve this accuracy.

The results of the calculations are displayed in Table 4.9.

Table 4.9 Predictor-Corrector solutions to the IVP (50) with $h = .1$

x_n	*Equations* (66) y_n	k	*Equations* (67) y_n	k	*Equations* (68) y_n	k	*Actual solution*
.0							1.000 00
.1	1.110 53	4					1.110 34
.2	1.243 21	4					1.242 81
.3	1.400 39	4					1.399 72
.4	1.584 64	4	1.583 65	2	1.583 65	2	1.583 65
.5	1.798 82	4	1.797 44	2	1.797 44	2	1.797 44
.6	2.046 06	4	2.044 24	2	2.044 24	2	2.044 24
.7	2.329 85	4	2.327 51	2	2.327 51	2	2.327 51
.8	2.654 05	4	2.651 08	2	2.651 08	2	2.651 08
.9	3.022 90	4	3.019 21	2	3.019 21	2	3.019 21
1.0	3.441 09	4	3.436 56	2	3.436 57	2	3.436 56

A numerical method for producing an approximate solution to an initial value problem is said to be convergent if, assuming there is no roundoff error, the numerical solution approaches the "true" solution as the stepsize approaches zero. All the numerical methods presented in this text are convergent. However, this does *not* mean that as the stepsize approaches zero, the numerical solution will *always* approach the "true" solution, since roundoff error will always be present. Sometimes the error of a numerical solution turns out to be larger than predicted by the local error estimate. And, furthermore as the stepsize is decreased, the error for a particular fixed value of the independent variable may become larger instead of smaller. This phenomenon is known as *numerical instability*. Numerical instability is a property of both the numerical method and the initial value problem. That is, a numerical method may be unstable for some initial value problems and stable for others. Numerical instability usually arises because a first order differential equation is approximated by a second or higher order difference equation. The approximating difference equation has two or more solutions: the fundamental solution which approximates the "true" solution of the initial value problem, and one or more parasitic solutions. The parasitic solutions are so named because they "feed" upon the errors (truncation and roundoff) of the numerical solution. If the parasitic solutions remain "small" relative to the fundamental solution, then the numerical method is stable, whereas if a parasitic solution becomes "large" relative to a fundamental solution, then the numerical method is unstable. For h sufficiently small, single-step methods do not exhibit any numerical instability for any initial value problems. On the other hand, multistep methods may be unstable for some initial value problems for a particular range of values of the stepsize or for all stepsizes. In practice one chooses a particular numerical method and produces numerical solutions for two or more reasonably "small" stepsizes. If the solutions produced are essentially the same, then the numerical method is probably stable for the problem under consideration and the results are probably also reasonably good. If the results are not similar, then one should reduce the stepsize further. If dissimilar results persist, then the numerical method is probably instable for the problem under consideration and a different numerical method should be employed.

EXAMPLE Use Euler's method ($y_{n+1} = y_n + hf_n$) and the midpoint rule ($y_{n+1} = y_{n-1} + 2hf_n$) with $h = .1$ to produce numerical solutions to the initial value problem

$$y' = -3y + 1; \qquad y(0) = 1 \tag{69}$$

on the interval [0, 2.5] and compare the results with the true solution.

We easily find the true solution of the linear initial value problem (69) to be

$$y = \frac{2e^{-3x} + 1}{3}.$$

Table 4.10 contains values of the true solution, Euler's solution, the difference between Euler's solution and the true solution, the midpoint rule solution, and the difference between the midpoint rule solution and the true solution. Notice that near the initial value $x_0 = 0$ the midpoint rule solution is more accurate than the Euler's method solution. This is due to the fact that the midpoint rule has a smaller truncation error. But notice that as x increases the error of the midpoint rule solution increases rapidly. This occurs because the parasitic solution is beginning to overwhelm the fundamental solution, so that for $x > .8$, the Euler's method solution is more accurate than the midpoint rule solution.

In summary, single-step methods, such as the fourth order Runge-Kutta method, have the advantages of being self-starting, numerically stable, and requiring a small amount of computer storage and the disadvantages of providing no error estimate and requiring multiple function evaluations per step. Multistep methods have the advantage of requiring only one function evaluation per step but have the disadvantages of requiring starting values, occasionally being numerically unstable, providing no error estimate, and requiring more computer storage than single-step methods. Predictor-corrector methods provide an error estimate at each step and require two or

Table 4.10 Solutions to the IVP (69)

x_n	True solution $y(x_n)$	Euler's method		Midpoint rule	
		y_n	*Difference*	y_n	*Difference*
.0	1.000 000	1.000 000	.000 000	1.000 000	.000 000
.2	.699 208	.870 000	.170 792	.703 673	.004 465
.4	.534 129	.596 300	.062 171	.540 668	.006 539
.6	.443 533	.462 187	.018 654	.452 303	.008 770
.8	.393 812	.396 472	.002 660	.406 768	.013 868
1.0	.366 525	.364 271	−.002 254	.387 669	.021 144
1.2	.351 549	.348 493	−.003 056	.388 130	.036 581
1.3	.346 828	.343 945	−.002 883	.299 686	−.047 142
1.4	.343 330	.340 762	−.002 568	.408 319	.064 989
1.5	.340 739	.338 533	−.002 206	.254 695	−.095 044
1.6	.338 820	.336 973	−.001 847	.455 502	.116 682
1.7	.337 398	.335 881	−.001 517	.181 394	−.156 004
1.8	.336 344	.355 117	−.001 227	.546 665	.210 321
1.9	.335 564	.334 582	−.000 982	.053 395	.282 169
2.0	.334 986	.334 207	−.000 779	.714 628	.379 642
2.1	.334 558	.333 945	−.000 613	−.175 381	−.509 934
2.2	.334 241	.333 762	−.000 479	1.019 856	.685 615
2.3	.334 005	.333 633	−.000 372	−.587 295	−.921 300
2.4	.333 831	.333 543	−.000 288	1.572 232	1.238 401
2.5	.333 702	.333 480	−.000 222	−1.330 634	−1.664 336

more function evaluations per step. The amount of computer storage and the numerical stability of predictor-corrector algorithms depend upon whether the formulas employed are single-step or multistep formulas.

Many people prefer to use a fourth or higher order single-step method such as Runge-Kutta to numerically solve simple initial value problems on a one-time basis over a "small" interval. Such a method is usually selected because it is self-starting and numerically stable but—most usually—because the user has a better understanding of and confidence in such a method. However, when a numerical method is to be used to solve the same or similar complex initial value problem many times or the solution is to be produced over a "large" interval, then some form of a multistep, predictor-corrector method should be chosen. Such a method of solution should be selected because it normally requires less computing time to produce a given accuracy and because an estimate of the error at each step is built into the method.

4.3.4 Numerical Case Studies

In this section we shall present some examples to illustrate the kinds of results one may obtain if a numerical solution to an initial value problem is generated without first analyzing the initial value problem itself. The purpose of this section is to point out the necessity of applying existence and uniqueness theorems to an initial value problem prior to generating any numerical solution.

EXAMPLE Use the single-step, multistep, and predictor-corrector methods of the previous sections with stepsize $h = .2$ and $h = .1$ to solve the initial value problem

$$y' = \tfrac{1}{2}y^3; \qquad y(-1) = \frac{1}{\sqrt{1.1}} \tag{70}$$

on the interval $[-1, 1]$.

Since $f(x, y) = \frac{1}{2}y^3$ is continuous, there exists a solution to the IVP (70) on some interval about $x = -1$. Since $f_y = \frac{3}{2}y^2$, f satisfies a Lipschitz condition with respect to y in any domain D which is bounded with respect to y. Consequently, the IVP (70) has a unique solution on some interval about $x = -1$, and this solution may be extended uniquely until the boundary of D is reached—in this case, until y becomes infinite.

Notice that if the initial value, (x_0, y_0), is such that $y_0 > 0$, then the solution to the differential equation $y' = \frac{1}{2}y^3$ will be a strictly monotone increasing function, since y' will always be positive. If, on the other hand, $y_0 < 0$, then y' will always be negative and the solution will be strictly monotone decreasing. If $y_0 = 0$, then the solution is $y(x) \equiv 0$.

Since the variables separate, the actual solution to the IVP (70) written as an integral equation is

$$2\int_{1/\sqrt{1.1}}^{y} y^{-3}\,dy = \int_{-1}^{x} dx.$$

Integrating and rearranging, we find

$$y^2 = \frac{1}{.1 - x}$$

or

$$y = \frac{\pm 1}{\sqrt{.1 - x}}. \tag{71}$$

Notice that (71) is not real for $x \geq .1$ and that the positive branch must be chosen to satisfy the initial condition $y(-1) = 1/\sqrt{1.1}$.

Each of the numerical methods produces a strictly monotone increasing solution to the IVP (70). It is of interest to record the last value of the independent variable for which each method successfully calculated an approximate solution value. This value of the independent variable should be near $x = .1$, since the actual solution $y(x)$ satisfies $\lim_{x \to .1-} y(x) = +\infty$. Table 4.11 contains the last values of x for which each of the numerical methods successfully calculated an approximate solution value for stepsizes $h = .2$ and $h = .1$. The results of this exercise indicate the unreliability of the numerical method when dealing with initial value problems similar to the given problem.

Table 4.11 Last x_n value for which y_n calculated for the IVP (70)

Method	Stepsize $h = .2$	Stepsize $h = .1$
Euler	>1.0	.8
Improved Euler	.6	.3
Fourth order Runge-Kutta	.2	.2
Adams-Bashforth		
$m = 1$	>1.0	.5
$m = 2$	.8	.5
$m = 3$	.8	.5
Nystrom		
$m = 1$	.8	.5
$m = 2$	.8	.5
$m = 3$	.8	.5
Milne's method	.8	.5
Predictor-corrector		
Equations (66)	−.2	.0
Equations (67)	.0	.0
Equations (68)	.0	.0

EXAMPLE Analyze and solve the initial value problem

$$y' = -\frac{x}{y}; \qquad y(-1) = .21 \tag{72}$$

on the interval $[-1, 3]$.

The function $f(x, y) = -x/y$ is defined and continuous except for $y = 0$, so there exists a solution to the IVP (72) on some interval about $x = -1$. Also since $f_y = x/y^2$ is defined except for $y = 0$, f satisfies a Lipschitz condition with respect to y in any domain which is bounded away from the x-axis ($y = 0$). Hence, the solution, y, of the IVP (72) can be extended uniquely until y becomes zero.

Since the variables separate, the solution to the IVP (72) is

$$\int_{.21}^{y} y\,dy = -\int_{-1}^{x} dx.$$

Integration leads to

$$\frac{y^2}{2} - \frac{(.21)^2}{2} = -\frac{x^2}{2} + \frac{1}{2},$$

which can be rewritten as

$$x^2 + y^2 = 1.0441. \tag{73}$$

The graph of this equation is a circle with center at the origin and radius $\sqrt{1.0441}$. Solving equation (73) for y we obtain the following explicit solution to the IVP (72):

$$y = \sqrt{1.0441 - x^2}.$$

Notice that the solution exists only for $|x| < \sqrt{1.0441}$, and the graph of the solution is a semicircle.

For instructive purposes we generated numerical solutions to the IVP (72) on the interval $[-1, 3]$. We employed all the methods discussed in the previous section with stepsizes $h = .2$ and $h = .1$. Figure 4.15 is a graph of the solution obtained using Euler's method with $h = .2$ and the improved Euler's method with $h = .1$. All other numerical methods except the fourth order Runge-Kutta method produced solutions with the same oscillatory behavior for $1 < x \leq 3$. For $h = .1$ the fourth order Runge-Kutta method produced large positive solution values for $1 < x \leq 3$. A graph of the Runge-Kutta solutions for $h = .2$ and $h = .1$ is shown in Figure 4.16.

EXAMPLE Analyze and solve the initial value problem

$$y' = 3xy^{1/3}; \qquad y(-1) = -1 \tag{74}$$

on the interval $[-1, 1]$.

Since $f(x, y) = 3xy^{1/3}$ is continuous for all values of x and y, a solution to the differential equation of (74) exists on some interval about x_0 for all (x_0, y_0). Hence, a solution to the IVP (74) exists and may be extended until x or y becomes infinite. Since $f_y = xy^{-2/3}$, f satisfies a Lipschitz condition with respect to y on any domain which is bounded away from the x-axis ($y = 0$) and f does not satisfy a Lipschitz condition with respect to y on any domain whose closure contains a point of the x-axis. Consequently, a solution y, of the IVP (74) can be continued uniquely at least until y becomes zero. Notice that $y = 0$ is a solution of the differential equation $y' = 3xy^{1/3}$.

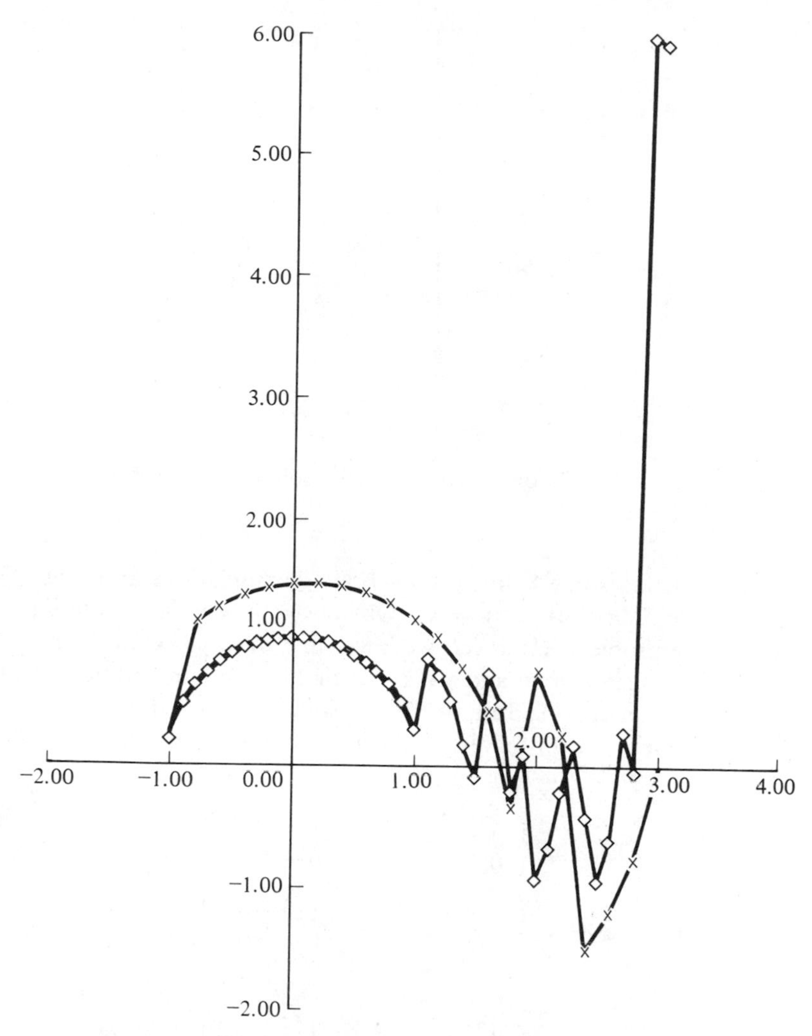

Figure 4.15 Solution to IVP $y' = -x/y$; $y(-1) = .21$. Euler's method $h = .2$, $\times$; improved Euler's method $h = .1$, $\diamond$.

Again the variables separate and we may write the solution of the IVP (74) as

$$\int_{-1}^{y} y^{-1/3}\, dy = 3 \int_{-1}^{x} x\, dx.$$

Hence,

$$\tfrac{3}{2}(y^{2/3} + 1) = \tfrac{3}{2}(x^2 + 1)$$

or

$$y^{2/3} = x^2.$$

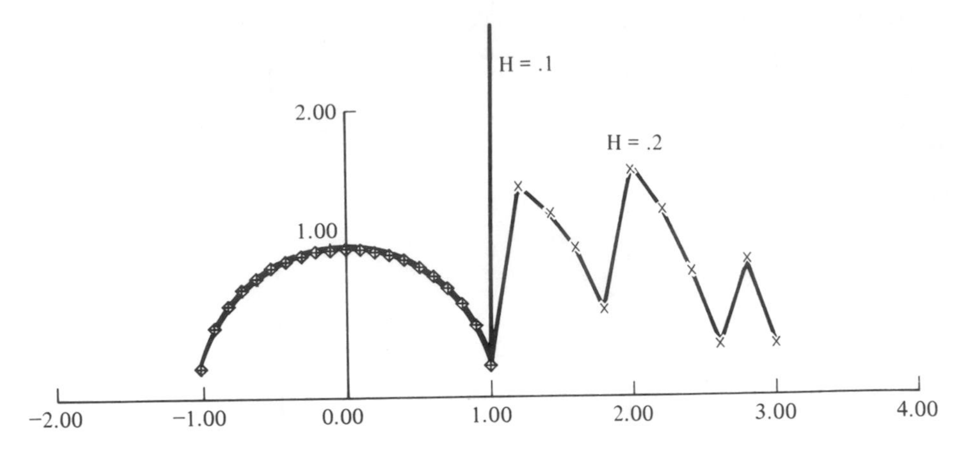

Figure 4.16 Fourth order Runge-Kutta solution to IVP $y' = -x/y$; $y(-1) = .21$.

It is tempting to do some "simple" algebraic manipulations and write *the* solution of the IVP (74) as $y = x^3$. Obviously, $y = x^3$ is a solution of the initial value problem. However, we know that once y becomes zero, the solution is no longer unique and we know that $y = 0$ is a solution of the differential equation of (74). So $y = x^3$ is *the* solution on the interval $[-1, 0]$, but for $x > 0$ there are an infinite number of solutions to the initial value problem. For example,

$$y_1(x) = \begin{cases} x^3, & -1 \le x \le 0 \\ 0, & 0 < x \le 1 \end{cases}$$

$$y_2(x) = \begin{cases} x^3, & -1 \le x \le 0 \\ -x^3, & 0 < x \le 1 \end{cases}$$

and

$$y_3(x) = \begin{cases} x^3, & -1 \le x \le 0 \\ 0, & 0 < x \le \frac{1}{2} \\ (x^2 - \frac{1}{4})^{3/2}, & \frac{1}{2} \le x \le 1 \end{cases}$$

are all solutions to the IVP (74).

The Adams-Bashforth solutions to the IVP (74) for $m = 1$ and $m = 2$ and a stepsize $h = .1$ are shown in Figure 4.17. The improved Euler method, the fourth order Runge-Kutta method, the Nystrom method with $m = 1$, and the predictor-corrector method of equations (66) all produce, for $h = .1$ and $h = .2$, numerical solutions similar to the Adams-Bashforth solution for $m = 1$. All other numerical methods discussed in the previous sections

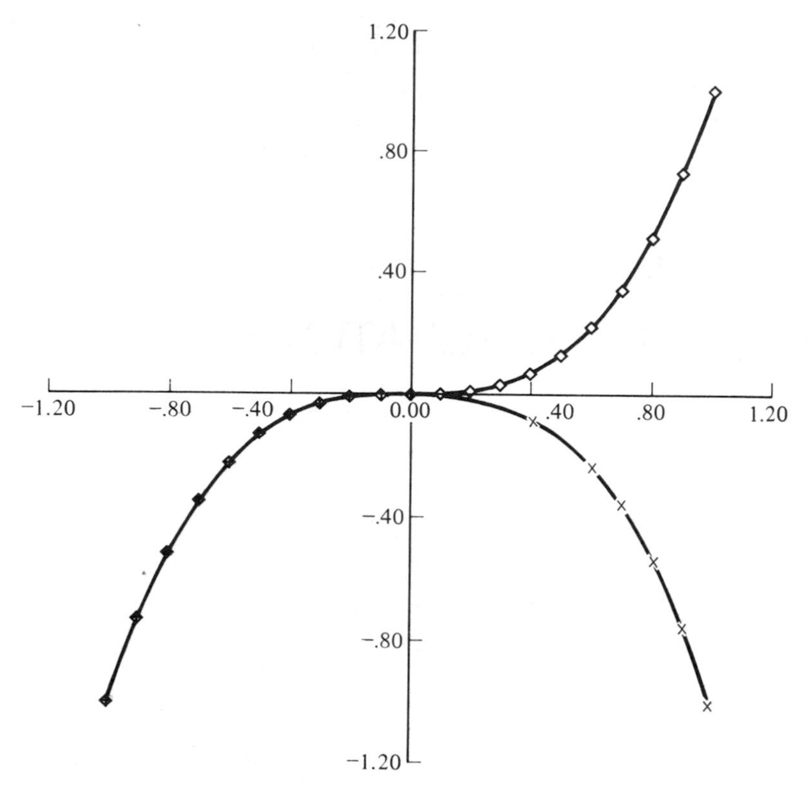

Figure 4.17 Solution to the IVP $y' = 3xy^{1/3}$; $y(-1) = -1$. Adams-Bashforth method $m = 1$, $\times$; $m = 2$, $\diamond$.

produce, for $h = .1$ and $h = .2$, solutions similar to the Adams-Bashforth solution with $m = 2$.

EXERCISES

1. Analyze the following initial value problems.
   (i) $y' = 2x/3y^2$; $y(-1) = 1$
   (ii) $y' = -3x^2/2y$; $y(-1) = 1$
   (iii) $y' = 2(xy)^{1/3}$; $y(-1) = 1$
   Use the fourth order Runge-Kutta method with stepsize $h = .2$ and $h = .1$ to solve these initial value problems on the interval $[-1, 1]$. Calculate the actual solutions and compare the results.

2. Use the Adams-Bashforth method with $m = 3$, the Nystrom method with $m = 3$, Milne's method, and the predictor-corrector methods (66), (67), and (68) to solve the initial value problems of Exercise 1 on the interval $[-1, 1]$. Use the fourth order Runge-Kutta solution values as starting values and stepsizes $h = .2$ and $h = .1$.

5

SYSTEMS OF FIRST ORDER EQUATIONS: THEORY

The purpose of this brief chapter is to show how the fundamental existence and uniqueness theorem for the scalar first order initial value problem $y' = f(x, y)$; $y(x_0) = c_0$ can be extended to a system of first order initial value problems. We then consider a linear system of equations in which all the coefficient functions are continuous on some interval I. Employing the stated existence and uniqueness theorem for systems of initial value problems, we prove that there exists a unique solution on the interval I to the initial value problem consisting of the linear system of equations with appropriate initial conditions. Next we show how an nth order differential equation may be written as a system of n first order equations. This equivalence and the existence and uniqueness theorem for a system of initial value problems yield an existence and uniqueness theorem for an nth order differential equation. Finally, applying this theorem to a linear nth order differential equation we are able to establish an existence and uniqueness theorem for a linear nth order differential equation with appropriate initial conditions. This is the theorem of immediate consequence, since Chapter 6 is devoted to the study of, and methods for obtaining solutions to, linear nth order differential equations.

Now let us consider the following system of n first order differential equations:

$$
\begin{aligned}
y_1' &= f_1(x, y_1, y_2, \ldots, y_n) \\
y_2' &= f_2(x, y_1, y_2, \ldots, y_n) \\
&\vdots \\
y_n' &= f_n(x, y_1, y_2, \ldots, y_n),
\end{aligned} \tag{1}
$$

where y_i $(i = 1, 2, \ldots, n)$ are real-valued functions of the independent variable x and f_i $(i = 1, 2, \ldots, n)$ are real-valued functions of $x, y_1, \ldots, y_n$. Systems such as (1) arise naturally from problems in which there is one independent variable and several dependent variables. For example, the equations of motion of a body in space may be expressed as a system of six first order differential equations. In this case, the independent variable, x, is time, y_1, y_3, and y_5 represent the position of the body in some three dimensional coordinate system and y_2, y_4, and y_6 represent the velocity of the body. Hence, the system might be

$$
\begin{aligned}
y_1' &= y_2 && = f_1(x, y_1, y_2, \ldots, y_6) \\
y_2' &= -\frac{y_1}{\sqrt{y_1^2 + y_3^2 + y_5^2}} && = f_2(x, y_1, y_2, \ldots, y_6) \\
y_3' &= y_4 && = f_3(x, y_1, y_2, \ldots, y_6) \\
y_4' &= -\frac{y_3}{\sqrt{y_1^2 + y_3^2 + y_5^2}} && = f_4(x, y_1, y_2, \ldots, y_6) \\
y_5' &= y_6 && = f_5(x, y_1, y_2, \ldots, y_6) \\
y_6' &= -\frac{y_5}{\sqrt{y_1^2 + y_3^2 + y_5^2}} && = f_6(x, y_1, y_2, \ldots, y_6),
\end{aligned}
$$

where the forces acting on the body—f_2, f_4, and f_6—are functions only of the position of the body—y_1, y_3, and y_5.

If the functions $y_1(x), y_2(x), \ldots, y_n(x)$ all have continuous first derivatives on an interval $I = [\alpha, \beta]$ and satisfy (1) on I, then the ordered n-tuple of functions

$$
\begin{pmatrix} y_1 \\ y_2 \\ \vdots \\ y_n \end{pmatrix}
$$

is called a *solution* of system (1) on I. For example,

$$
\begin{pmatrix} \sin x \\ \cos x \end{pmatrix}
$$

is a solution of the system

$$
\begin{aligned}
y_1' &= y_2 \\
y_2' &= -y_1
\end{aligned} \tag{2}
$$

on the interval $(-\infty, \infty)$.

For our purposes a *vector*, $\mathbf{v}$, will be an ordered n-triple of real numbers or real-valued functions and we will represent $\mathbf{v}$ by $(v_1, v_2, \ldots, v_n)$ or

$$
\begin{pmatrix} v_1 \\ v_2 \\ \vdots \\ v_n \end{pmatrix}
$$

The *Euclidean norm* of a vector is $\|\mathbf{v}\| = (v_1^2 + v_2^2 + \cdots + v_n^2)^{1/2}$ and the triangle inequality $\|\mathbf{u} + \mathbf{v}\| \leq \|\mathbf{u}\| + \|\mathbf{v}\|$ holds. The *inner product* of two vectors $\mathbf{u}$ and $\mathbf{v}$ is defined as $\mathbf{u} \cdot \mathbf{v} = u_1 v_1 + u_2 v_2 + \cdots + u_n v_n$ and satisfies Schwarz's inequality $|\mathbf{u} \cdot \mathbf{v}| \leq \|\mathbf{u}\| \, \|\mathbf{v}\|$. Let

$$
\mathbf{y} = \begin{pmatrix} y_1 \\ y_2 \\ \vdots \\ y_n \end{pmatrix}
$$

be a vector of real-valued functions. Then the vector

$$
\mathbf{y}' = \begin{pmatrix} y_1' \\ y_2' \\ \vdots \\ y_n' \end{pmatrix}
$$

and the vector

$$
\int_a^b \mathbf{y} \, dx = \begin{pmatrix} \int_a^b y_1 \, dx \\ \int_a^b y_2 \, dx \\ \vdots \\ \int_a^b y_n \, dx \end{pmatrix}
$$

Let

$$
\mathbf{f}(x, \mathbf{y}) = \begin{pmatrix} f_1(x, \mathbf{y}) \\ f_2(x, \mathbf{y}) \\ \vdots \\ f_n(x, \mathbf{y}) \end{pmatrix}
$$

Then system (1) may be represented more concisely by the vector differential equation.

$$\mathbf{y}' = \mathbf{f}(x, \mathbf{y}), \tag{3}$$

and a *solution* of (3) on an interval I is a vector of functions which has a continuous first derivative and satisfies (3) on I.

5.1 THE INITIAL VALUE PROBLEM: $\mathbf{y}' = \mathbf{f}(x, \mathbf{y}); \mathbf{y}(x_0) = \mathbf{c}_0$

The vector initial value problem which corresponds to the scalar initial value problem $y' = f(x, y); y(x_0) = c_0$ is

$$\mathbf{y}' = \mathbf{f}(x, \mathbf{y}); \qquad \mathbf{y}(x_0) = \mathbf{c}_0, \tag{4}$$

where $\mathbf{c}_0$ is a constant vector—a vector with components which are real numbers. Hence, an initial value problem for a system of equations is to find a solution of (3) subject to the conditions $\mathbf{y}(x_0) = \mathbf{c}_0$.

EXAMPLE Solve the system (2) subject to the initial conditions

$$\mathbf{y}(0) = \begin{pmatrix} 2 \\ 3 \end{pmatrix}.$$

The general solution of (2) is

$$\mathbf{y} = A\begin{pmatrix} \sin x \\ \cos x \end{pmatrix} + B\begin{pmatrix} \cos x \\ -\sin x \end{pmatrix} = \begin{pmatrix} A \sin x + B \cos x \\ A \cos x - B \sin x \end{pmatrix} = \begin{pmatrix} y_1 \\ y_2 \end{pmatrix},$$

where A and B are arbitrary constants. (We will show how to find the general solution of a linear system of equations with constant coefficients in a later chapter.) In order to satisfy the initial conditions, A and B must satisfy

$$\mathbf{y}(0) = \begin{pmatrix} 2 \\ 3 \end{pmatrix} = \begin{pmatrix} B \\ A \end{pmatrix}.$$

Hence $A = 3$ and $B = 2$ and the solution of this vector initial value problem is

$$\mathbf{y} = \begin{pmatrix} 3 \sin x + 2 \cos x \\ 3 \cos x - 2 \sin x \end{pmatrix}.$$

5.1.1 Existence and Uniqueness Theorems

One proceeds in a manner similar to that employed for the scalar initial value problem to establish conditions under which the vector initial value problem (4) is well-posed. Consequently, in what follows we will dispense with many of the details.

First, one proves the following theorem relating the vector initial value problem to a vector integral equation.

Theorem 5.1 Let $\mathbf{f}(x, \mathbf{y})$ be a vector whose components, $f_1, f_2, \ldots, f_n$, are all continuous functions of $x, y_1, y_2, \ldots, y_n$. That is, let $\mathbf{f}(x, \mathbf{y}) \in \mathscr{C}$. Then $\mathbf{y}$ is a solution of the IVP (4) on an interval I about x_0 if and only if $\mathbf{y}$ satisfies the integral equation

$$\mathbf{y}(x) = \mathbf{c}_0 + \int_{x_0}^{x} \mathbf{f}(t, \mathbf{y}(t))\, dt \tag{5}$$

for all $x \in I$.

The proof of this theorem is similar to the proof of Theorem 4.1 and will be omitted.

A vector-valued function $\mathbf{f}(x, \mathbf{y})$ satisfies a *Lipschitz condition with respect to* $\mathbf{y}$ *in a domain* D of $(x, \mathbf{y})$-space if and only if there exists a positive constant M such that $(x, \mathbf{y}_1), (x, \mathbf{y}_2) \in D$ implies $\|\mathbf{f}(x, \mathbf{y}_1) - \mathbf{f}(x, \mathbf{y}_2)\| \leq M\|\mathbf{y}_1 - \mathbf{y}_2\|$.

Theorem 5.2 Let $\mathbf{f}(x, \mathbf{y}) \in \mathscr{C}$ on some domain D of $(x, \mathbf{y})$-space and let $\mathbf{f}(x, \mathbf{y})$ satisfy a Lipschitz condition with respect to $\mathbf{y}$ in D. Let $(x_0, \mathbf{c}_0) \in D$. Then there exists a unique solution to the IVP (4) on some closed interval about x_0.

To prove this theorem one proceeds as in the scalar case—see Theorem 4.2. First one shows how to appropriately choose a closed interval I with center x_0. Then, for $x \in I$, one defines the following sequence of vectors:

$$\mathbf{y}_0(x) = \mathbf{c}_0$$

$$\mathbf{y}_n(x) = \mathbf{c}_0 + \int_{x_0}^{x} \mathbf{f}(t, \mathbf{y}_{n-1}(t))\, dt, \qquad n = 1, 2, \ldots.$$

This sequence is shown to be well defined on I and to converge uniformly to a vector $\mathbf{y}^*$ which is defined and continuous on I and which satisfies the IE (5) on I. Consequently, $\mathbf{y}^*$ satisfies the IVP (4) on I. Finally, the solution $\mathbf{y}^*$ is shown to be unique on I.

The following theorem regarding the unique extension of the solution can be proven.

Theorem 5.3 Under the hypotheses of Theorem 5.2, the solution $\mathbf{y}^*$ of the vector IVP (4) can be extended uniquely until the boundary of D is reached.

Lemma Let D be a bounded convex domain and let $\mathbf{f}(x, \mathbf{y}) \in \mathscr{C}$ and have continuous first partial derivatives with respect to $y_1, y_2, \ldots, y_n$ on the closure of D. Then f satisfies a Lipschitz condition with respect to $\mathbf{y}$ on D and a suitable Lipschitz constant is $M = n \sup_{\substack{D \\ 1 \le i,j \le n}} |\partial f_i/\partial y_j|$.

The proof of this lemma is similar to the proof of Theorem 4.3. However, the proof requires the use and understanding of the inner product of two vectors and Schwarz's inequality and, therefore, will be omitted. We can deduce the following theorem from the lemma above and Theorem 5.2.

Theorem 5.4 If $\mathbf{f}(x, \mathbf{y}) \in \mathscr{C}$ and has continuous first partial derivatives with respect to $y_1, y_2, \ldots, y_n$ on the closure of a domain D, then the vector IVP $\mathbf{y}' = \mathbf{f}(x, \mathbf{y}); \mathbf{y}(x_0) = \mathbf{c}_0$ where $(x_0, \mathbf{c}_0) \in D$ has a unique solution on some interval containing x_0.

5.1.2 Linear Systems

If each component f_i $(i = 1, 2, \ldots, n)$ of $\mathbf{f}(x, \mathbf{y})$ is linear, then the system $\mathbf{y}' = \mathbf{f}(x, \mathbf{y})$ is called a *linear* system. Consider the general linear system

$$
\begin{aligned}
y_1' &= a_{11}(x)y_1 + a_{12}(x)y_2 + \cdots + a_{1n}(x)y_n + b_1(x) = f_1(x, \mathbf{y}) \\
y_2' &= a_{21}(x)y_1 + a_{22}(x)y_2 + \cdots + a_{2n}(x)y_n + b_2(x) = f_2(x, \mathbf{y}) \\
&\vdots \qquad\qquad\qquad\qquad\qquad\qquad\qquad\qquad\qquad\qquad\quad \vdots \\
y_n' &= a_{n1}(x)y_1 + a_{n2}(x)y_2 + \cdots + a_{nn}(x)y_n + b_n(x) = f_n(x, \mathbf{y}),
\end{aligned} \tag{6}
$$

where $a_{ij}(x)$ and $b_i(x)$ $(i, j = 1, 2, \ldots, n)$ are all continuous on some closed interval $I = [\alpha, \beta]$ which contains x_0. If $b_i(x) \equiv 0$ for $i = 1, 2, \ldots, n$, the system is called *homogeneous*; otherwise, it is called *nonhomogeneous.*

Since $\partial f_i/\partial y_j = a_{ij}(x)$ and $b_i(x)$ are continuous for $x \in I$, $\mathbf{f}(x, \mathbf{y}) \in \mathscr{C}$, and has continuous first partial derivatives with respect to $y_1, y_2, \ldots, y_n$ for $x \in I$. Let $D = \{(x, \mathbf{y}) \mid \alpha < x < \beta \text{ and } \|\mathbf{y}\| < \infty\}$. Since I is closed and $a_{ij}(x)$ is continuous,

$$K = \sup_{\substack{D \\ 1 \le i,j \le n}} \left|\frac{\partial f_i}{\partial y_j}\right| = \max_{\substack{I \\ 1 \le i,j \le n}} |a_{ij}(x)| < \infty.$$

Applying Theorems 5.3 and 5.4 to the initial value problem consisting of the system (6) and the initial condition $(x_0, \mathbf{c}_0)$, we obtain the following theorem:

Theorem 5.5 The linear system (6) has a unique solution on I, $\mathbf{y}(x)$, which satisfies $\mathbf{y}(x_0) = \mathbf{c}_0$.

5.2 *n*th ORDER DIFFERENTIAL EQUATIONS

There is a close relationship between *n*th order differential equations which are written in normal form and a system of *n* first order differential equations. We wish to exhibit this relationship and then exploit it. Consider the *n*th order differential equation.

$$y^{(n)} = f(x, y, y^{(1)}, \ldots, y^{(n-1)}). \tag{7}$$

Let $y_1 = y$, $y_2 = y^{(1)}, \ldots, y_n = y^{(n-1)}$. Then the *n*th order differential equation (7) may be rewritten as

$$\begin{aligned} y_1' &= y_2 &&= f_1(x, \mathbf{y}) \\ y_2' &= y_3 &&= f_2(x, \mathbf{y}) \\ &\vdots && \vdots \\ y_{n-1}' &= y_n &&= f_{n-1}(x, \mathbf{y}) \\ y_n' &= f(x, y_1, y_2, \ldots, y_n) &&= f_n(x, \mathbf{y}). \end{aligned} \tag{8}$$

Hence, the *n*th order differential equation (7) is equivalent to the system of *n* first order differential equations (8). Observe that (8) is a special case of (3) $\mathbf{y}' = \mathbf{f}(x, \mathbf{y})$. Recall that the initial value problem associated with (3) is

$$\mathbf{y}' = f(x, \mathbf{y}); \qquad \mathbf{y}(x_0) = \mathbf{c}_0. \tag{9}$$

Making use of the correspondence between (7) and (8), we see that the initial conditions for (7) which are equivalent to the initial conditions $\mathbf{y}(x_0) = \mathbf{c}_0$ for (8) are $y_1(x_0) = y(x_0) = c_1$, $y_2(x_0) = y^{(1)}(x_0) = c_2, \ldots, y_n(x_0) = y^{(n-1)}(x_0) = c_n$. Hence, the initial value problem associated with (7) which is equivalent to (9) is

$$\begin{gathered} y^{(n)} = f(x, y, y^{(1)}, \ldots, y^{(n-1)}); \\ y(x_0) = c_1,\ y^{(1)}(x_0) = c_2, \ldots, y^{(n-1)}(x_0) = c_n. \end{gathered} \tag{10}$$

Since $f_1(x, \mathbf{y}) = y_2$, $f_2(x, \mathbf{y}) = y_3, \ldots, f_{n-1}(x, \mathbf{y}) = y_n$ are all continuous in any domain D of $(x, \mathbf{y})$-space and since for $i = 1, 2, \ldots, n-1$ and $j = 1, 2, \ldots, n$

$$\frac{\partial f_i}{\partial y_j} = \begin{cases} 0, & j \neq i+1 \\ 1, & j = i+1, \end{cases}$$

system (8) satisfies the hypotheses of Theorem 5.2, if $f(x, y_1, y_2, \ldots, y_n) \in \mathscr{C}$ and satisfies a Lipschitz condition with respect to $\mathbf{y}$ on some domain D. Hence, we have the following theorem for the IVP (10).

Theorem 5.6 If $f(x, y, y^{(1)}, \ldots, y^{(n-1)}) \in \mathscr{C}$ on some bounded domain D of $(x, y, y^{(1)}, \ldots, y^{(n-1)})$-space, if f satisfies a Lipschitz condition with respect to $y, y^{(1)}, \ldots, y^{(n-1)}$ on D and if $(x_0, c_1, \ldots, c_n) \in D$, then there exists a unique solution to the IVP (10) on some interval about x_0 and this solution can be extended uniquely until the boundary of D is reached.

The hypotheses of the following theorem—which are sufficient conditions for the hypotheses of Theorem 5.6—are usually easier to verify than those of Theorem 5.6.

Theorem 5.7 If $f(x, y, y^{(1)}, \ldots, y^{(n-1)}) \in \mathscr{C}$ and has continuous first partial derivatives with respect to $y, y^{(1)}, \ldots, y^{(n-1)}$ on some bounded domain D and if $(x_0, c_1, \ldots, c_n) \in D$, then there exists a unique solution to the IVP (10) on some interval about x_0 and this solution can be extended uniquely until the boundary of D is reached.

If $f(x, y, y^{(1)}, \ldots, y^{(n-1)})$ is linear, then the IVP (10) has the form

$$
\begin{aligned}
&y^{(n)} = a_1(x)y + a_2(x)y^{(1)} + \cdots + a_n(x)y^{(n-1)} + b(x); \\
&y(x_0) = c_1, \qquad y^{(1)}(x_0) = c_2, \qquad \ldots, \qquad y^{(n-1)}(x_0) = c_n
\end{aligned} \tag{11}
$$

and the equivalent vector IVP (9) has the form

$$
\begin{aligned}
y_1' &= y_2 \\
y_2' &= y_3 \\
&\vdots \\
y_{n-1}' &= y_n \\
y_n' &= a_1(x)y_1 + a_2(x)y_2 + \cdots + a_n(x)y_n + b(x); \qquad \mathbf{y}(x_0) = \mathbf{c}_0.
\end{aligned} \tag{12}
$$

If $a_1(x), a_2(x), \ldots, a_n(x)$, and $b(x)$ are all continuous on some interval I, then (12) and consequently (11) satisfies the hypotheses of Theorem 5.5. Thus, we have the following theorem for linear nth order differential equations.

Theorem 5.8 If $a_1(x), a_2(x), \ldots, a_n(x)$, and $b(x)$ are all continuous on some interval I containing x_0, then (11) has a unique solution on I.

EXERCISES

1. Verify that (x, x^2) is the solution of the vector initial value problem

$$
\begin{aligned}
y_1' &= 1; & y_1(-1) &= -1 \\
y_2' &= 2y_1; & y_2(-1) &= 1.
\end{aligned}
$$

2. Verify that (e^x, e^x) is the solution of the vector initial value problem

$$
\begin{aligned}
y_1' &= y_2; & y_1(0) &= 1 \\
y_2' &= y_1; & y_2(0) &= 1.
\end{aligned}
$$

3. Let $D = \{(x, y_1, y_2) \mid x^2 + y_1^2 + y_2^2 < 2\}$.

   (a) Show that $\mathbf{f}(x, \mathbf{y}) = \begin{pmatrix} f_1(x, \mathbf{y}) \\ f_2(x, \mathbf{y}) \end{pmatrix} = \begin{pmatrix} 1 \\ 2y_1 \end{pmatrix}$

   of Exercise 1 satisfies a Lipschitz condition with respect to $\mathbf{y}$ on D. What is a suitable value for the Lipschitz constant on D?

   (b) Show that $\mathbf{f}(x, \mathbf{y})$ of Exercise 2 satisfies a Lipschitz condition with respect to $\mathbf{y}$ on D and find a Lipschitz constant for $\mathbf{f}$ on D.

So the solutions to the IVPs of Exercises 1 and 2 are unique on D. Notice that if the domain D is enlarged, the Lipschitz constant remains the same. Since the Lipschitz constant is independent of the domain, the solutions exist for all real x.

4. (a) Verify that $(1/x, x^2)$ is the solution to the vector initial value problem

$$\begin{aligned} y_1' &= -1/y_2; \qquad & y_1(1) &= 1 \\ y_2' &= 2y_1y_2; \qquad & y_2(1) &= 1. \end{aligned}$$

   (b) Let $D = \{(x, y_1, y_2) \mid |x| < 100, |y_1| < 100, \text{ and } \frac{1}{100} < y_2 < 100\}$. Show that $\mathbf{f}(x, \mathbf{y})$ satisfies a Lipschitz condition with respect to $\mathbf{y}$ on D and a suitable Lipschitz constant for $\mathbf{f}$ on D.

   (c) What is the maximum interval on which the solution exists?

5. Write the system of first order equations which is equivalent to each of the following differential equations.
   (a) $y'' + xy' + x^2y = x^3$
   (b) $y''' + xy'y'' - x(y')^2 = 2xy$

6. On what interval does each of the following initial value problems have a solution?
   (a) $y'' + y'/x - y/(x - 2) = 1/(x^2 + 1)$; $y(1) = 1$, $y'(1) = 3$
   (b) $y'' - xy' + y/x = \sin x$; $y(1) = 1$, $y'(1) = 2$

6

LINEAR DIFFERENTIAL EQUATIONS

6.1 GENERAL nth ORDER LINEAR EQUATIONS

An nth order linear differential equation is any equation of the form

$$(1) \qquad f_0(x)y^{(n)} + f_1(x)y^{(n-1)} + \cdots + f_{n-1}(x)y^{(1)} + f_n(x)y = g(x),$$

where $f_0(x) \not\equiv 0$. If $g(x) \equiv 0$, then (1) is said to be *homogeneous*, whereas if $g(x) \not\equiv 0$, (1) is said to be *nonhomogeneous*. Dividing (1) by $f_0(x)$ and letting $a_i(x) = f_i(x)/f_0(x)$ and $b(x) = g(x)/f_0(x)$ we deduce the following theorem regarding (1) from Theorem 5.8.

Theorem 6.1 Let $x_0 \in (\alpha, \beta) = I$ and let $c_0, c_1, \ldots, c_{n-1}$ be arbitrary real constants. If $f_0(x), f_1(x), \ldots, f_n(x)$, and $g(x)$ are continuous on I and $f_0(x) \neq 0$ for all $x \in I$, then there exists a unique solution $y(x)$ of (1) defined on I which satisfies $y(x_0) = c_0, y'(x_0) = c_1, \ldots, y^{(n-1)}(x_0) = c_{n-1}$.

We have the following superposition theorem for any two solutions of an nth order linear homogeneous differential equation.

Theorem 6.2 If $y_1(x)$ and $y_2(x)$ are solutions of the linear homogeneous equation

(2) $$f_0(x)y^{(n)} + f_1(x)y^{(n-1)} + \cdots + f_{n-1}(x)y^{(1)} + f_n(x)y = 0,$$

then $y_3 = c_1y_1 + c_2y_2$ where c_1 and c_2 are arbitrary constants is a solution of (2).

Proof

$$f_0(x)y_3^{(n)} + \cdots + f_n(x)y_3 = c_1(f_0(x)y_1^{(n)} + \cdots + f_n(x)y_1) + c_2(f_0(x)y_2^{(n)} + \cdots + f_n(x)y_2) = c_1 \cdot 0 + c_2 \cdot 0 = 0.$$

We can easily extend this theorem and show that if $y_1, y_2, \ldots, y_m$ are solutions of (2), then any linear combination, $y = \sum_{i=1}^{m} c_i y_i$ where c_i are any constants, is a solution of (2). Hence, any linear combination of solutions of an nth order linear homogeneous differential equation is a solution of that equation.

Consider a set of functions $\{y_1, y_2, \ldots, y_n\}$ which are defined on an interval I. This set is *linearly dependent on I* if there exist constants $c_1, c_2, \ldots, c_n$ not all zero such that $c_1y_1(x) + c_2y_2(x) + \cdots + c_ny_n(x) = 0$ for all $x \in I$. Otherwise, the set is said to be *linearly independent on I*. That is, the set is linearly independent on I if $c_1y_1(x) + c_2y_2(x) + \cdots + c_ny_n(x) = 0$ for all $x \in I$ implies that $c_1 = c_2 = \cdots = c_n = 0$.

Let $y_1, y_2, \ldots, y_n$ be differentiable at least $(n - 1)$ times for all $x \in I$. The *Wronskian* of $y_1, y_2, \ldots, y_n$ on I is the function defined by

$$W(y_1, y_2, \ldots, y_n, x) = \begin{vmatrix} y_1 & y_2 & \cdots & y_n \\ y_1^{(1)} & y_2^{(1)} & \cdots & y_n^{(1)} \\ \vdots & \vdots & & \vdots \\ y_1^{(n-1)} & y_2^{(n-1)} & \cdots & y_n^{(n-1)} \end{vmatrix}.$$

The Wronskian is named in honor of the Polish mathematician H. Wronski (1778–1853).

Theorem 6.3 If the Wronskian is not zero for any $x \in I$, then the functions $y_1, y_2, \ldots, y_n$ are linearly independent on I.

Proof Consider the following system of n linear equations in the n unknowns $c_1, c_2, \ldots, c_n$

$$\begin{aligned} c_1y_1 + c_2y_2 + \cdots + c_ny_n &= 0 \\ c_1y_1^{(1)} + c_2y_2^{(1)} + \cdots + c_ny_n^{(1)} &= 0 \\ &\vdots \\ c_1y_1^{(n-1)} + c_2y_2^{(n-1)} + \cdots + c_ny_n^{(n-1)} &= 0. \end{aligned}$$

This system of equations may be written in matrix-vector notation as

$$\begin{pmatrix} y_1 & y_2 & \cdots & y_n \\ y_1^{(1)} & y_2^{(1)} & \cdots & y_n^{(1)} \\ \vdots & \vdots & & \vdots \\ y_1^{(n-1)} & y_2^{(n-1)} & \cdots & y_n^{(n-1)} \end{pmatrix} \begin{pmatrix} c_1 \\ c_2 \\ \vdots \\ c_n \end{pmatrix} = \begin{pmatrix} 0 \\ 0 \\ \vdots \\ 0 \end{pmatrix}. \tag{3}$$

It follows from a theorem of linear algebra that if the determinant of the matrix in the matrix-vector equation (3) is nonzero, then the vector of unknowns must be the zero vector. Hence, if the Wronskian of $y_1, y_2, \ldots, y_n$—the determinant of the matrix of (3)—is not zero for any $x \in I$, then the solution of (3) is $\mathbf{c} = \mathbf{0}$. Since the first equation of the system of equations is the linear combination which one checks to determine linear independence, we see that the set $\{y_1, y_2, \ldots, y_n\}$ is linearly independent on I.

Thus, if $W(y_1, y_2, \ldots, y_n, x) \neq 0$ for any $x \in I$, $y_1, y_2, \ldots, y_n$ are linearly independent on I. What about the converse theorem? That is, if $y_1, y_2, \ldots, y_n$ are linearly independent on I, is the Wronskian nonzero on I? Unfortunately, the answer is no, as the following example shows.

EXAMPLE Let $y_1(x) = x^3$ and $y_2(x) = |x|^3$. These functions are linearly independent on $(-2, 2)$, but $W(y_1, y_2, x) = 0$ for $x \in (-2, 2)$ and $x \neq 0$ as the following computations show. In order to see that y_1 and y_2 are linearly independent on $(-2, 2)$, we assume that they are linearly dependent on $(-2, 2)$. That is, we assume that there exist constants c_1 and c_2 not both zero such that $c_1 y_1(x) + c_2 y_2(x) = 0$ for all $x \in (-2, 2)$. We find for $x = -1$, $-c_1 + c_2 = 0$ and for $x = +1$, $c_1 + c_2 = 0$, from which we conclude that $c_1 = c_2 = 0$, which contradicts our assumption. Hence, $y_1(x)$ and $y_2(x)$ are linearly independent on $(-2, 2)$. And since

$$\frac{d|x|^3}{dx} = 3|x|^2 \frac{d|x|}{dx} = 3|x|^2 \frac{|x|}{x} = \begin{cases} -3x^2, & x < 0 \\ 3x^2, & x > 0, \end{cases}$$

we have for $x \neq 0$,

$$W(y_1, y_2, x) = \begin{vmatrix} x^3 & |x|^3 \\ 3x^2 & 3|x|^2 \dfrac{|x|}{x} \end{vmatrix} = 3x^3|x|^2 \frac{|x|}{x} - 3x^2|x|^3 = 0.$$

If in addition to being linearly independent on I, the functions $y_1, y_2, \ldots, y_n$ are all analytic on I—that is, if the functions all have a Taylor series expansion which is valid on the interval I, it can be shown that the Wronskian does not vanish on I. Or as we shall show if in addition to being linearly independent on I the functions $y_1, y_2, \ldots, y_n$ satisfy the same nth order linear homogeneous differential equation, then the Wronskian does not vanish on I. First, we prove the following theorem for second order linear homogeneous differential equations.

Theorem 6.4 Let $f_0(x)$, $f_1(x)$, and $f_2(x)$ be continuous on some interval I and suppose $f_0(x) \neq 0$ for all $x \in I$. If y_1 and y_2 are linearly independent solutions on I of the second order linear homogeneous equation,

$$f_0(x)y'' + f_1(x)y' + f_2(x)y = 0, \tag{4}$$

then the Wronskian of y_1 and y_2 does not vanish on I [that is, $W(y_1, y_2, x) \neq 0$ for any $x \in I$].

Proof Suppose that y_1 and y_2 are linearly independent on I and satisfy (4) on I. Also assume that for some $x_0 \in I$, $W(y_1, y_2, x_0) = 0$. Hence,

$$\begin{vmatrix} y_1(x_0) & y_2(x_0) \\ y_1'(x_0) & y_2'(x_0) \end{vmatrix} = 0,$$

from which we deduce that

$$\begin{pmatrix} y_1(x_0) \\ y_1'(x_0) \end{pmatrix} = k\begin{pmatrix} y_2(x_0) \\ y_2'(x_0) \end{pmatrix} \tag{5}$$

for some k. Since y_1 and y_2 are solutions of (4) on I, so is any linear combination of y_1 and y_2. In particular, $Y(x) = y_1(x) - ky_2(x)$ is a solution of (4) on I. In view of (5) we have

$$Y(x_0) = 0 \quad \text{and} \quad Y'(x_0) = 0. \tag{6}$$

But by Theorem 6.1 there exists one and only one solution of (4) on I which satisfies the initial conditions (6). Obviously, that solution is the zero solution. Hence, $Y(x) \equiv 0$ on I. Therefore, y_1 and y_2 are linearly dependent on I, which is a contradiction. Consequently, if y_1 and y_2 are linearly independent solutions of (4) on I, their Wronskian does not vanish on I.

It follows from Theorems 6.3 and 6.4 that y_1 and y_2 are linearly independent solutions of (4) on I if and only if y_1 and y_2 satisfy (4) on I and $W(y_1, y_2, x) \neq 0$ for all $x \in I$. Thus, we have a reasonably good method for determining whether two solutions to (4) on I are linearly independent on I or not. However, we can improve upon this result. Let y_1 and y_2 be any solutions—not necessarily linearly independent solutions—of (4) on some interval I. Then

$$W(y_1, y_2, x) = \begin{vmatrix} y_1(x) & y_2(x) \\ y_1'(x) & y_2'(x) \end{vmatrix} = y_1(x)y_2'(x) - y_1'(x)y_2(x)$$

and differentiating, we find

$$W'(y_1, y_2, x) = \frac{d}{dx}(y_1y_2' - y_1'y_2)$$
$$= y_1'y_2' + y_1y_2'' - y_1''y_2 - y_1'y_2' = y_1y_2'' - y_1''y_2.$$

Since y_1 and y_2 satisfy (4),

$$y_1'' = -\frac{f_1}{f_0}y_1' - \frac{f_2}{f_0}y_1 \quad \text{and} \quad y_2'' = -\frac{f_1}{f_0}y_2' - \frac{f_2}{f_0}y_2.$$

So

$$W'(y_1, y_2, x) = y_1\left(-\frac{f_1}{f_0}y_2' - \frac{f_2}{f_0}y_2\right) - y_2\left(-\frac{f_1}{f_0}y_1' - \frac{f_2}{f_0}y_1\right)$$
$$= -\frac{f_1}{f_0}(y_1y_2' - y_1'y_2) = -\frac{f_1}{f_0}W(y_1, y_2, x).$$

Thus, the Wronskian satisfies the first order linear differential equation

$$W' + \frac{f_1}{f_0}W = 0.$$

The solution of this equation is

$$W(y_1, y_2, x) = W(y_1, y_2, x_0)\exp\left\{\int_{x_0}^{x}\frac{f_1(t)}{f_0(t)}\,dt\right\},$$

where x_0 is any point in the interval I. [This result was obtained in 1827 by the Norwegian mathematician N. H. Abel (1802–1829) and is known as Abel's formula.] Since the exponential function is always positive, the Wronskian of any two solutions y_1 and y_2 of (4) is either positive on the entire interval I, negative on the entire interval I, or zero on the entire interval I. And, of course, whether the Wronskian is positive, negative, or zero on the entire interval can be determined by calculating the value of the Wronskian at any convenient point x_0 in the interval I. So two solutions y_1 and y_2 of (4) are linearly independent on an interval I if and only if y_1 and y_2 satisfy (4) on I and $W(y_1, y_2, x_0) \neq 0$ for some $x_0 \in I$. This result can be generalized to nth order linear homogeneous equations. Consequently, we have the following important theorem.

Theorem 6.5 If $f_0(x), f_1(x), \ldots, f_n(x)$ are continuous on some interval I and $f_0(x) \neq 0$ for all $x \in I$, then $y_1, y_2, \ldots, y_n$ are linearly independent solutions of the nth order linear homogeneous differential equation

$$f_0(x)y^{(n)} + f_1(x)y^{(n-1)} + \cdots + f_n(x)y = 0 \tag{7}$$

if and only if $y_1, y_2, \ldots, y_n$ satisfy (7) on I and $W(y_1, y_2, \ldots, y_n, x) \neq 0$ for some $x \in I$.

We have seen that if $y_1, y_2, \ldots, y_m$ are linearly independent solutions of (7) on I, then any linear combination, $y = \sum_{i=1}^{m} c_i y_i$ where the c_i's are real constants, is also a solution of (7) on I. We need to establish (a) that there are n linearly independent solutions to (7) and (b) that any solution of (7) may be written as a linear combination of a set of n linear independent solutions. Thus, in vector space parlance, the solution space of (7) is an n-dimensional function space and any set of n linearly independent solutions of (7) form a basis for the solution space.

Theorem 6.6 There exist n linearly independent solutions of (7) on I.

Proof Let $x_0 \in I$. By Theorem 6.1 there exists n unique solutions $y_1(x)$, $y_2(x), \ldots, y_n(x)$ defined on I which satisfy the following n sets of initial conditions, respectively:

$$(8) \qquad y_i^{(m)}(x_0) = \begin{cases} 1, & m = i - 1 \\ 0, & m \neq i - 1, \end{cases} \qquad m = 0, 1, \ldots, n-1; i = 1, 2, \ldots, n.$$

Suppose that there exist n real constants $c_1, c_2, \ldots, c_n$ such that

$$c_1 y_1(x) + c_2 y_2(x) + \cdots + c_n y_n(x) \equiv 0 \quad \text{for all } x \in I.$$

By repeated differentiation we find

$$\begin{array}{llll} c_1 y_1'(x) & + c_2 y_2'(x) & + \cdots + c_n y_n'(x) & \equiv 0 \quad \text{for all } x \in I \\ \vdots & & \vdots & \vdots \\ c_1 y_1^{(n-1)}(x) & + c_2 y_2^{(n-1)}(x) & + \cdots + c_n y_n^{(n-1)}(x) & \equiv 0 \quad \text{for all } x \in I. \end{array}$$

In particular, these identities must hold for $x = x_0$. Setting $x = x_0$ in the equations above and substituting from (8) we find that $c_1 = c_2 = \cdots = c_n = 0$. Therefore, the set of functions $\{y_1, y_2, \ldots, y_n\}$ which are the solutions of (7) which satisfy the initial conditions (8) are linearly independent on I.

So now we know that there are at least n linearly independent solutions to (7) on I. The following theorem shows that there are at most n linearly independent solutions to (7) on I and provides a representation for every solution of (7) in terms of any n linearly independent solutions. The theorem does not say that there is only one set of n linearly independent solutions—the set of Theorem 6.6—but that the maximum number of members in any solution set which is linearly independent on I is n.

Theorem 6.7 If $y_1, y_2, \ldots, y_n$ are linearly independent solutions of (7) on I, then every solution y of (7) may be written as $y = \sum_{i=1}^{n} c_i y_i$, where the c_i's are suitably chosen constants.

Proof Let y be any solution of (7) and let $x_0 \in I$. Consider $z(x) = \sum_{i=1}^n c_i y_i(x)$. By an extension of Theorem 6.2 $z(x)$ is a solution to (7) on I. If we can choose the c_i's so that

$$z(x_0) = y(x_0), \quad z^{(1)}(x_0) = y^{(1)}(x_0), \quad \ldots, \quad z^{(n-1)}(x_0) = y^{(n-1)}(x_0), \tag{9}$$

then $z(x) \equiv y(x)$ for all $x \in I$ since by Theorem 6.1 there is only one solution of (7) which satisfies (9). Differentiating $z(x)$ repeatedly and evaluating each derivative at x_0, we obtain the following system of n equations in the n unknowns c_i:

$$\begin{aligned}
\sum_{i=1}^n c_i y_i(x_0) &= y(x_0) \\
\sum_{i=1}^n c_i y_i^{(1)}(x_0) &= y^{(1)}(x_0) \\
&\vdots \\
\sum_{i=1}^n c_i y_i^{(n-1)}(x_0) &= y^{(n-1)}(x_0).
\end{aligned}$$

Or, in matrix-vector notation,

$$\begin{pmatrix} y_1(x_0) & y_2(x_0) & \cdots & y_n(x_0) \\ y_1^{(1)}(x_0) & y_2^{(1)}(x_0) & \cdots & y_n^{(1)}(x_0) \\ \vdots & \vdots & & \vdots \\ y_1^{(n-1)}(x_0) & y_2^{(n-1)}(x_0) & \cdots & y_n^{(n-1)}(x_0) \end{pmatrix} \begin{pmatrix} c_1 \\ c_2 \\ \vdots \\ c_n \end{pmatrix} = \begin{pmatrix} y(x_0) \\ y^{(1)}(x_0) \\ \vdots \\ y^{(n-1)}(x_0) \end{pmatrix}. \tag{10}$$

We know from linear algebra that this system of equations has a unique solution if and only if the determinant of the square matrix in (10) is nonzero. We know that the determinant of the square matrix in (10) is nonzero, since it is the Wronskian of $y_1, y_2, \ldots, y_n$ evaluated at x_0, which is nonzero because the functions $y_1, y_2, \ldots, y_n$ are linearly independent on I. So there exists a unique solution $(c_1, c_2, \ldots, c_n)$ to (10), and consequently $y \equiv z = \sum_{i=1}^n c_i y_i$ on I.

The linear combination $y = \sum_{i=1}^n c_i y_i$, where the c_i's are arbitrary constants is called the *general solution* of the homogeneous equation (7).

Now let us consider the general nth order nonhomogeneous linear differential equation

$$f_0(x)y^{(n)} + f_1(x)y^{(n-1)} + \cdots + f_{n-1}(x)y^{(1)} + f_n(x)y = g(x), \tag{11}$$

where $f_0(x) \neq 0$ and $g(x) \not\equiv 0$ for x in some interval I. The *associated homogeneous equation* is

$$f_0(x)y^{(n)} + f_1(x)y^{(n-1)} + \cdots + f_{n-1}(x)y^{(1)} + f_n(x)y = 0. \tag{12}$$

Theorem 6.8 If y_p is any solution of (11) and $y_1, y_2, \ldots, y_n$ are linearly independent solutions of (12), then every solution of (11) has the form

$$y = c_1y_1 + c_2y_2 + \cdots + c_ny_n + y_p = y_c + y_p, \tag{13}$$

where $c_1, c_2, \ldots, c_n$ are suitably chosen constants.

Proof Let $z(x)$ be any solution of (11). We must show that the constants $c_1, c_2, \ldots, c_n$ can be chosen so that $z = y_c + y_p$. Since z and y_p are solutions of the nonhomogeneous equation (11), $z - y_p$ is a solution of the homogeneous equation (12). By Theorem 6.7 there exist constants $c_1, c_2, \ldots, c_n$ such that

$$z - y_p = c_1y_1 + c_2y_2 + \cdots + c_ny_n = y_c.$$

Hence,

$$z = y_c + y_p$$

for suitably chosen constants $c_1, c_2, \ldots, c_n$.

The function y_p in equation (13) is called a *particular solution* of the nonhomogeneous equation (11). The linear combination y_c in equation (13), which is the general solution of the homogeneous equation (12), is called the *complementary function* for (11). And equation (13) where $c_1, c_2, \ldots, c_n$ are arbitrary constants is known as the *general solution* of the nonhomogeneous equation (11).

EXERCISES

1. Verify that e^x and e^{-x} are solutions of the differential equation $y'' - y = 0$. Why are $\sinh x = (e^x - e^{-x})/2$ and $\cosh x = (e^x + e^{-x})/2$ also solutions?

2. Verify that e^{ix} and e^{-ix} are solutions of the differential equation $y'' + y = 0$. Why are $\sin x = (e^{ix} - e^{-ix})/2i$ and $\cos x = (e^{ix} + e^{-ix})/2$ also solutions?

3. Verify that x, x^{-2}, and $c_1x + c_2x^{-2}$, where c_1 and c_2 are arbitrary constants, are solutions of the differential equation $x^2y'' + 2xy' - 2y = 0$ for $x > 0$.

4. Verify that the functions 1 and x^2 are linearly independent and are solutions of the differential equation $2yy'' - (y')^2 = 0$. Why is $c_1 + c_2x^2$ not a solution of the differential equation for all arbitrary constants c_1 and c_2?

5. Use the Wronskian to show that the following sets of functions are linearly independent.

(a) $1, x, x^2$ (b) $\sin x, \cos x$

(c) $\sin x, \sin 2x$ (d) e^x, e^{2x}, xe^{2x}

6. Show that the following sets of functions are linearly dependent by finding constants $c_1, c_2, \ldots, c_n$ not all zero such that $c_1y_1 + c_2y_2 + \cdots + c_ny_n = 0$.

(a) $1, x, 3x - 4$ (b) $1, \sin^2 x, \cos^2 x$

(c) $x, e^x, xe^x, (x + 2)e^x$

6.2 HOMOGENEOUS LINEAR EQUATIONS WITH CONSTANT COEFFICIENTS

We now turn our attention to the simplest linear differential equations —those with constant coefficients. These equations have the general form

$$a_0y^{(n)} + a_1y^{(n-1)} + \cdots + a_ny = 0, \tag{14}$$

where $a_0, a_1, \ldots, a_n$ are real constants and $a_0 \neq 0$. Theorem 6.6 tells us that there exists a set of n linearly independent solutions to (14) which are all defined on $(-\infty, \infty)$, say $\{y_1, y_2, \ldots, y_n\}$, and Theorem 6.7 tells us that any solution y of (14) which is defined on $(-\infty, \infty)$ can be written as the linear combination

$$y = \sum_{i=1}^{n} c_iy_i, \tag{15}$$

where the c_i's are properly chosen constants. The linear combination (15) where the c_i's are arbitrary is called the *general solution* of (14). Our objective then is to find a set of n linearly independent solutions to (14) and hence the general solution.

Daniel Bernoulli and Euler both knew how to solve second order homogeneous linear equations with constant coefficients prior to 1740. Euler published his results first in 1743. The general method for effecting a solution of (14) is to assume that there exists a solution of the form $y = e^{rx}$, where r is an unknown real constant. Differentiating n times, substituting into (14) and rearranging leads to

$$e^{rx}(a_0r^n + a_1r^{n-1} + \cdots + a_{n-1}r + a_n) = 0. \tag{16}$$

Clearly, if r_i is a root of the polynomial

$$a_0r^n + a_1r^{n-1} + \cdots + a_{n-1}r + a_n = 0, \tag{17}$$

then $y = e^{r_ix}$ is a solution of the differential equation (14). Equation (17) is called the *auxiliary* or *characteristic equation*. The method for constructing a set of n linear independent solutions to (14) depends upon the relationship of the roots of the auxiliary equation (17). We shall consider all the possibilities in the following paragraphs.

Distinct real roots If the roots $r_1, r_2, \ldots, r_n$ of equation (17) are all real and no two roots are equal, then the functions $y_1 = e^{r_1 x}, y_2 = e^{r_2 x}, \ldots, y_n = e^{r_n x}$ form a linearly independent set of real-valued solutions to (14) on the interval $(-\infty, \infty)$ and the general solution to (14) on $(-\infty, \infty)$ is $y = \sum_{i=1}^{n} c_i e^{r_i x}$, where the c_i's are arbitrary constants. Clearly, the y_i's are solutions to (14). All we need to do is verify that they are linearly independent. We may do so by showing that their Wronskian is nonzero for some $x_0 \in (-\infty, \infty)$.

$$W(y_1, y_2, \ldots, y_n, x) = \begin{vmatrix} e^{r_1 x} & e^{r_2 x} & \cdots & e^{r_n x} \\ r_1 e^{r_1 x} & r_2 e^{r_2 x} & & r_n e^{r_n x} \\ \vdots & \vdots & & \vdots \\ r_1^{n-1} e^{r_1 x} & r_2^{n-1} e^{r_2 x} & \cdots & r_n^{n-1} e^{r_n x} \end{vmatrix}.$$

So a convenient choice for x_0 is $x_0 = 0$. Making this choice, we find

$$\begin{aligned} W(y_1, y_2, \ldots, y_n, 0) &= \begin{vmatrix} 1 & 1 & \cdots & 1 \\ r_1 & r_2 & & r_n \\ \vdots & \vdots & & \vdots \\ r_1^{n-1} & r_2^{n-1} & & r_n^{n-1} \end{vmatrix} \\ &= [(r_2 - r_1)(r_3 - r_1)\cdots(r_n - r_1)] \\ &\quad \cdot [(r_3 - r_2)\cdots(r_n - r_2)]\cdots[(r_n - r_{n-1})] \neq 0 \end{aligned}$$

since the roots are distinct. The last determinant in the calculations above is known as the *Vandermonde determinant* and its value is well known.

EXAMPLE Find the general solution of

$$2y''' - y'' - 2y' + y = 0.$$

The auxiliary equation

$$2r^3 - r^2 - 2r + 1 = 0$$

has distinct real roots $-1, 1, \frac{1}{2}$. Hence, the general solution is

$$y = c_1 e^{-x} + c_2 e^{x} + c_3 e^{x/2}.$$

Repeated real roots Consider the differential equation

$$y''' - 6y'' + 12y' - 8y = 0. \tag{18}$$

The auxiliary equation for (18) is

$$r^3 - 6r^2 + 12r - 8 = 0.$$

The roots of this equation are $r_1 = r_2 = r_3 = 2$. Thus, one solution of (18) is $y_1 = e^{r_1 x} = e^{2x}$. However, $y_2 = e^{r_2 x} = e^{2x} = y_1$ and $y_3 = e^{r_3 x} = e^{2x} = y_1$. Consequently, y_2 and y_3 are identical to y_1, and therefore the set $\{y_1, y_2, y_3\}$ is not linearly independent. So when some roots of the auxiliary equation are real and equal, the technique of the previous paragraph will not suffice.

If r_i is a real root of the auxiliary equation of multiplicity $k \geq 2$, then $y_1 = e^{r_i x}$ will be one member of the linearly independent solution set; however, we must find $k - 1$ other linearly independent solutions corresponding to the root r_i. In order to find the other $k - 1$ linearly independent solutions corresponding to r_i, we assume that there are $k - 1$ solutions of the form $y_{l+1} = x^l e^{r_i x}$, where $1 \leq l \leq k - 1$. We then verify that y_{l+1} satisfies the differential equation and that the set $\{y_1, y_2, \ldots, y_k\}$ is linearly independent.

Returning to our example we must verify that $y_2 = xe^{2x}$ and $y_3 = x^2e^{2x}$ satisfy equation (18) and that $y_1 = e^{2x}$, y_2, and y_3 are linearly independent. Calculating the first, second, and third derivatives of y_2 and y_3, we find

$$\begin{aligned} y_2' &= (1 + 2x)e^{2x}; & y_3' &= (2x + 2x^2)e^{2x} \\ y_2'' &= (4 + 4x)e^{2x}; & y_3'' &= (2 + 8x + 4x^2)e^{2x} \\ y_2''' &= (12 + 8x)e^{2x}; & y_3''' &= (12 + 24x + 8x^2)e^{2x}. \end{aligned}$$

Substituting y_2 into (18), we obtain

$$[(12 + 8x) - 6(4 + 4x) + 12(1 + 2x) - 8x]e^{2x} = 0 \cdot e^{2x} = 0.$$

Hence, y_2 satisfies (15). Substituting y_3 into (18) results in

$$[(12 + 24x + 8x^2) - 6(2 + 8x + 4x^2) + 12(2x + 2x^2) - 8x^2]e^{2x} = 0 \cdot e^{2x} = 0.$$

Hence, y_3 satisfies (18). In order to show that y_1, y_2, and y_3 are linearly independent, we examine their Wronskian at $x = 0$.

$$W(y_1, y_2, y_3, 0) = \begin{vmatrix} y_1(0) & y_2(0) & y_3(0) \\ y_1'(0) & y_2'(0) & y_3'(0) \\ y_1''(0) & y_2''(0) & y_3''(0) \end{vmatrix} = \begin{vmatrix} 1 & 0 & 0 \\ 2 & 1 & 0 \\ 4 & 4 & 2 \end{vmatrix} = 2 \neq 0.$$

Hence, y_1, y_2, and y_3 are linearly independent.

Using differential operator notation we can show in general that if r_i is a root of (14) of multiplicity k, then $y_{l+1} = x^l e^{r_i x}$, where $l = 0, 1, \ldots, k - 1$ are k linearly independent solutions of (14). Let D denote differentiation with respect to x. That is, $D \equiv d/dx$. Hence, $Dy = dy/dx = y' = y^{(1)}$. In general, $D^m = d^m/dx^m$, where m is a positive integer and $D^m y = y^{(m)}$. Hence, (14) may be written as

$$(19) \qquad (a_0D^n + a_1D^{n-1} + \cdots + a_{n-1}D + a_n)y = p(D)y = 0$$

and the auxiliary equation is $p(r) = 0$. If r_i is a root of p of multiplicity k, then p has $(r - r_i)^k$ as a factor and (19) has $(D - r_i)^k$ as a factor. That is, $p(D)y = q(D)(D - r_i)^k y$.

Let v be a function of x. Then

$$(D - r_i)(ve^{r_ix}) = D(ve^{r_ix}) - r_ive^{r_ix} = e^{r_ix}Dv$$

$$(D - r_i)^2(ve^{r_ix}) = (D - r_i)[(D - r_i)(ve^{r_ix})] = (D - r_i)(e^{r_ix}Dv) = e^{r_ix}D^2v.$$

And, in general,

$$(D - r_i)^k(ve^{r_ix}) = e^{r_ix}D^kv.$$

If $v(x) = x^l$ and $l < k$, then $D^kx^l = 0$. Hence, for $l < k$, $y_{l+1} = x^le^{r_ix}$ satisfies $(D - r_i)^k(y_{l+1}) = 0$, and consequently y_{l+1} satisfies $p(D)y_{l+1} = q(D)(D - r_i)^k y_{l+1} = 0$. The y_{l+1}'s associated with r_i are linearly independent, since $1, x, x^2, \ldots, x^{k-1}$ are linearly independent—which the reader should already know or verify—and the y_{l+1}'s are linearly independent of all other solutions y_j which are associated with other roots r_j of the auxiliary equation.

EXAMPLE Find the general solution of the equation

$$y^{(4)} - 2y^{(2)} + y = 0.$$

The roots of the auxiliary equation,

$$r^4 - 2r^2 + 1 = 0,$$

are $-1, -1, 1, 1$. Therefore, the general solution is

$$y = c_1e^{-x} + c_2xe^{-x} + c_3e^x + c_4xe^x.$$

EXAMPLE Find the solution of the initial value problem

$$y^{(4)} - y^{(2)} = 0; \; y(0) = 1, \; y^{(1)}(0) = -1, \; y^{(2)}(0) = 2, \; y^{(3)}(0) = 1.$$

The auxiliary equation

$$r^4 - r^2 = 0$$

has roots $-1, 0, 0, 1$. So the general solution of the given differential equation is

$$y = c_1e^{-x} + c_2 + c_3x + c_4e^x.$$

The four initial conditions yield the following system of four linear equations in the unknowns c_1, c_2, c_3, and c_4.

$$\begin{aligned} y(0) &= 1 = c_1 + c_2 + c_4 \\ y^{(1)}(0) &= -1 = -c_1 + c_3 + c_4 \\ y^{(2)}(0) &= 2 = c_1 + c_4 \\ y^{(3)}(0) &= 1 = -c_1 + c_4. \end{aligned}$$

Solving this system of equations we find that $c_1 = \frac{1}{2}$, $c_2 = -1$, $c_3 = -2$, and $c_4 = \frac{3}{2}$. Thus, the solution of the initial value problem is

$$y = \tfrac{1}{2}e^{-x} - 1 - 2x + \tfrac{3}{2}e^{x}.$$

Complex roots Again consider the linear homogeneous differential (14) and (19) $(a_0D^n + a_1D^{n-1} + \cdots + a_{n-1}D + a_n)y = p(D)y = 0$. If $r_1 = a + bi$, where a and b are real and $b \neq 0$, is a root of the auxiliary equation $p(r) = 0$, then $r_2 = a - bi$ is also a root of the auxiliary equation, since the coefficients $a_0, a_1, \ldots, a_n$ of the auxiliary equation are real. That is, complex roots of the auxiliary equation occur in conjugate pairs. Corresponding to the roots r_1 and r_2 of $p(r) = 0$ are two linearly independent solutions of the differential equation $p(D)y = 0$, namely $y_1^* = e^{r_1x}$ and $y_2^* = e^{r_2x}$. Using *Euler's formula*:

$$e^{i\theta} = \cos\theta + i\sin\theta$$

and the identities $\cos(-x) = \cos x$ and $\sin(-x) = -\sin x$, we find

$$\text{(20a)} \qquad y_1^* = e^{r_1x} = e^{(a+ib)x} = e^{ax}e^{ibx} = e^{ax}(\cos bx + i\sin bx)$$

and

$$\text{(20b)} \qquad y_2^* = e^{r_2x} = e^{(a-ib)x} = e^{ax}e^{-ibx} = e^{ax}(\cos bx - i\sin bx).$$

Thus, the solutions y_1^* and y_2^* are complex solutions of (19). Because the differential equation (19) is real, we would prefer to have two linearly independent real solutions which correspond to r_1 and r_2, if possible. Since y_1^* and y_2^* are solutions to the differential equation, so is any linear combination

$$\text{(21)} \qquad y = c_1^*y_1^* + c_2^*y_2^*,$$

where c_1^* and c_2^* are any arbitrary complex numbers. From (20) and (21) we find

$$\begin{aligned} y &= c_1^*e^{ax}(\cos bx + i\sin bx) + c_2^*e^{ax}(\cos bx - i\sin bx) \\ &= (c_1^* + c_2^*)e^{ax}\cos bx + i(c_1^* - c_2^*)e^{ax}\sin bx. \end{aligned}$$

Choosing $c_1^* = c_2^* = \frac{1}{2}$ we obtain the real solution

$$\text{(22a)} \qquad y_1 = e^{ax}\cos bx.$$

Choosing $c_1^* = -c_2^* = -i/2$ we obtain the second real solution,

$$\text{(22b)} \qquad y_2 = e^{ax}\sin bx.$$

These two real solutions are linearly independent since

$$W(y_1, y_2, 0) = \begin{vmatrix} y_1(0) & y_2(0) \\ y_1'(0) & y_2'(0) \end{vmatrix} = \begin{vmatrix} 1 & 0 \\ a & b \end{vmatrix} = b \neq 0.$$

Hence, two linearly independent real solutions of the differential equation $p(D)y = 0$ which correspond to the complex conjugate roots r_1 and r_2 are y_1 and y_2.

EXAMPLE Find the general solution of the equation

$$y^{(3)} + y^{(1)} + 10y = 0.$$

The auxiliary equation

$$r^3 + r + 10 = 0$$

has $r_1 = -2$ as a real root. Factoring we find

$$r^3 + r + 10 = (r + 2)(r^2 - 2r + 5) = 0.$$

Using the quadratic formula we find the two complex roots to be $1 \pm 2i$. Therefore, the general solution is

$$y = c_1e^{-2x} + c_2e^x \cos 2x + c_3e^x \sin 2x.$$

When the auxiliary equation has repeated complex roots, linearly independent real solutions are obtained in a manner analogous to the repeated-real-root case. For example, if $r = a + bi$ is a root of the auxiliary equation of multiplicity 3, then $r = a - bi$ is a root of multiplicity 3 and the six linearly independent real solutions of the differential equation associated with these roots are

$$y_{2k-1} = x^{k-1}e^{ax} \cos bx, \qquad y_{2k} = x^{k-1}e^{ax} \sin bx, \qquad k = 1, 2, 3.$$

EXAMPLE Solve the differential equation

$$y^{(4)} + 8y^{(3)} + 26y^{(2)} + 40y^{(1)} + 25y = 0.$$

The roots of the auxiliary equation are $-2 + i$, $-2 + i$, $-2 - i$, and $-2 - i$. So the general solution is

$$y = (c_1 + c_2x)e^{-2x} \cos x + (c_3 + c_4x)e^{-2x} \sin x.$$

EXERCISES

Find the general solution of the following differential equations.

1. $y'' - 4y = 0$
2. $y'' + 4y = 0$
3. $y'' + 4y' + 4y = 0$
4. $y'' + 4y' - 4y = 0$
5. $4y'' + 4y' - 3y = 0$
6. $y'' - 2y' + 2y = 0$

7. $y''' - 4y'' + 6y' - 4y = 0$

8. $y'''' - 2y''' + y'' = 0$

9. $y'''' - 16y = 0$

10. $y'''' + 16y = 0$

11. $y'''' - 4y''' + 8y'' - 8y' + 4y = 0$

Find the solution of the following initial value problems.

12. $y'' + 2y' - 3y = 0;\ y(0) = 3,\ y'(0) = -1$

13. $y'' - 2y' + 3y = 0;\ y(0) = 1,\ y'(0) = 1 + \sqrt{2}$

14. $y'''' + y'' = 0;\ y(0) = 2,\ y'(0) = -1,\ y''(0) = 1,\ y'''(0) = -1$

6.3 NONHOMOGENEOUS LINEAR EQUATIONS WITH CONSTANT COEFFICIENTS

In this section we shall present the *method of undetermined coefficients* and the *Laplace transform method.* These methods are often used to solve differential equations of the form

$$a_0 y^{(n)} + a_1 y^{(n-1)} + \cdots + a_{n-1} y^{(1)} + a_n y = p(D)y = g(x), \tag{23}$$

where $a_0, a_1, \ldots, a_n$ are real constants, $a_0 \neq 0$, and $g(x) \not\equiv 0$. When the nonhomogeneous term, $g(x)$, is itself a solution of some homogeneous linear differential equation with constant coefficients, the method of undetermined coefficients is an effective procedure for finding a particular solution of (23). Thus, the method of undetermined coefficients will provide a particular solution to (23) when $g(x)$ is an exponential function, a sine or cosine function, a polynomial function or the sums and products of such functions. In general, $g(x)$ will not be a solution of any homogeneous linear differential equation with constant coefficients, and the method of undetermined coefficients can not be used to solve (23). In the event that $g(x)$ has a Laplace transform, the solution of (23) may be obtained using the Laplace transform method.

6.3.1 Undetermined Coefficients

From Theorem 6.8 we know that the general solution of (23) has the form $y = y_c + y_p$, where y_c is the general solution of the homogeneous equation associated with (23), namely, $p(D)y = 0$, and y_p is any particular solution of (23). Using the methods of the preceding section we can obtain y_c. Thus, our problem reduces to one of finding y_p.

Suppose that $g(x)$ satisfies some homogeneous linear differential equation with constant coefficients. That is, suppose there exists a polynomial q with constant coefficients such that $q(D)g(x) = 0$. In general, a differential operator $q(D)$ is said to *annihilate* a function $g(x)$ if $q(D)g(x) = 0$. Operating on the nonhomogeneous equation (23) $p(D)y = g(x)$ with $q(D)$, a polynomial annihilator of $g(x)$, we find $q(D)y(D)y = q(D)g(x) = 0$. Since this equation is homogeneous, we can again find by the methods of the preceding section the general solution, y_G. Since y_G is the general solution of $q(D)p(D)y = 0$ and since

$$q(D)p(D)(y_c + y_p) = q(D)(p(D)y_c + p(D)y_p) = q(D)(0 + g(x)) = 0,$$

we can determine y_p by deleting y_c from y_G and determining the remaining constants so that $p(D)(y_G - y_c) = p(D)y_p = g(x)$.

EXAMPLE Find the general solution of the nonhomogeneous equation

$$y'' - y = 3x + 2 \sin x.$$

The associated homogeneous equation is $y'' - y = (D^2 - 1)y = p(D)y = 0$ and the auxiliary equation $p(r) = 0$ has roots -1 and 1. Hence,

$$y_c = c_1e^{-x} + c_2e^{x}.$$

In this example $g(x) = 3x + 2 \sin x$. The term $3x$ will be a solution of any homogeneous equation which has a double root of zero in the auxiliary equation and the term $2 \sin x$ will be a solution of any homogeneous equation which has complex conjugate roots $\pm i$ in the auxiliary equation. Hence, the differential operator, $q(D)$, of smallest degree which will annihilate $g(x)$ is $q(D) = D^2(D^2 + 1)$. Since the roots of $q(r)p(r) = 0$ are $-1, 1, 0, 0, \pm i$, the general solution of $q(D)p(D)y = 0$ is

$$y_G = c_1e^{-x} + c_2e^{x} + c_3 + c_4x + c_5 \cos x + c_6 \sin x.$$

Eliminating y_c we see that

$$y_p = c_3 + c_4x + c_5 \cos x + c_6 \sin x$$

and the constants must be chosen so that $p(D)y_p = g(x)$. Differentiating, we obtain

$$y_p' = c_4 - c_5 \sin x + c_6 \cos x$$
$$y_p'' = -c_5 \cos x - c_6 \sin x.$$

Substituting into the nonhomogeneous equation, $y'' - y = g(x)$, we find that c_3, c_4, c_5, and c_6 must satisfy

$$-2c_5 \cos x - 2c_6 \sin x - c_3 - c_4x = 3x + 2 \sin x.$$

Hence, $c_3 = 0$, $c_4 = -3$, $c_5 = 0$, and $c_6 = -1$, and therefore

$$y_p = -3x - \sin x.$$

The general solution is $y = y_c + y_p = c_1e^{-x} + c_2e^x - 3x - \sin x$.

EXAMPLE Solve the differential equation

$$y'' - 3y' + 2y = xe^{2x}.$$

The auxiliary equation of the associated homogeneous equation $p(r) = r^2 - 3r + 2 = 0$ has roots 1 and 2, so the complementary solution is

$$y_c = c_1e^x + c_2e^{2x}.$$

The function $g(x) = xe^{2x}$ can be annihilated by any operator that has roots 2, 2. Hence, $q(D) = (D - 2)^2$ and $q(r)p(r) = 0$ has roots 1, 2, 2, 2. Consequently,

$$y_G = c_1e^x + c_2e^{2x} + c_3xe^{2x} + c_4x^2e^{2x}$$

and

$$y_p = y_G - y_c = c_3xe^{2x} + c_4x^2e^{2x} = (c_3x + c_4x^2)e^{2x},$$

where c_3 and c_4 must be chosen so that $p(D)y_p = g(x)$. Differentiating, we find

$$y_p' = (c_3 + 2c_3x + 2c_4x + 2c_4x^2)e^{2x}$$

and

$$y_p'' = (4c_3 + 2c_4 + 4c_3x + 8c_4x + 4c_4x^2)e^{2x}.$$

Substituting into the differential equation, rearranging, and dividing by e^{2x} yields $(c_3 + 2c_4) + 2c_4x = x$. Consequently, $c_3 = -1$ and $c_4 = \frac{1}{2}$ and

$$y_p = (-x + \tfrac{1}{2}x^2)e^{2x}.$$

Hence, the general solution is

$$y = y_c + y_p = c_1e^x + c_2e^{2x} - xe^{2x} + \tfrac{1}{2}x^2e^{2x}.$$

When $g(x) = g_1(x) + g_2(x) + \cdots + g_m(x)$, it is often advisable to find m particular solutions y_{p_i} of the m nonhomogeneous differential equations

$$p(D)y = g_i(x), \qquad i = 1, 2, \ldots, m.$$

Consider $y = \sum_{i=1}^m y_{p_i}$. Since the differential equation is linear, we have

$$p(D)y = p(D)\sum_{i=1}^m y_{p_i} = \sum_{i=1}^m p(D)y_{p_i} = \sum_{i=1}^m g_i(x) = g(x).$$

Thus, if y_{p_i} is a particular solution of $p(D)y = g_i(x)$ for $i = 1, 2, \ldots, m$, then $y_p = \sum_{i=1}^m y_{p_i}$ is a particular solution of

$$p(D)y = \sum_{i=1}^m g_i(x) = g(x).$$

EXAMPLE Find the solution of the initial value problem

$$y^{(3)} + 9y^{(1)} = 4x + e^{-x} + 2\cos x - \sin 3x;$$
$$y(0) = -\tfrac{1}{60}, \quad y^{(1)}(0) = -\tfrac{1}{180}, \quad y^{(2)}(0) = \tfrac{1}{90}.$$

The roots of the auxiliary equation of the associated homogeneous equation, $p(r) = r^3 + 9r = 0$, are 0, $-3i$, and $+3i$. Consequently, the complementary solution is

$$y_c = c_1 + c_2 \cos 3x + c_3 \sin 3x.$$

Next we seek particular solutions of the nonhomogeneous equations

$$y^{(3)} + 9y^{(1)} = 4x \tag{24a}$$
$$y^{(3)} + 9y^{(1)} = e^{-x} \tag{24b}$$
$$y^{(3)} + 9y^{(1)} = 2\cos x \tag{24c}$$
$$y^{(3)} + 9y^{(1)} = -\sin 3x. \tag{24d}$$

Since zero is a root of $p(r) = 0$, and since a homogeneous differential equation must have an auxiliary equation with two zero roots to annihilate $4x$, we try for a particular solution to (24a) of the form

$$y_{p_1} = Ax^2 + Bx.$$

Differentiating, we find

$$y_{p_1}^{(1)} = 2Ax + B$$
$$y_{p_1}^{(2)} = 2A$$
$$y_{p_1}^{(3)} = 0.$$

Substituting into (24a), we obtain

$$18Ax + 9B = 4x.$$

Hence, $A = \frac{2}{9}$ and $B = 0$ and

$$y_{p_1} = \tfrac{2}{9}x^2.$$

We did not add a constant C to y_{p_1} since zero is a root of the associated homogeneous equation.

Since -1 is not a root of $p(r) = 0$, we try to find a particular solution of (24b) of the form

$$y_{p_2} = Ae^{-x}.$$

Repeated differentiation yields

$$y_{p_2}^{(1)} = -Ae^{-x}$$
$$y_{p_2}^{(2)} = Ae^{-x}$$
$$y_{p_2}^{(3)} = -Ae^{-x}.$$

Substitution into (24b) results in

$$-10Ae^{-x} = e^{-x}.$$

Therefore, $A = -\frac{1}{10}$ and

$$y_{p_2} = -\tfrac{1}{10}e^{-x}.$$

We seek a particular solution of (24c) of the form

$$y_{p_3} = A\cos x + B\sin x$$

since $\pm i$ is not a root of $p(r) = 0$ and since a real homogeneous differential equation must have an auxiliary equation with roots $\pm i$ in order to annihilate the nonhomogeneous term $2\cos x$. Differentiating and substituting into (24c) we find

$$-8A\sin x + 8B\cos x = 2\cos x.$$

Consequently, $A = 0$, $B = \frac{1}{4}$, and

$$y_{p_3} = \tfrac{1}{4}\sin x.$$

Since $\pm 3i$ is a root of $p(r) = 0$ and since a real homogeneous differential equation that annihilates $\sin 3x$ must also have roots $\pm 3i$, we seek a particular solution of (24d) of the form

$$y_{p_4} = Ax\cos 3x + Bx\sin 3x.$$

Differentiation yields

$$\begin{aligned} y_{p_4}^{(1)} &= A\cos 3x + 3(B - A)x\sin 3x + B\sin 3x \\ y_{p_4}^{(2)} &= -3B\cos 3x + 9(B - A)x\cos 3x + 3(B - 2A)\sin 3x \\ y_{p_4}^{(3)} &= 9(2B - 3A)\cos 3x - 27(B - A)x\sin 3x + 9B\sin 3x. \end{aligned}$$

Substituting into (24d) and rearranging, we find

$$18(B - A)\cos 3x + 18B\sin 3x = -\sin 3x.$$

Hence, $A = B = -\frac{1}{18}$ and

$$y_{p_4} = -\tfrac{1}{18}x(\cos 3x + \sin 3x).$$

The general solution is

$$\begin{aligned} y = c_1 + c_2\cos 3x + c_3\sin 3x + \tfrac{2}{9}x^2 - \tfrac{1}{10}e^{-x} \\ + \tfrac{1}{4}\sin x - \tfrac{1}{18}x(\cos 3x + \sin 3x). \end{aligned}$$

To find the solution of the initial value problem, we must satisfy the initial conditions. Differentiating, we find

$$y^{(1)} = -3c_2 \sin 3x + 3c_3 \cos 3x + \tfrac{4}{9}x + \tfrac{1}{10}e^{-x} + \tfrac{1}{4}\cos x$$
$$-\tfrac{1}{18}[(3x + 1)\cos 3x + (1 - 3x)\sin 3x]$$

and

$$y^{(2)} = -9c_2 \cos 3x - 9c_3 \sin 3x + \tfrac{4}{9} - \tfrac{1}{10}e^{-x} - \tfrac{1}{4}\cos x$$
$$-\tfrac{1}{6}[(2 - 3x)\cos 3x - (3x + 2)\sin 3x].$$

To satisfy the initial conditions c_1, c_2, and c_3 must simultaneously satisfy the equations

$$y(0) = -\tfrac{1}{60} = c_1 + c_2 - \tfrac{1}{10}$$
$$y^{(1)}(0) = -\tfrac{1}{180} = 3c_3 + \tfrac{1}{10} + \tfrac{1}{4} - \tfrac{1}{18}$$
$$y^{(2)}(0) = \tfrac{1}{90} = -9c_2 + \tfrac{4}{9} - \tfrac{1}{10} - \tfrac{1}{4} - \tfrac{1}{3}.$$

Solving we find $c_1 = \tfrac{1}{9}$, $c_2 = -\tfrac{1}{36}$, $c_3 = -\tfrac{1}{10}$.

One must remember that it is the coefficients of the general solution of nonhomogeneous equation, $y = y_c + y_p$, and not the coefficients of the general solution of the associated homogeneous equation, y_c, which must be chosen to satisfy the initial conditions.

EXERCISES

Find the general solution of the following differential equations.

1. $y'' - 4y = 2x - 3\cos x + e^{-x}$
2. $y'' - 4y = 3xe^{2x}$
3. $y'' + 4y = 3\sin 2x + x^2 - 1$
4. $y^{(4)} - 2y^{(3)} + y^{(2)} = 3x^2 + 2x - e^x$
5. $y^{(3)} - 4y^{(2)} + 6y^{(1)} - 4y = x + e^x \cos x$

Find the solution of the following initial value problems.

6. $y'' + 4y' + 4y = 4e^{2x}$; $y(0) = 1$; $y'(0) = -1$
7. $y'' + 3y' = 6x + 5$; $y(0) = 0$, $y'(0) = 4$
8. $y'' - 2y' + y = xe^x + e^{-x}$; $y(0) = 1$, $y'(0) = 1$
9. $y'' + 9y = x - 2\cos 3x$; $y(\pi/2) = -2 + 2\pi/9$, $y'(\pi/2) = -\tfrac{23}{9}$
10. $y'' + 4y' + 5y = 1 - 5x + 8(\cos x - \sin x)$; $y(0) = 0$, $y'(0) = 7$

6.3.2 Laplace Transforms

Let $f(x)$ be a function defined on the interval $[0, +\infty)$. The *Laplace transform* of f is

$$L[f] = \int_0^\infty e^{-sx} f(x)\, dx = F(s). \tag{26}$$

The Laplace transform was named in honor of P. S. Laplace (1749–1827), the famous French mathematician and astronomer. The Laplace transform of f exists only when the improper integral of (26) exists. It can be shown that if the integral converges for some fixed value s_0, it will converge for all $s > s_0$. For some functions, f, the Laplace transform exists for all real s.

EXAMPLE Find the Laplace transform of $f(x) = e^{ax}$.

$$L[e^{ax}] = \int_0^\infty e^{-sx}e^{ax}\,dx = \frac{-1}{s-a}e^{-(s-a)x}\Big|_0^\infty = \frac{1}{s-a} \quad \text{provided } s > s_0 = a.$$

If $s \leq a$, $L[e^{ax}]$ does not exist. Letting $a = 0$, we get

$$L[1] = \frac{1}{s} \quad \text{for } s > 0.$$

EXAMPLE Calculate the Laplace transform of $f(x) = x^n$, where n is a positive integer.

$$L[x^n] = \int_0^\infty e^{-sx}x^n\,dx.$$

Recall the following formula for integration by parts

$$\int_a^b u\,dv = uv\big|_a^b - \int_a^b v\,du.$$

Letting $u = x^n$ and $dv = e^{-sx}\,dx$ and differentiating and integrating, we find $du = nx^{n-1}\,dx$ and $v = -(1/s)e^{-sx}$. Hence,

$$L[x^n] = -\frac{x^n}{s}e^{-sx}\Big|_0^\infty + \frac{n}{s}\int_0^\infty e^{-sx}x^{n-1}\,dx = \frac{n}{s}L[x^{n-1}].$$

If $n = 1$, we have

$$L[x] = \frac{1}{s}L[1] = \frac{1}{s^2} \quad \text{for } s > 0.$$

If $n > 1$, we repeatedly use integration by parts to obtain

$$L[x^n] = \frac{n(n-1)}{s^2}L[x^{n-2}]$$

$$\vdots$$

$$L[x^n] = \frac{n!}{s^n}L[1] = \frac{n!}{s^{n+1}} \quad \text{for } s > 0.$$

EXAMPLE Find $L[\sin bx] = \int_0^\infty e^{-sx} \sin bx\, dx$.

Again we use integration by parts. This time we let $u = \sin bx$ and $dv = e^{-sx}\, dx$ and find $du = b \cos bx\, dx$ and $v = -e^{-sx}/s$. Consequently,

$$L[\sin bx] = \frac{-1}{s} \sin bx\, e^{-sx}\Big|_0^\infty + \frac{b}{s}\int_0^\infty e^{-sx} \cos bx\, dx$$

$$= \frac{b}{s}\int_0^\infty e^{-sx} \cos bx\, dx \quad \text{provided that } s > 0.$$

Now let $u = \cos bx$ and $dv = e^{-sx}\, dx$. Then $du = -b \sin bx\, dx$ and $v = -e^{-sx}/s$. Upon integration by parts, we find

$$L[\sin bx] = \frac{b}{s}\left\{\frac{-1}{s} \cos bx\, e^{-sx}\Big|_0^\infty - \frac{b}{s}\int_0^\infty e^{-sx} \sin bx\, dx\right\}$$

$$= \frac{b}{s^2} - \frac{b^2}{s^2} L[\sin bx] \quad \text{provided that } s > 0.$$

Solving for $L[\sin bx]$ yields

$$L[\sin bx] = \frac{b/s^2}{1 + b^2/s^2} = \frac{b}{s^2 + b^2} \quad \text{for } s > 0.$$

Not all functions have a Laplace transform, as the following example illustrates.

EXAMPLE Show that $L[e^{x^2}] = \int_0^\infty e^{-sx}e^{x^2}\, dx$ does not exist.

Clearly the integral does not exist if $s \leq 0$. Suppose that the integral exists for some $s > 0$. Then

$$L[e^{x^2}] = \int_0^\infty e^{-sx}e^{x^2}\, dx = \int_0^{2s} e^{x(x-s)}\, dx + \int_{2s}^\infty e^{x(x-s)}\, dx.$$

The first integral on the right-hand side of the last equality is positive since the integrand is positive for all real x and s. For $x \geq 2s$, $x - s \geq s$. So for $x \geq 2s$, $e^{x(x-s)} \geq e^{sx}$, and therefore the second integral satisfies the inequality

$$\int_{2s}^\infty e^{x(x-s)}\, dx \geq \int_{2s}^\infty e^{sx}\, dx = \infty.$$

Thus, for any $s > 0$, we have

$$L[e^{x^2}] = \int_0^{2s} e^{x(x-s)}\, dx + \int_{2s}^\infty e^{x(x-s)}\, dx \geq \int_{2s}^\infty e^{sx}\, dx = \infty.$$

Consequently, $L[e^{x^2}]$ does not exist.

Linear property The Laplace transform is a linear operator. That is, if f_1 and f_2 are functions that have Laplace transforms for $s > s_1$ and $s > s_2$, respectively, and if c_1 and c_2 are constants, then

$$L[c_1 f_1(x) + c_2 f_2(x)] = c_1 L[f_1(x)] + c_2 L[f_2(x)] \quad \text{for } s > \max(s_1, s_2).$$

Proof Let $s > \max(s_1, s_2)$. Then, by definition,

$$\begin{aligned} L[c_1 f_1(x) + c_2 f_2(x)] &= \int_0^\infty e^{-sx}[c_1 f_1(x) + c_2 f_2(x)]\, dx \\ &= c_1 \int_0^\infty e^{-sx} f_1(x)\, dx + c_2 \int_0^\infty e^{-sx} f_2(x)\, dx \\ &= c_1 L[f_1(x)] + c_2 L[f_2(x)]. \end{aligned}$$

EXAMPLE Calculate $L[2x^2 + 3]$.
Using the linear property of the Laplace transform, we find

$$L[2x^2 + 3] = 2L[x^2] + 3L[1] = 2\frac{2}{s^3} + 3\frac{1}{s} = \frac{4}{s^3} + \frac{3}{s} \quad \text{for } s > 0.$$

EXAMPLE Calculate $L[\sinh bx]$.
Since $\sinh bx = \frac{1}{2}e^{bx} - \frac{1}{2}e^{-bx}$, we obtain

$$\begin{aligned} L[\sinh bx] &= L[\tfrac{1}{2}e^{bx} - \tfrac{1}{2}e^{-bx}] \\ &= \tfrac{1}{2}L[e^{bx}] - \tfrac{1}{2}L[e^{-bx}] \\ &= \frac{1}{2}\frac{1}{s-b} - \frac{1}{2}\frac{1}{s+b} \\ &= \frac{b}{s^2 - b^2} \quad \text{for } s > |b|. \end{aligned}$$

Translation property If $L[f(x)] = F(s)$ for $s > s_0$, then $L[e^{ax}f(x)] = F(s - a)$ for $s > s_0 + a$.

Proof By definition and hypothesis,

$$L[f(x)] = \int_0^\infty e^{-sx} f(x)\, dx = F(s) \quad \text{for } s > s_0.$$

So

$$\begin{aligned} L[e^{ax}f(x)] &= \int_0^\infty e^{-sx} e^{ax} f(x)\, dx \\ &= \int_0^\infty e^{-(s-a)x} f(x)\, dx \\ &= F(s - a) \quad \text{for } s - a > s_0. \end{aligned}$$

EXAMPLE Calculate $L[x^n e^{ax}]$ where n is a positive integer.

We have already found that

$$L[x^n] = \frac{n!}{s^{n+1}} = F(s) \quad \text{for } s > 0.$$

So using the translation property, we get

$$L[x^n e^{ax}] = F(s - a) = \frac{n!}{(s - a)^{n+1}} \quad \text{for } s > a.$$

A function $f(x)$ is of *exponential order a as $x \to +\infty$* if and only if there exists positive constants M and x_0 and a constant a such that

$$|f(x)| < Me^{ax} \quad \text{for } x \geq x_0.$$

Thus, a function which is of exponential order $a > 0$ as $x \to +\infty$ may become infinite as $x \to +\infty$, but it may not become infinite more rapidly than Me^{ax}. It follows from this definition that all bounded functions are of exponential order 0 as $x \to +\infty$. Also if $a < b$ and f is of exponential order a as $x \to +\infty$, then f is of exponential order b as $x \to +\infty$, since $a < b$ implies that $Me^{ax} < Me^{bx}$. It should be noted that there are functions which are not of exponential order a as $x \to +\infty$ for any a. For instance, e^{x^2} is not of exponential order a as $x \to +\infty$ for any a. As a matter of fact, it can be shown that for any positive constants a and M, no matter how large, there exists an x_0—which depends upon M and a—such that $e^{x^2} > Me^{ax}$ for all $x > x_0$.

A function $f(x)$ is *piecewise continuous on a finite interval* $[a, b]$ if and only if (i) $f(x)$ is continuous on $[a, b]$ except at a finite number of points, (ii) the limits

$$f(a^+) = \lim_{x \to a^+} f(x) \quad \text{and} \quad f(b^-) = \lim_{x \to b^-} f(x),$$

both exist and are finite, and (iii) if $c \in (a, b)$ is a point of discontinuity of f, then the following limits exist and are finite:

$$f(c^-) = \lim_{x \to c^-} f(x) \quad \text{and} \quad f(c^+) = \lim_{x \to c^+} f(x).$$

When the limits in (iii) are equal, f is said to have a *removable discontinuity at c* and when the limits in (iii) are unequal, f is said to have a *jump discontinuity at c*. The function graphed in Figure 6.1 is piecewise continuous on $[a, b]$ and has a removable discontinuity at c_1 and jump discontinuities at c_2 and c_3. If $f(x)$ is piecewise continuous on a finite interval $[a, b]$ and is continuous except possibly at the points

$$a = a_1 < a_2 < \cdots < a_n = b,$$

then f is integrable on $[a, b]$ and

$$\int_a^b f(x)\,dx = \int_{a_1}^{a_2} f(x)\,dx + \int_{a_2}^{a_3} f(x)\,dx + \cdots + \int_{a_{n-1}}^{a_n} f(x)\,dx.$$

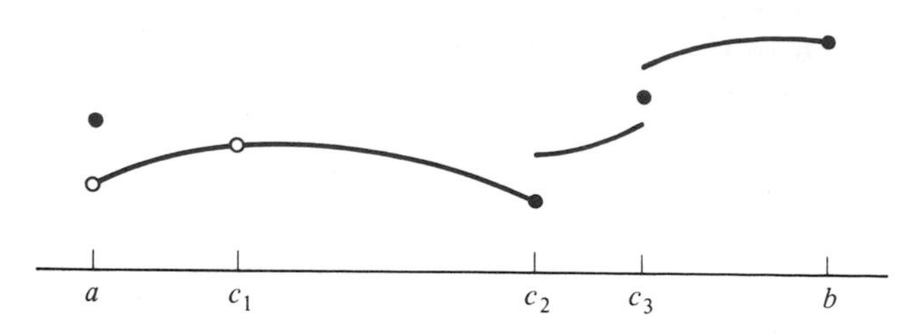

Figure 6.1 Graph of a piecewise continuous function.

The following theorem provides sufficient conditions for the existence of a Laplace transform of a function $f(x)$. These conditions are not necessary conditions for the existence of a Laplace transform, as we shall show in the example following the theorem.

Theorem 6.9 If $f(x)$ is piecewise continuous on $[0, b]$ for all finite $b > 0$ and if $f(x)$ is of exponential order a as $x \to +\infty$, then the Laplace transform of $f(x)$ exists for $s > a$.

Proof Since $f(x)$ is assumed to be of exponential order a as $x \to +\infty$, there exist positive constants M and x_0 such that $|f(x)| < Me^{ax}$ for $x > x_0$, we find

$$L[f(x)] = \int_0^\infty e^{-sx}f(x)\,dx = \int_0^{x_0} e^{-sx}f(x)\,dx + \int_{x_0}^\infty e^{-sx}f(x)\,dx.$$

The first integral on the right-hand side of this equality exists, since $f(x)$ is piecewise continuous on $[0, x_0]$, which implies that $e^{-sx}f(x)$ is piecewise continuous on $[x, x_0]$, which ensures integrability on $[0, x_0]$. The second integral on the right-hand side of the last equality exists for $s > a$, since

$$\left|\int_{x_0}^\infty e^{-sx}f(x)\,dx\right| \le \int_{x_0}^\infty e^{-sx}|f(x)|\,dx < M\int_{x_0}^\infty e^{-sx}e^{ax}\,dx$$

$$= \frac{M}{a-s}\,e^{(a-s)x}\Big|_0^\infty = \frac{M}{s-a}\,e^{(a-s)x_0} \quad \text{for } s > a.$$

Table 6.1 contains the Laplace transforms of several continuous functions that are of exponential order a as $x \to +\infty$.

EXAMPLE The function $f(x) = 1/\sqrt{x}$ is of exponential order 0 as $x \to +\infty$ but is not piecewise continuous on $[0, b]$ for any $b > 0$ since $\lim\limits_{x \to 0+} f(x) = +\infty$. Thus $f(x) = 1/\sqrt{x}$ does not satisfy the first hypothesis of Theorem 6.9; however, the Laplace transform exists, as the following calculations show:

$$L\left[\frac{1}{\sqrt{x}}\right] = \int_0^\infty e^{-sx}x^{-1/2}\,dx = \frac{1}{\sqrt{s}}\int_0^\infty e^{-t}t^{-1/2}\,dt = \sqrt{\frac{\pi}{s}}\,.$$

Table 6.1 Laplace transforms

$f(x)$	$L[f] = F(s)$
e^{ax}	$\frac{1}{s-a}$, $s > a$
1	$\frac{1}{s}$, $s > 0$
x^n, n a positive integer	$\frac{n!}{s^{n+1}}$, $s > 0$
$\sin bx$	$\frac{b}{s^2 + b^2}$, $s > 0$
$\cos bx$	$\frac{s}{s^2 + b^2}$, $s > 0$
$\sinh bx$	$\frac{b}{s^2 - b^2}$, $s > \lvert b \rvert$
$\cosh bx$	$\frac{s}{s^2 - b^2}$, $s > \lvert b \rvert$
$x^n e^{ax}$, n a positive integer	$\frac{n!}{(s-a)^{n+1}}$, $s > a$
$e^{ax} \sin bx$	$\frac{b}{(s-a)^2 + b^2}$, $s > a$
$e^{ax} \cos bx$	$\frac{s-a}{(s-a)^2 + b^2}$, $s > a$
$e^{ax} \sinh bx$	$\frac{b}{(s-a)^2 - b^2}$, $s > \lvert b \rvert + a$
$e^{ax} \cosh bx$	$\frac{s}{(s-a)^2 - b^2}$, $s > \lvert b \rvert + a$

To obtain this result we made the change of variable $t = sx$ and found the value of the last definite integral in a table of integrals.

Theorem 6.10 If $f(x)$ and $g(x)$ are defined and piecewise continuous on $[0, b]$ for all finite $b > 0$ and of exponential order a as $x \to +\infty$ and if $L[f] = L[g]$, then $f(x) = g(x)$ at all points $x \in [0, \infty)$ where f and g are both continuous.

Proof Since f and g are piecewise continuous on $[0, b]$ for all finite $b > 0$ and of exponential order a as $x \to +\infty$, $f(x) - g(x)$ is piecewise continuous on $[0, b]$ for all finite $b > 0$ and of exponential order a as $x \to +\infty$. Consequently, $L[f - g]$ exists and since the Laplace transform is linear, we have $L[f - g] = L[f] - L[g] = 0$. Thus,

$$L[f - g] = \int_0^\infty e^{-sx}[f(x) - g(x)]\, dx = 0.$$

It can be shown that the integrand $e^{-sx}[f(x) - g(x)] = 0$ on $[0, \infty)$ except at a countable number of points which must be points of discontinuity of the integrand. Hence, $f(x) = g(x)$ for all points $x \in [0, \infty)$ which are not points of discontinuity of f or g.

EXAMPLE Find a function $f(x)$ whose Laplace transform is $F(s) = 2/s(s + 1)$. Using the method of partial fractions, we find

$$L[f] = \frac{2}{s(s+1)} = \frac{2}{s} - \frac{2}{s+1} = 2L[1] - 2L[e^{-x}] = L[2 - 2e^{-x}].$$

Hence, by Theorem 6.10, $f(x) = 2 - 2e^{-x}$, where f is continuous.

EXERCISES

1. Let a and b be real constants and let n be a positive integer. Use the definition to calculate the Laplace transform of each of the following functions.
   (a) $\cos bx$ (b) $\cosh bx$
   (c) $x \sin bx$ (d) $x \cos bx$

2. Find $L[f]$ for

$$f(x) = \begin{cases} 0, & 0 \le x < 3. \\ 1, & 3 \le x \end{cases}$$

3. Find $L[g]$ for

$$g(x) = \begin{cases} 1 - x, & 0 \le x < 1. \\ x - 1, & 1 \le x \end{cases}$$

4. Find $L[h]$ for

$$h(x) = \begin{cases} x, & 0 \le x < 2 \\ -x + 4, & 2 \le x < 4. \\ 0, & 4 \le x \end{cases}$$

5. Find $L[k]$ for

$$k(x) = \begin{cases} 0, & 0 \le x < 1 \\ 1, & 1 \le x < 2. \\ 0, & 2 \le x \end{cases}$$

6. Use Table 6.1 and the linear property of Laplace transforms to find the following transforms.
   (a) $L[5]$ (b) $L[e]$
   (c) $L[3x - 2]$ (d) $L[e^{-x}(2x + 1)]$
   (e) $L[e^{2x} \sin 3x - 2 \cos x]$ (f) $L[e^{3x+2}]$

7. Use the translation property to calculate the following Laplace transforms.

(a) $L[e^{ax} \sin bx]$ (b) $L[e^{ax} \cos bx]$
(c) $L[e^{ax} \sinh bx]$ (d) $L[e^{ax} \cosh bx]$

8. Show that if f and g are of exponential order a as $x \to +\infty$, then $f - g$ is of exponential order a as $x \to +\infty$.

9. For each of the following functions $F(s)$, find a function $f(x)$ such that $L[f] = F(s)$.
(a) $3/s^3$ (b) $4/(s + 2)^2$
(c) $-2s/(s^2 + 3)$ (d) $1/s^2(s + 1)$
(e) $(s - 1)/(s^2 - 2s + 5)$ (f) $2/(s^2 - 2s + 5)$
(g) $-4/s(s^2 + 1)$ (h) $(2s + 5)/(s^2 + 2s + 2)$
(i) $\dfrac{1}{s^2} + \dfrac{2}{s^2 - 1}$ (j) $\dfrac{3s}{s^2 - 4s + 3}$

6.3.2.1 Application of Laplace transforms to differential equations

The Laplace transform method for solving linear differential equations—homogeneous or nonhomogeneous—is a three-step process. First, the differential equation is transformed by the Laplace transform into an algebraic equation in s. The unknown of the algebraic equation—the transform of the solution of the differential equation $L[y]$—is solved for by algebraic manipulations. And finally the solution of the algebraic equation is transformed back by the inverse Laplace transform and the result is the solution of the original differential equation.

The Laplace transform method immediately yields the solution of nonhomogeneous differential equations and initial value problems. That is, one does not have to (1) find the general solution of the associated homogeneous equation, (2) find a particular solution to the nonhomogeneous equation, and (3) add these solutions to get the general solution. In addition, if the problem is an initial value problem, the initial conditions are incorporated in the transforming equations. Hence, one does not have to find the general solution and then choose the constants to satisfy the initial conditions.

Let us formally calculate the Laplace transform of the derivative of the function $y(x)$. Integrating by parts, we obtain

$$L[y'] = \int_0^\infty e^{-sx} y'(x)\, dx = ye^{-sx}\Big|_0^\infty + s\int_0^\infty e^{-sx} y(x)\, dx.$$

If y is the exponential order a as $x \to +\infty$, then for $s > a$, $ye^{-sx} \to 0$ as $x \to +\infty$ and, therefore,

$$L[y'] = -y(0) + sL[y] \quad \text{for } s > a.$$

Next, we formally calculate the Laplace transform of the second derivative of the function $y(x)$. Again using integration by parts, we find

$$L[y''] = \int_0^\infty e^{-sx}y''(x)\,dx = y'e^{-sx}|_0^\infty + s\int_0^\infty e^{-sx}y'(x)\,dx.$$

If y' is of exponential order a as $x \to +\infty$, then for $s > a$, $y'e^{-sx} \to 0$ as $x \to +\infty$, and

$$L[y''] = -y'(0) + sL[y'] = -y'(0) - sy(0) + s^2L[y] \quad \text{for } s > a.$$

By induction we obtain the following general formula

$$(27) \qquad L[y^{(n)}] = -y^{(n-1)}(0) - sy^{(n-2)}(0) - \cdots - s^{n-1}y(0) + s^nL[y],$$

which is valid when $y, y^{(1)}, \ldots, y^{(n-1)}$ are all of exponential order a as $x \to +\infty$. Since the solutions of homogeneous linear differential equations with constant coefficients and all their derivatives are of exponential order a for some constant a as $x \to +\infty$, equation (27) is valid for (19) $p(D)y = 0$. If $g(x)$ has a Laplace transform—in particular, if $g(x)$ is of exponential order a as $x \to +\infty$, then equation (27) can be applied to the nonhomogeneous linear differential equation with constant coefficients (23) $p(D)y = g(x)$.

EXAMPLE Use the Laplace transform method to find the general solution of the homogeneous linear differential equation $y'' + 4y = 0$.

Taking the Laplace transform, we find

$$\begin{aligned} L[y'' + 4y] &= L[0]; \\ L[y''] + 4L[y] &= 0; \\ -y'(0) - sy(0) + s^2L[y] + 4L[y] &= 0; \\ -y'(0) - sy(0) + (s^2 + 4)L[y] &= 0. \end{aligned}$$

Solving for $L[y]$ and letting $y(0) = A$ and $y'(0) = B$, where A and B are arbitrary real constants, we obtain

$$\begin{aligned} L[y] &= \frac{B + As}{s^2 + 4} = -\frac{B}{2}\frac{2}{s^2 + 4} + A\frac{s}{s^2 + 4} \\ &= -\frac{B}{2}L[\sin 2x] + AL[\cos 2x] \\ &= L\left[-\frac{B}{2}\sin 2x + A\cos 2x\right]. \end{aligned}$$

Hence, the general solution is

$$y = C\sin 2x + A\cos 2x,$$

where A and $C = -B/2$ are arbitrary real constants.

EXAMPLE Use the Laplace transform method to find the general solution of the nonhomogeneous linear differential equation $y'' + y' - 2y = x^2 - 1$.

Taking the Laplace transform of the differential equation, we get

$$L[y'' + y' - 2y] = L[x^2 - 1];$$
$$L[y''] + L[y'] - 2L[y] = L[x^2] - L[1];$$
$$-y'(0) - sy(0) + s^2L[y] - y(0) + sL[y] - 2L[y] = \frac{2}{s^3} - \frac{1}{s}.$$

Letting $y(0) = A$ and $y'(0) = B$ and solving for $L[y]$ yields

$$L[y] = \left(A(s + 1) + B + \frac{2}{s^3} - \frac{1}{s}\right)\Big/(s^2 + s - 2)$$
$$= \frac{A(s + 1) + B}{(s + 2)(s - 1)} + \frac{2}{s^3(s + 2)(s - 1)} - \frac{1}{s(s + 2)(s - 1)}.$$

Expanding the right-hand side of this equation by partial fractions and combining like terms, we find

$$L[y] = \frac{c_1}{s + 2} + \frac{c_2}{s - 1} - \frac{1}{4s} + \frac{1}{2s^2} - \frac{1}{s^3},$$

where $c_1 = (4A - 4B - 1)/12$ and $c_2 = (2A + B + 1)/3$ are arbitrary real constants since A and B are arbitrary real constants. Hence,

$$L[y] = c_1L[e^{-2x}] + c_2L[e^x] - \tfrac{1}{4}L[1] - \tfrac{1}{2}L[x] - \tfrac{1}{2}L[x^2]$$
$$= L[c_1e^{-2x} + c_2e^x - \tfrac{1}{4} - \tfrac{1}{2}x - \tfrac{1}{2}x^2].$$

Therefore, the general solution is

$$y = c_1e^{-2x} + c_2e^x - \tfrac{1}{4} - \tfrac{1}{2}x - \tfrac{1}{2}x^2.$$

EXAMPLE Find the solution of the initial value problem

$$y'' - 4y' + 5y = 2e^x - \sin x; \qquad y(0) = 1, \quad y'(0) = -1$$

by the Laplace transform method.

The Laplace transform of the differential equation is

$$-y'(0) - sy(0) + s^2L[y] + 4y(0) - 4sL[y] + 5L[y] = \frac{2}{s - 1} - \frac{1}{s^2 + 1}.$$

Substituting the initial conditions and solving for $L[y]$, we get

$$L[y] = \left(s - 5 + \frac{2}{s - 1} - \frac{1}{s^2 + 1}\right)\Big/(s^2 - 4s + 5)$$
$$= \frac{s^4 - 6s^3 + 8s^2 - 7s + 8}{(s - 1)(s^2 + 1)(s^2 - 4s + 5)}.$$

Using the method of partial fractions, we find

$$L[y] = \frac{1}{s-1} - \frac{1}{8}\frac{1}{s^2+1} - \frac{1}{8}\frac{s}{s^2+1} - \frac{19}{8}\frac{1}{s^2-4s+5} + \frac{1}{8}\frac{s}{s^2-4s+5}$$

$$= \frac{1}{s-1} - \frac{1}{8}\frac{1}{s^2+1} - \frac{1}{8}\frac{s}{s^2+1} - \frac{17}{8}\frac{1}{(s-2)^2+1} + \frac{1}{8}\frac{s-2}{(s-2)^2+1}$$

$$= L[e^x] - \tfrac{1}{8}L[\sin x] - \tfrac{1}{8}L[\cos x] - \tfrac{17}{8}L[e^{2x}\sin x] + \tfrac{1}{8}L[e^{2x}\cos x].$$

So the solution of the initial value problem is

$$y = e^x - \tfrac{1}{8}(\sin x + \cos x + 17e^{2x}\sin x - e^{2x}\cos x).$$

EXERCISES

Use the Laplace transform method to find the general solution of the following differential equations.

1. $y' - y = 0$

2. $y'' - 2y' + 5y = 0$

3. $y' + 2y = 4$

4. $y'' - 9y = 2\sin 3x$

5. $y'' + 9y = 2\sin 3x$

6. $y'' + y' - 2y = xe^x - 3x^2$

7. $y^{(4)} - 2y^{(3)} + y^{(2)} = xe^x - 3x^2$

Find the solution of each of the following initial value problems by using the Laplace transform method.

8. $y' = e^x;\ y(0) = -1$

9. $y' - y = 2e^x;\ y(0) = 1$

10. $y'' - 9y = x + 2;\ y(0) = -1,\ y'(0) = 1$

11. $y'' + 9y = x + 2;\ y(0) = -1,\ y'(0) = 1$

12. $y'' - y' + 6y = -2\sin 3x;\ y(0) = 0,\ y'(0) = -1$

13. $y'' - 2y' + 2y = 1 - x^2;\ y(0) = 1,\ y'(0) = 0$

14. $y''' + 3y'' + 2y' = x + \cos x;\ y(0) = 1,\ y'(0) = -1,\ y''(0) = 2$

6.3.2.2 Convolution and the Laplace transform

One can often write a Laplace transform $H(s)$ as the product of two Laplace transforms $F(s)$ and $G(s)$ in such a way that both $F(s)$ and $G(s)$ are the Laplace transforms of known functions, say $f(x)$ and $g(x)$, respectively. That is, one sometimes encounters the situation $H(s) = F(s)G(s)$, where $F(s) = L[f(x)]$ and $G(s) = L[g(x)]$. Hence, $H(s) = L[f(x)]L[g(x)]$. Momentarily, we might expect that $H(s) = L[f(x)]L[g(x)] = L[f(x)g(x)]$. Stated verbally, we might anticipate that the Laplace transform distributes over the multiplication of functions. However, we know from calculus that the integral of the product of two functions is not equal to the product of the integrals of the two functions. And since the Laplace transform is defined in terms of an integral, we should not expect the Laplace transform to distribute over the multiplication of functions. The following example illustrates that

$$L[f(x)]L[g(x)] \neq L[f(x)g(x)].$$

Let $f(x) = x$ and let $g(x) = e^x$. Then

$$L[f(x)]L[g(x)] = L[x]L[e^x] = \frac{1}{s^2}\frac{1}{s-1}$$

and

$$L[f(x)g(x)] = L[xe^x] = \frac{1}{(s-1)^2}\cdot$$

Clearly,

$$L[f(x)g(x)] \neq L[f(x)]L[g(x)] \quad \text{for any real } s.$$

Two questions should now come to mind: "Is there any function $h(x)$ such that $L[h(x)] = H(s) = L[f(x)]L[g(x)]$?" and "If so, how is $h(x)$ related to $f(x)$ and $g(x)$?" We shall now answer these questions. The *convolution of $f(x)$ and $g(x)$* is defined by

$$f(x) * g(x) = \int_0^x f(x - \xi)g(\xi)\, d\xi. \tag{28}$$

Making the change of variable $\eta = x - \xi$ in the integral appearing in equation (28), we see that

$$\begin{aligned} f(x) * g(x) &= \int_0^x f(x - \xi)g(\xi)\, d\xi = -\int_x^0 f(\eta)g(x - \eta)\, d\eta \\ &= \int_0^x g(x - \eta)f(\eta)\, d\eta = g(x) * f(x). \end{aligned}$$

Thus, we have shown that the convolution operator is commutative. Indeed, the convolution operator has many of the same properties as ordinary multiplication. For instance,

$$\begin{aligned} f(x) * (g_1(x) + g_2(x)) &= f(x) * g_1(x) + f(x) * g_2(x), \\ f(x) * (g(x) * h(x)) &= (f(x) * g(x)) * h(x), \\ f(x) * 0 &= 0. \end{aligned}$$

Consequently, the convolution operator may be thought of as a "generalized multiplication" operator. However, the convolution operator does not have some of the properties of ordinary multiplication. For example, it is not true for all functions $f(x)$ that $f(x) * 1 = f(x)$.

Suppose that $f(x)$ and $g(x)$ both have a Laplace transform for $s > a$. That is, suppose

$$L[f] = \int_0^\infty e^{-sx} f(x)\, dx$$

and

$$L[g] = \int_0^\infty e^{-sx} g(x)\, dx$$

both exist for $s > a$. By definition,

$$L[f(x) * g(x)] = \int_0^\infty e^{-sx} \left[\int_0^x f(x - \xi) g(\xi)\, d\xi \right] dx \tag{29}$$

provided that both integrals exist. The domain and order of integration of equation (29) is as shown in Figure 6.2(a). Assuming that the order of integration can be interchanged, we see that

$$L[f(x) * g(x)] = \int_0^\infty \left[\int_\xi^\infty e^{-sx} f(x - \xi)\, dx \right] g(\xi)\, d\xi. \tag{30}$$

See Figure 6.2(b). Now in the innermost integral in equation (30), we make the change of variable $\eta = x - \xi$ and thereby find

$$\begin{aligned} L[f(x) * g(x)] &= \int_0^\infty \left[\int_0^\infty e^{-s(\xi+\eta)} f(\eta)\, d\eta \right] g(\xi)\, d\xi \\ &= \int_0^\infty e^{-s\eta} f(\eta)\, d\eta \int_0^\infty e^{-s\xi} g(\xi)\, d\xi \\ &= L[f(x)]L[g(x)]. \end{aligned}$$

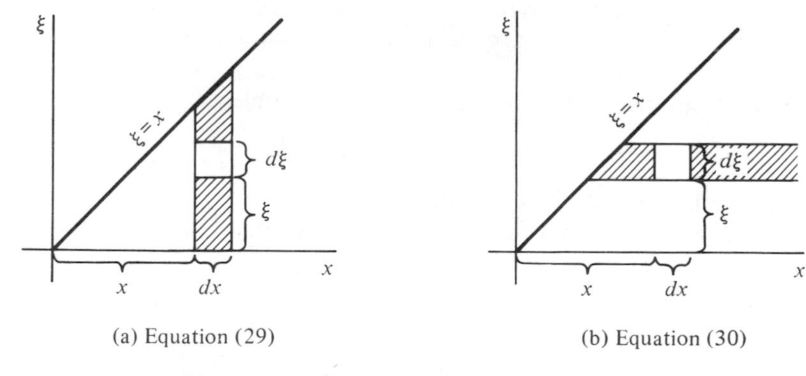

Figure 6.2 Domain and order of integration.

Thus, we have proven the following theorem.

Theorem 6.11 If $L[f(x)]$ and $L[g(x)]$ both exist for $s > a$, then

$$L[f(x)]L[g(x)] = L[f(x) * g(x)] \quad \text{for } s > a.$$

EXAMPLE Find a function $h(x)$ whose Laplace transform is $H(s) = 1/s^2(s+1)$.

We can write

$$H(s) = \frac{1}{s^2}\frac{1}{s+1} = F(s)G(s),$$

where

$$F(s) = \frac{1}{s^2} = L[x]$$

and

$$G(s) = \frac{1}{s+1} = L[e^{-x}].$$

By Theorem 6.11 and the commutative property of the convolution operator, we know that

$$h(x) = x * e^{-x} = \int_0^x (x - \xi)e^{-\xi}\,d\xi$$

and

$$h(x) = e^{-x} * x = \int_0^x e^{-(x-\xi)}\xi\,d\xi.$$

Evaluating the second integral, which is slightly simpler, we find

$$\begin{aligned} h(x) &= e^{-x}\int_0^x \xi e^{\xi}\,d\xi = e^{-x}[e^{\xi}(\xi - 1)]\Big|_0^x \\ &= e^{-x}[e^x(x-1) + 1] = x - 1 + e^{-x}. \end{aligned}$$

EXAMPLE Find the solution to the initial value problem

$$y' + y = x; \qquad y(0) = 0$$

by using the Laplace transform method.

Taking the Laplace transform of the given differential equation, we obtain

$$\begin{aligned} L[y' + y] &= L[x]; \\ L[y'] + L[y] &= L[x]; \\ -y(0) + sL[y] + L[y] &= \frac{1}{s^2}\,. \end{aligned}$$

Imposing the initial condition $y(0) = 0$ and combining like terms, we get

$$(s + 1)L[y] = \frac{1}{s^2}.$$

Solving for $L[y]$, we find that the Laplace transform of the solution, $y(x)$, of the initial value problem satisfies

$$L[y] = \frac{1}{s^2(s + 1)}.$$

Recalling the results of the previous example, we see that

$$y(x) = x - 1 + e^{-x}.$$

EXERCISES

For each of the following functions $H(s)$, use the convolution theorem to find a function $h(x)$ such that $L[h] = H(s)$.

1. $1/s(s^2 + 9)$

2. $1/(s + 1)(s - 2)$

3. $1/(s + 1)(s - 2)^2$

4. $1/s(s^2 - 2s + 5)$

5. $s/(s - 1)(s^2 + 4)$

6. $1/s^2(s^2 - 4)$

Find the solution of the following initial value problems by using the Laplace transform method and the convolution theorem.

7. $y' - 2y = 6$; $y(0) = 2$

8. $y' + y = e^x$; $y(0) = \frac{5}{2}$

9. $y'' + 9y = 1$; $y(0) = 0$, $y'(0) = 0$ (*Hint:* See Exercise 1.)

10. $y'' + 9y = 18e^{3x}$; $y(0) = -1$, $y'(0) = 6$

11. $y'' - y' - 2y = 0$; $y(0) = 0$, $y'(0) = 3$ (*Hint:* See Exercise 2.)

12. $y'' - y' - 2y = x^2$; $y(0) = \frac{11}{4}$, $y'(0) = \frac{1}{2}$

13. $y'' - 2y' + y = 2 \sin x$; $y(0) = -2$, $y'(0) = 0$

14. $y''' - y'' + 4y' - 4y = 0$; $y(0) = 0$, $y'(0) = 5$, $y''(0) = 5$ (*Hint:* See Exercise 5.)

15. Show that for any continuous function $f(x)$ and any constant $a \neq 0$, the solution of the initial value problem

$$y'' + a^2 y = f(x); \qquad y(0) = 0, \quad y'(0) = 0$$

is

$$y(x) = \frac{1}{a} \sin ax * f(x) = \frac{1}{a} \int_0^x \sin a(x - \xi) f(\xi)\, d\xi$$

$$= \frac{\sin ax \int_0^x \cos a\xi f(\xi)\, d\xi - \cos ax \int_0^x \sin a\xi f(\xi)\, d\xi}{a}.$$

16. Show that for any continuous function $f(x)$ and any constant $a \neq 0$ the solution of the initial value problem

$$y'' - a^2y = f(x); \qquad y(0) = 0, \quad y'(0) = 0$$

is

$$y(x) = \frac{1}{2a}e^{ax} * f(x) - \frac{1}{2a}e^{-ax} * f(x)$$

$$= \frac{e^{ax}\int_0^x e^{-a\xi}f(\xi)\,d\xi - e^{-ax}\int_0^x e^{a\xi}f(\xi)\,d\xi}{2a}.$$

17. Show that for any continuous function $f(x)$ and any constant $a \neq 0$ the solution of the initial value problem

$$y'' - 2ay' + a^2y = f(x); \qquad y(0) = 0, \quad y'(0) = 0$$

is

$$y(x) = xe^{ax} * f(x) = \int_0^x (x - \xi)e^{a(x-\xi)}f(\xi)\,d\xi$$

$$= e^{ax}\left(x\int_0^x e^{-a\xi}f(\xi)\,d\xi - \int_0^x \xi e^{-a\xi}f(\xi)\,d\xi\right).$$

6.3.2.3 The unit function and time-delayed functions

The Laplace transform method is useful not only for solving nonhomogeneous linear differential equations and initial value problems with constant coefficients when the nonhomogeneous function—which is also called the forcing function—is itself a solution to a homogeneous linear differential equation with constant coefficients but also when the forcing function is discontinuous or an impulse and yet has a Laplace transform. Forcing functions which are discontinuous or an impulse frequently occur in electrical and mechanical systems. In this section we shall consider forcing functions which are discontinuous and in the next section we shall consider forcing functions which are impulses.

One of the simplest functions which has a jump discontinuity of 1 at $x = c \geq 0$ is the *unit step function* or the *Heaviside function*, $u(x - c)$, which is defined as follows:

$$u(x - c) = \begin{cases} 0, & x < c \\ 1, & x \geq c \end{cases}$$

Other discontinuous step functions can be written as a linear combination of unit step functions. For instance,

$$h(x) = \begin{cases} 0, & x < 1 \\ 2, & 1 \leq x < 2 \\ 1, & 2 \leq x \end{cases} \tag{31}$$

can be written as $h(x) = 2u(x - 1) - u(x - 2)$. Thus, the unit step function is the basic function for constructing other step functions. We easily calculate the Laplace transform of the unit step function as follows:

$$L[u(x - c)] = \int_0^\infty e^{-sx}u(x - c)\,dx = \int_c^\infty e^{-sx}\,dx = \frac{e^{-cs}}{s} \quad \text{for } s > 0.$$

Using the linear property of Laplace transforms, we will easily be able to calculate the Laplace transform of any particular step function once it is written as a linear combination of unit step functions. For example, the Laplace transform of $h(x) = 2u(x - 1) - u(x - 2)$ is calculated as follows:

$$\begin{aligned} L[h(x)] &= L[2u(x - 1) - u(x - 2)] \\ &= 2L[u(x - 1)] - L[u(x - 2)] \\ &= \frac{2e^{-s}}{s} - \frac{e^{-2s}}{s} \quad \text{for } s > 0. \end{aligned}$$

Suppose that two identical sensing devices are placed at points A and B which are separated from one another by a relatively large distance. Further suppose that at time $x = 0$ a signal is sent from point A. Let the signal received by the sensing device at point A be the function $f(x)$ shown in Figure 6.3(a). Assuming that no distortion or attenuation has occurred, the signal received by the sensing device at point B will be the function

$$g(x) = \begin{cases} 0, & 0 \le x < c \\ f(x - c), & c \le x, \end{cases}$$

where $c > 0$ is the time it takes the signal to travel from point A to point B. The function $g(x)$ is shown in Figure 6.3(b). The function $g(x)$ is called the *c-time delay function of* $f(x)$. That is, the function $g(x)$ is the function $f(x)$ delayed by c units of time. One encounters time-delayed functions in many physical circumstances. In most physical situations the function $f(x)$ is

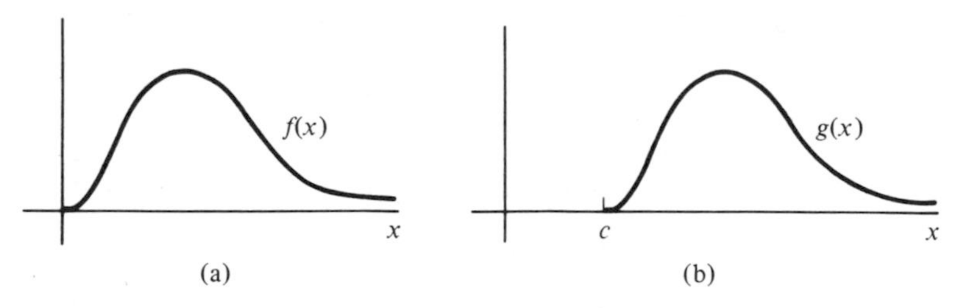

Figure 6.3 The function $f(x)$ and the associated c-time delay function $g(x)$.

defined only for $x \geq 0$. We have defined the c-time delay function $g(x)$ of $f(x)$ so that $g(x) = 0$ for $0 \leq x < c$, since no disturbance at point B is caused by the signal sent from point A during the interval of time $[0, c)$. Notice that $g(x)$ can be defined more concisely by

$$g(x) = u(x - c)f(x - c).$$

The following theorem provides us with the relationship between the Laplace transform of a function $f(x)$ and the Laplace transform of the c-time delay function of $f(x)$, $u(x - c)f(x - c)$.

Theorem 6.12 If $L[f(x)]$ exists for $s > a \geq 0$ and if $c > 0$, then

$$L[u(x - c)f(x - c)] = e^{-cs}L[f(x)] \quad \text{for } s > a.$$

Proof By definition,

$$\begin{aligned} L[u(x - c)f(x - c)] &= \int_0^\infty e^{-sx}u(x - c)f(x - c)\,dx \\ &= \int_c^\infty e^{-sx}f(x - c)\,dx. \end{aligned}$$

Making the change of variable $\xi = x - c$, we find

$$\begin{aligned} L[u(x - c)f(x - c)] &= \int_0^\infty e^{-s(\xi + c)}f(\xi)\,d\xi \\ &= e^{-cs}\int_0^\infty e^{-s\xi}f(\xi)\,d\xi \\ &= e^{-cs}L[f(x)] \quad \text{for } s > a. \end{aligned}$$

EXAMPLE Calculate the Laplace transform of

$$f(x) = \begin{cases} \cos x, & 0 \leq x < \pi \\ 0, & \pi \leq x. \end{cases}$$

Once we discover that

$$f(x) = \cos x - u(x - \pi)\cos x = \cos x + u(x - \pi)\cos(x - \pi),$$

we easily find

$$\begin{aligned} L[f(x)] &= L[\cos x] + L[u(x - \pi)\cos(x - \pi)] \\ &= L[\cos x] + e^{-\pi s}L[\cos x] \\ &= \frac{(1 + e^{-\pi s})s}{s^2 + 1}. \end{aligned}$$

EXAMPLE Find a function $f(x)$ whose Laplace transform is $e^{-\pi s}/(s^2 + 4)$.

Since $2/(s^2 + 4)$ is the Laplace transform of $\sin 2x$ and since the factor $e^{-\pi s}$ indicates that the function $\sin 2x$ should be delayed π units, we calculate the Laplace transform of $\frac{1}{2}u(x - \pi) \sin 2(x - \pi)$ and find

$$L[\tfrac{1}{2}u(x - \pi) \sin 2(x - \pi)] = \tfrac{1}{2}e^{-\pi s}L[\sin 2x] = \frac{e^{-\pi s}}{s^2 + 4}.$$

Consequently, $f(x) = \frac{1}{2}u(x - \pi) \sin 2(x - \pi)$. Noting that $\sin (2x - 2\pi) = \sin 2x$, we can write $f(x)$ more conventionally as

$$f(x) = \begin{cases} 0, & 0 \leq x < \pi \\ \frac{1}{2} \sin 2x, & \pi \leq x. \end{cases}$$

EXAMPLE Find the solution of the initial value problem

$$y'' + y = 2u(x - 1) - u(x - 2); \qquad y(0) = 0, \quad y'(0) = 1. \tag{32}$$

The forcing function in this instance is the step function $h(x)$ of equation (31). This function is discontinuous at $x = 1$ and $x = 2$ and is not the solution of any linear homogeneous differential equation with constant coefficients. So here we have our first example of an initial value problem which we can not easily solve using the method of undetermined coefficients (Section 6.3.1). If we were to use the method of undetermined coefficients to solve this initial value problem, we would have to consider the differential equation of equation (32) to be three distinct differential equations defined, respectively, on the intervals $(-\infty, 1]$, $[1, 2]$, $[2, \infty)$. Then we would have to require that (i) the solution y_1 on the interval $(-\infty, 1]$ satisfy the given initial conditions, (ii) the solution y_2 on the interval $[1, 2]$ satisfy $y_2(1) = y_1(1)$ and $y_2'(1) = y_1'(1)$, and (iii) the solution y_3 on the interval $[2, \infty)$ satisfy $y_3(2) = y_2(2)$ and $y_3'(2)$ and $y_2'(2)$. In this manner we could obtain a solution,

$$y(x) = \begin{cases} y_1(x), & -\infty < x \leq 1 \\ y_2(x), & 1 \leq x \leq 2, \\ y_3(x), & 2 \leq x < \infty \end{cases}$$

of the given initial value problem which would be valid on $(-\infty, \infty)$. That is, y and y' would be continuous on $(-\infty, \infty)$—in particular, at $x = 1$ and $x = 2$—and y would satisfy the differential equation and initial conditions of equation (32). Hence, in order to solve the initial value problem (32) using the method of undetermined coefficients, one would in effect have to solve the following three initial value problems in succession:

$$y_1' + y_1 = 0; \qquad y_1(0) = 0, \qquad y_1'(0) = 1 \quad \text{for } -\infty < x \leq 1 \tag{33a}$$

$$y_2' + y_2 = 2; \qquad y_2(1) = y_1(1), \quad y_2'(1) = y_1'(1) \quad \text{for } 1 \leq x \leq 2 \tag{33b}$$

$$y_3' + y_3 = 1; \qquad y_3(2) = y_2(2), \quad y_3'(2) = y_2'(2) \quad \text{for } 2 \leq x < \infty. \tag{33c}$$

The advantage of the Laplace transform method is that the solution of equation (32) can be obtained with one application of the method—not three separate applications. In addition, it should be noted that the solution obtained by the Laplace transform method will also simultaneously satisfy equations (33). And, therefore, these equations can serve to check the validity of the solution.

Taking the Laplace transform of the differential equation of equation (32), we get

$$L[y''] + L[y] = 2L[u(x-1)] - L[u(x-2)];$$

$$-y'(0) - sy(0) + s^2L[y] + L[y] = \frac{2e^{-s}}{s} - \frac{e^{-2s}}{s}.$$

Imposing the initial conditions of equation (32) and solving for $L[y]$, we find

$$L[y] = \frac{2e^{-s}/s - e^{-2s}/s + 1}{s^2+1}$$

$$= \frac{2e^{-s}}{s(s^2+1)} - \frac{e^{-2s}}{s(s^2+1)} + \frac{1}{s^2+1}.$$

The function

$$\frac{1}{s(s^2+1)} = \frac{1}{s} - \frac{s}{s^2+1}$$

is the Laplace transform of

$$f(x) = 1 - \cos x.$$

Hence,

$$L[y] = 2e^{-s}L[1 - \cos x] - e^{-2s}L[1 - \cos x] + L[\sin x].$$

Applying Theorem 6.12, we see that

$$L[y] = L[2u(x-1)(1-\cos(x-1))] + L[-u(x-2)(1-\cos(x-2))] + L[\sin x]$$

$$= L[2u(x-1)(1-\cos(x-1)) - u(x-2)(1-\cos(x-2)) + \sin x].$$

Hence,

$$y(x) = 2u(x-1)[1-\cos(x-1)] - u(x-2)[1-\cos(x-2)] + \sin x.$$

Or, writing $y(x)$ in the more usual way, we have

$$y(x) = \begin{cases} \sin x, & -\infty < x \le 1 \\ \sin x + 2 - 2\cos(x-1), & 1 \le x \le 2 \\ \sin x + 1 - 2\cos(x-1) + \cos(x-2), & 2 \le x < \infty. \end{cases}$$

The reader should verify that this solution simultaneously satisfies equations (33).

EXERCISES

1. Express each of the following functions in terms of unit step functions.

(a) $f_1(x) = \begin{cases} 2, & 0 \le x < 1 \\ 1, & 1 \le x \end{cases}$

(b) $f_2(x) = \begin{cases} 1, & 2 \le x < 4 \\ 0, & \text{otherwise} \end{cases}$

(c) $f_3(x) = \begin{cases} 0, & 0 \le x < 1 \\ (x-1)^2, & 1 \le x \end{cases}$

(d) $f_4(x) = \begin{cases} 0, & 0 \le x < 1 \\ x^2 - 2x + 3, & 1 \le x \end{cases}$

(e) $f_5(x) = \begin{cases} 0, & 0 \le x < \pi \\ \sin 3(x - \pi), & \pi \le x \end{cases}$

(f) $f_6(x) = \begin{cases} x, & 0 \le x < 1 \\ 1, & 1 \le x \end{cases}$

(g) $f_7(x) = \begin{cases} x, & 0 \le x < 1 \\ 0, & 1 \le x \end{cases}$

2. Find the Laplace transform of the functions of Exercise 1.

3. For each of the following functions $F(s)$, find a function $f(x)$ such that $L[f(x)] = F(s)$.

(a) $e^{-s}/(s + 2)$
(b) $(1 - e^{-2s})/s^2$
(c) $se^{-\pi s}/(s^2 + 9)$
(d) $se^{-\pi s}/(s^2 - 9)$
(e) $e^{-2s}/(s^2 + 2s + 2)$
(f) $se^{-3s}/(s^2 + 2s + 2)$
(g) $e^{-3s}/(s^2 + 2s - 3)$
(h) $e^{-s}/(s^2 - 2s + 1)$

4. Solve the following initial value problems. The functions $f_i(x)$ are as defined in Exercise 1.

(a) $y' + 2y = f_1(x);\ y(0) = 1$
(b) $y'' - y' - 2y = f_2(x);\ y(0) = 0,\ y'(0) = 1$
(c) $y'' - 2y' = f_3(x);\ y(0) = 1,\ y'(0) = 0$
(d) $y'' - 2y' + y = f_4(x);\ y(0) = 0,\ y'(0) = 1$
(e) $y'' + 4y = f_5(x);\ y(0) = 1,\ y'(0) = 1$
(f) $y'' - 4y = f_6(x);\ y(0) = 0,\ y'(0) = 0$
(g) $y'' - 4y' + 5y = f_7(x);\ y(0) = 1,\ y'(0) = 0$

6.3.2.4 Impulse functions

A force which is of relative large magnitude and which acts on a system for a relative short period of time is called an *impulse force.* A golf club striking a golf ball, a hammer striking a mass suspended on a spring, and a voltage source connected to an electrical circuit for a short interval of time are all examples of impulse forces. A function $f(x)$ which represents an impulse force is naturally called an *impulse function.* Since impulse functions represent impulse forces, impulse functions are zero except for a short interval of time. Suppose that $f(x)$ is an impulse function and $f(x) = 0$ except for $x_1 < x < x_2$. Then the integral

$$I_f = \int_{-\infty}^{\infty} f(x)\,dx = \int_{x_1}^{x_2} f(x)\,dx$$

is called the total impulse of the function $f(x)$ and represents the total force imparted to the system.

Let us consider the set of impulse step functions $d_\epsilon(x)$ defined by

$$d_\epsilon(x) = \begin{cases} 1/\epsilon, & 0 < x < \epsilon \\ 0, & x \leq 0 \text{ or } x \geq \epsilon, \end{cases}$$

where ϵ is a small positive constant. A graph of one function $d_\epsilon(x)$ is shown in Figure 6.4. Notice that for all $\epsilon > 0$, the total impulse of $d_\epsilon(x)$ is

$$I_{d_\epsilon} = \int_{-\infty}^{\infty} d_\epsilon(x)\,dx = \int_0^{\epsilon} \frac{1}{\epsilon}\,dx = 1.$$

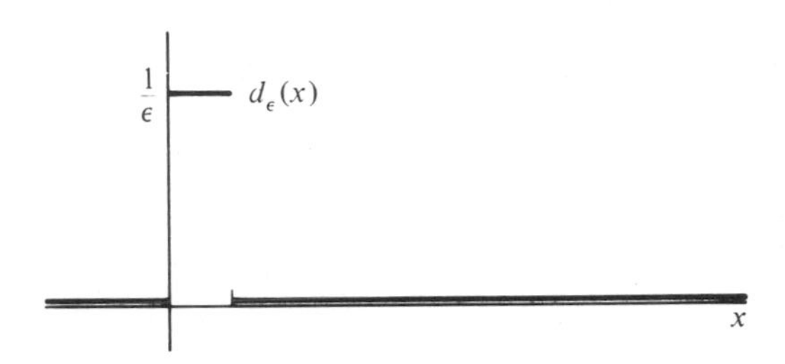

Figure 6.4 Graph of the impulse step function $d_\epsilon(x)$.

We would like to define an idealized impulse function $\delta(x)$ to be the limit as $\epsilon \to 0^+$ of the impulse step functions $d_\epsilon(x)$ and we would also like for the integral of the idealized step function $\delta(x)$ to be the limit as $\epsilon \to 0^+$ of I_{d_ϵ}. Thus, we define

$$\delta(x) = \lim_{\epsilon \to 0^+} d_\epsilon(x) = \begin{cases} \infty, & x = 0 \\ 0, & x \neq 0. \end{cases}$$

Notice that $\delta(x)$, which is called the *Dirac delta function*, in honor of P. A. M. Dirac (1902–), is not a function in the usual sense but is a "generalized function." Laurent Schwartz developed a mathematical theory called *distribution theory* in the early 1950s. Within the framework of this theory the Dirac delta function is an acceptable generalized function and is integrable with

$$\int_{-\infty}^{\infty} \delta(x)\,dx = \lim_{\epsilon \to 0^+} \int_{-\infty}^{\infty} d_\epsilon(x)\,dx = 1. \tag{34}$$

The first integral of equation (34) is not the Riemann integral with which we are familiar but the generalized integral of distribution theory. Because of the property of the Dirac delta function exhibited in equation (34), the "function" $\delta(x)$ is also often referred to as the *unit impulse function.* Following the notation developed earlier for translating functions, we have

$$\delta(x - c) = \begin{cases} \infty, & x = c \\ 0, & x \neq c \end{cases}$$

and

$$\int_{-\infty}^{\infty} \delta(x - c)\, dx = 1.$$

Let $f(x)$ be a function which is continuous on some interval about $x = c$. We shall define the generalized integral of the product of $\delta(x - c)$ and $f(x)$ as follows:

$$\begin{aligned} \int_{-\infty}^{\infty} \delta(x - c)f(x)\, dx &= \lim_{\epsilon \to 0+} \int_{-\infty}^{\infty} d_\epsilon(x - c)f(x)\, dx \\ &= \lim_{\epsilon \to 0+} \int_{c}^{c+\epsilon} \frac{1}{\epsilon} f(x)\, dx \\ &= \lim_{\epsilon \to 0+} \frac{1}{\epsilon} f(\xi)(c + \epsilon - c), \end{aligned}$$

where $c < \xi < c + \epsilon$. The last equality was obtained by using the mean value theorem for integrals. Since $c < \xi < c + \epsilon$ and $f(x)$ is assumed to be continuous on some interval about $x = c$, we have $\lim_{\epsilon \to 0+} f(\xi) = f(c)$. Consequently, in the context of distribution theory we have for any function $f(x)$ which is continuous at $x = c$,

$$\int_{-\infty}^{\infty} \delta(x - c)f(x)\, dx = f(c).$$

Therefore, letting $f(x) = e^{-sx}$, we have some justification for defining the Laplace transform of the Dirac delta function $\delta(x - c)$, where $c > 0$, to be

$$L[\delta(x - c)] = \int_{0}^{\infty} e^{-sx}\delta(x - c)\, dx = e^{-sc}.$$

In what follows we will operate with the Dirac delta function, $\delta(x - c)$, as though it were an ordinary function and we will use the properties discussed and developed in the preceding paragraphs even though the development of these properties was not mathematically rigorous. It is perhaps comforting to know that in another context all these operations and properties can and have been proven rigorously.

EXAMPLE Solve the initial value problem

$$y'' + 4y' + 5y = \delta(x - 1); \qquad y(0) = 0, \quad y'(0) = 0.$$

This equation could represent the damped motion of a mass on a spring which is initially at rest but which is set in motion at time $x = 1$ by a unit impulse force. Taking the Laplace transform of the given differential equation and imposing the initial conditions, we find

$$L[y'' + 4y' + 5y] = L[\delta(x - 1)];$$
$$L[y''] + 4L[y'] + 5L[y] = e^{-s};$$
$$-y'(0) - sy(0) + s^2L[y] - 4y(0) + 4sL[y] + 5L[y] = e^{-s};$$
$$(s^2 + 4s + 5)L[y] = e^{-s}.$$

Solving for $L[y]$, we get

$$L[y] = \frac{e^{-s}}{s^2 + 4s + 5} = \frac{e^{-s}}{(s + 2)^2 + 1} = e^{-s}L[e^{-2x} \sin x].$$

Applying Theorem 6.12, we see that

$$L[y] = e^{-s}L[e^{-2x} \sin x] = L[u(x - 1)e^{-2(x-1)} \sin (x - 1)].$$

Hence, the solution of the given initial value problem is

$$y(x) = u(x - 1)e^{-2(x-1)} \sin (x - 1) = \begin{cases} 0, & 0 \le x \le 1 \\ e^{-2(x-1)} \sin (x - 1), & 1 < x. \end{cases}$$

EXERCISES

Find the solution of the following initial value problems by using Laplace transforms.

1. $y' + 3y = \delta(x - 2); \ y(0) = 0$
2. $y' - 3y = \delta(x - 1) + 2u(x - 2); \ y(0) = 0$
3. $y'' + 9y = \delta(x - \pi) - \delta(x - 3\pi); \ y(0) = 0, \ y'(0) = 0$
4. $y'' - 2y' + y = 2\delta(x - 1); \ y(0) = 0, \ y'(0) = 1$
5. $y'' - 2y' + 5y = \cos x + \delta(x - \pi); \ y(0) = 1, \ y'(0) = 0$
6. $y'' + 4y = \delta(x - \pi) \cos x; \ y(0) = 0, \ y'(0) = 1$
7. $y'' + a^2y = \delta(x - \pi)f(x); \ y(0) = 0, \ y'(0) = 0$
   where a is a real constant and $f(x)$ is a function that is continuous on some interval about the point $x = \pi$.

6.4 HOMOGENEOUS LINEAR EQUATIONS WITH NONCONSTANT COEFFICIENTS

In this section we shall examine one method for obtaining the general solution of the nth order linear homogeneous differential equation

$$f_0(x)y^{(n)} + f_1(x)y^{(n-1)} + \cdots + f_n(x)y = 0, \tag{35}$$

where $f_0(x), f_1(x), \ldots, f_n(x)$ are assumed to be continuous—but not necessarily constant—on some interval I and $f_0(x) \neq 0$ for $x \in I$. The method of solution which we shall present is due to Jean D'Alembert (1717–1783) and is known as *D'Alembert's method* or *the method of reduction of order.*

First, let us consider the second order homogeneous linear differential equation

$$y'' + p(x)y' + q(x)y = 0, \tag{36}$$

where $p(x)$ and $q(x)$ are assumed to be continuous functions on some interval I. Suppose that $y_1(x) \not\equiv 0$ is known to be a solution of (36). Since (36) is linear, we know that $cy_1(x)$ is also a solution for any constant c. The following question should naturally come to mind: Can we find any nonconstant function $v(x)$ such that $v(x)y_1(x)$ is also a solution of (36)? To answer this question, we set

$$y(x) = v(x)y_1(x).$$

Differentiating we find

$$y' = vy_1' + v'y_1$$
$$y'' = vy_1'' + 2v'y_1' + v''y_1.$$

Substituting into (36) and rearranging, we obtain

$$v(y_1'' + py_1' + qy_1) + v'(2y_1' + py_1) + v''y_1 = 0. \tag{37}$$

Since y_1 is a solution of (36), (37) reduces to

$$v'(2y_1' + py_1) + v''y_1 = 0.$$

Letting $w = v'$, we obtain the following linear equation in w:

$$y_1w' + (2y_1' + py_1)w = 0. \tag{38}$$

Multiplying (38) by the integrating factor

$$\exp \int^x \left\{\frac{2y_1'}{y_1} + p\right\} ds = y_1^2 \exp\left\{\int^x p\, ds\right\}$$

and integrating we find

$$wy_1^2 \exp\left\{\int^x p\, ds\right\} = k_1,$$

where k_1 is an arbitrary constant. Hence,

$$v' = w = k_1 y_1^{-2} \exp\left\{-\int^x p\, ds\right\},$$

and consequently,

$$(39) \qquad y(x) = y_1(x)v(x) = y_1(x)\left[k_1 \int^x y_1^{-2}(t) \exp\left\{-\int^t p(s)\, ds\right\} dt + k_2\right]$$
$$= k_1 y_1(x)u(x) + k_2 y_1(x),$$

where k_2 is an arbitrary constant. Since k_1 and k_2 are arbitrary constants and since $u(x)$ is not a constant function, $y_2(x) = y_1(x)u(x)$ is a solution of (36) which is linearly independent of y_1. One should not memorize equation (39) but should remember that if y_1 is a solution of (36), then there is a linearly independent solution of (36) which has the form $y_2 = v(x)y_1(x)$ and that $v(x)$ can be determined explicitly by the procedure above.

EXAMPLE Verify that $y_1 = x$ is a solution of

$$(40) \qquad 2x^2y'' + xy' - y = 0$$

and find a second linearly independent solution.

Substituting $y_1 = x$, $y_1' = 1$, and $y_1'' = 0$ into (40) yields $2x^2 \cdot 0 + x \cdot 1 - x = 0$. Hence, y_1 is a solution of the given equation. To obtain a second linearly independent solution, we let

$$y(x) = v(x)y_1(x) = v(x)x.$$

Differentiating, we find

$$y'(x) = v(x) + v'(x)x$$
$$y''(x) = 2v'(x) + v''(x)x.$$

Substituting into (40) we obtain

$$2x^2[2v'(x) + v''(x)x] + x[v(x) + v'(x)x] - v(x)x = 0.$$

Rearranging and dividing by $2x^3$, we find that v satisfies the differential equation

$$v''(x) + \frac{5}{2x}v'(x) = 0.$$

Letting $w = v'$ and multiplying by $\exp\left(\frac{5}{2}\int^x ds/s\right) = x^{5/2}$ we obtain

$$w'x^{5/2} + \tfrac{5}{2}x^{3/2}w = d(wx^{5/2}) = 0.$$

Integrating and solving for w, we find

$$w = k_1 x^{-5/2}.$$

Hence,

$$v = k_1 \int^x s^{-5/2}\, ds = -\tfrac{2}{3} k_1 x^{-3/2} + k_2.$$

Choosing $k_1 = 1$ and $k_2 = 0$, we see that

$$y_2 = v y_1 = (-\tfrac{2}{3} x^{-3/2}) x = -\tfrac{2}{3} x^{-1/2}$$

is a solution of (40) which is linearly independent of y_1. Notice that y_2 is defined and real only for $x > 0$.

Now let us consider the general nth order linear homogeneous equation (35). If $y_1(x)$ is known to be a solution of (35), then we can attempt to find additional linearly independent solutions of the form $y(x) = v(x)y_1(x)$, where $v(x)$ is a nonconstant function. Substituting into (35) we see that v must satisfy

$$f_0(x)(v(x)y_1(x))^{(n)} + f_1(x)(v(x)y_1(x))^{(n-1)} + \cdots + f_{n-1}(x)(v(x)y_1(x))^{(1)} + f_n(x)v(x)y_1(x) = 0.$$

Or

$$f_0(v^{(n)}y_1 + \cdots + vy_1^{(n)}) + f_1(v^{(n-1)}y_1 + \cdots + vy_1^{(n-1)}) + \cdots + f_{n-1}(v^{(1)}y_1 + vy_1^{(1)}) + f_n v y_1 = 0$$

Rearranging this nth order differential equation in v, we obtain

$$f_0 y_1 v^{(n)} + (nf_0 y_1^{(1)} + f_1 y_1)v^{(n-1)} + \cdots + (nf_0 y_1^{(n-1)} + (n-1)f_1 y_1^{(n-2)} + \cdots + f_{n-1}y_1)v^{(1)} + (f_0 y_1^{(n)} + f_1 y_1^{(n-1)} + \cdots + f_{n-1}y_1^{(1)} + f_n y_1)v = 0.$$

Since y_1 satisfies (35), the coefficient of v in the last equation is zero. Letting $w = v'$ we obtain a linear homogeneous differential equation of order $n - 1$. If we can find $n - 1$ linearly independent solutions $w_1, w_2, \ldots, w_{n-1}$ to this equation, then y_1, $y_2 = v_1 y_1$, $y_3 = v_2 y_1, \ldots, y_n = v_{n-1} y_1$, where $v_i = \int^x w_i$ for $i = 1, 2, \ldots, n - 1$ will be linearly independent solutions of (35). If at least one solution w_1 of the reduced equation of order $n - 1$ can be found, then one may again apply the method of reduction of order and obtain a differential equation of order $n - 2$. Perhaps this process can be continued until one reduces the equation to order 1. Once this is accomplished, a solution to the first order equation can be found and by repeated integration one may obtain n linearly independent solutions to (35). For $n \geq 3$ the reduced equation of order $n - 1$ is generally more difficult to solve than the original differential equation of order n. Hence, the method of reduction of order is usually only applied to linear differential equations of second order. So when $f_0(x), f_1(x), \ldots, f_n(x)$ are not all constant functions and $n \geq 3$, it is often necessary to employ numerical techniques or power series methods (Chapter 8) to solve (35).

EXERCISES

In Exercises 1–8, verify that the given function y_1 satisfies the given differential equation, use the method of reduction of order to find a linearly independent solution, and determine the range of x for which the general solution is valid.

1. $x^2y'' - 2y = 0,\ y_1 = x^2$

2. $x^2y'' + 2xy' - 2y = 0,\ y_1 = x$

3. $x^2y'' - xy' + y = 0,\ y_1 = x$

4. $x^2y'' + xy' + y = 0,\ y_1 = \cos(\ln x)\ x > 0.$

5. $(x - 1)y'' - xy' + y = 0,\ y_1 = e^x$

6. $(2x + 1)y'' - 4(x + 1)y' + 4y = 0,\ y_1 = e^{2x}$

7. $x^2y'' + xy' + (x^2 - \frac{1}{4})y = 0,\ y_1 = x^{-1/2}\cos x$

8. $(1 - x^2)y'' - 2xy' + 2y = 0,\ y_1 = x$

9. Find the general solution of

$$x^3y''' - 6x^2y'' + x(x^2 + 18)y' - 2(x^2 + 12)y = 0$$

given that there is a solution of the form x^n.

6.5 NONHOMOGENEOUS LINEAR EQUATIONS WITH NONCONSTANT COEFFICIENTS

We shall now turn our attention to a general method for determining a particular solution of the nth order linear nonhomogeneous differential equation

$$f_0(x)y^{(n)} + f_1(x)y^{(n-1)} + \cdots + f_n(x)y = g(x), \tag{41}$$

where $f_0(x), f_1(x), \ldots, f_n(x)$, and $g(x)$ are all assumed to be continuous on an interval I, $f_0(x) \neq 0$ on I, and $g(x) \not\equiv 0$ on I. The method which we shall consider is called the *method of variation of parameters* or *Lagrange's method* in honor of J. L. Lagrange (1736–1813), who employed this method in 1774.

The basic assumption of this method is that the general solution of the associated homogeneous equation

$$f_0(x)y^{(n)} + f_1(x)y^{(n-1)} + \cdots + f_n(x)y = 0 \tag{42}$$

is known. One should be warned that the general solution of (42) may be very difficult or impossible to obtain. However, when the coefficient functions $f_0(x), f_1(x), \ldots, f_n(x)$ are all constant functions, we can find the general solution of (42). All that is required of $g(x)$ in order to find a particular solution using the method of variation of parameters is that $g(x)$ be continuous on I. Thus, this method is more general than the method of undetermined coefficients, since the method of undetermined coefficients requires that $f_0(x)$, $f_1(x), \ldots, f_n(x)$ be constant and that $g(x)$ be a solution of some linear homogeneous differential equation with constant coefficients.

Again, let us first consider the second order nonhomogeneous linear equation

$$y'' + p(x)y' + q(x)y = g(x) \tag{43}$$

and the associated homogeneous equation

$$y'' + p(x)y' + q(x)y = 0 \tag{44}$$

where $p(x)$, $q(x)$, and $g(x)$ are continuous on some interval I. Suppose that y_1 and y_2 are linearly independent solutions of (44). Then the complementary function for (43) is $y_c = c_1y_1 + c_2y_2$, where c_1 and c_2 are arbitrary constants. In the method of reduction of order once we had found one solution, say y_1, of (44), we knew that cy_1 was also a solution for any constant c, since (44) was linear. Yet we tried and succeeded in finding a second linearly independent solution of the form $y = v(x)y_1$. Likewise, in the method of variation of parameters we seek a particular solution of (43) of the form

$$y_p = v_1(x)y_1(x) + v_2(x)y_2(x), \tag{45}$$

where v_1 and v_2 are functions which are to be determined. Notice that we replace the arbitrary constants c_1 and c_2 of the general solution of the associated homogeneous equation (44) with unknown functions v_1 and v_2 and try to find a particular solution of (43) of form (45). In order for the two unknown functions v_1 and v_2 of (45) to be uniquely determined except for an arbitrary constant, one must specify two conditions. So far only one condition has been specified—namely, that (45) satisfy (43). Therefore, we may impose a second condition v_1 and v_2. We will select this condition so that the calculations will be simplified. Differentiating (45), we find

$$y_p' = (v_1'y_1 + v_2'y_2) + (v_1y_1' + v_2y_2').$$

If we were to differentiate this expression for y_p' as it stands, we would obtain an expression for y_p'' which involves v_1'' and v_2''. Substituting y_p, y_p', and the resulting expression for y_p'' into (43), we would obtain a differential equation which involved v_1'' and v_2''. To obtain a differential equation that involves only first derivatives of v_1 and v_2 after substituting into (43), we impose the following second condition on v_1 and v_2:

$$v_1'y_1 + v_2'y_2 = 0. \tag{46}$$

Hence, y_p' reduces to

$$y_p' = v_1y_1' + v_2y_2' \tag{47}$$

and differentiation yields

$$y_p'' = v_1'y_1' + v_2'y_2' + v_1y_1'' + v_2y_2''. \tag{48}$$

Substituting (45), (47), and (48) into (43) and rearranging, we see that v_1 and v_2 must satisfy

$$v_1(y_1'' + py_1' + qy_1) + v_2(y_2'' + py_2' + qy_2) + v_1'y_1' + v_2'y_2' = g(x).$$

Since y_1 and y_2 are solutions of (44), this condition reduces to

$$v_1'y_1' + v_2'y_2' = g(x). \tag{49}$$

Thus, the two equations which v_1 and v_2 must satisfy for (45) to be a particular solution of (43) are (46) and (49). Rewriting this system of equations in matrix-vector notation we see that v_1 and v_2 must satisfy

$$\begin{pmatrix} y_1 & y_2 \\ y_1' & y_2' \end{pmatrix}\begin{pmatrix} v_1' \\ v_2' \end{pmatrix} = \begin{pmatrix} 0 \\ g(x) \end{pmatrix}. \tag{50}$$

Since y_1 and y_2 are linearly independent solutions of (44), the determinant of the matrix on the left-hand side of (50)—which is the Wronskian $W(y, y_2)$—is nonzero; and, therefore, the solution of (50) is

$$v_1' = \frac{\begin{vmatrix} 0 & y_2 \\ g(x) & y_2' \end{vmatrix}}{W(y_1, y_2)} = \frac{-y_2g(x)}{W(y_1, y_2)} \tag{51a}$$

$$v_2' = \frac{\begin{vmatrix} y_1 & 0 \\ y_1' & g(x) \end{vmatrix}}{W(y_1, y_2)} = \frac{y_1g(x)}{W(y_1, y_2)}\,. \tag{51b}$$

Integration yields v_1 and v_2, and substitution into (45) provides a particular solution of (43). Again one should not memorize equations (51a) and (51b) but should remember the solution procedure outlined in this section or equation (50).

EXAMPLE Find the general solution of $y'' + 4y' + 3y = \sin e^x$.

The complementary function is

$$y_c = c_1e^{-x} + c_2e^{-3x}.$$

So we seek a particular solution of the form

(52) $$y_p = v_1e^{-x} + v_2e^{-3x}.$$

Hence,

$$y_p' = v_1'e^{-x} + v_2'e^{-3x} - v_1e^{-x} - 3v_2e^{-3x}.$$

Setting

(53) $$v_1'e^{-x} + v_2'e^{-3x} = 0$$

and then differentiating, we obtain

$$y_p'' = -v_1'e^{-x} - 3v_2'e^{-3x} + v_1e^{-x} + 9v_2e^{-3x}.$$

Substituting y_p, y_p', and y_p'' into the given differential equation and simplifying, we get the equation

(54) $$-v_1'e^{-x} - 3v_2'e^{-3x} = \sin e^x.$$

Adding (53) and (54) and solving for v_2' results in

$$v_2' = -\tfrac{1}{2}e^{3x}\sin e^x.$$

Integrating by parts twice, we find

$$\begin{aligned} v_2 &= -\tfrac{1}{2}\int^x e^{2x}(e^x \sin e^x)\,dx \\ &= \tfrac{1}{2}e^{2x}\cos x - \int^x e^{2x}\cos e^x\,dx \\ &= \tfrac{1}{2}e^{2x}\cos x - e^x\sin e^x + \int^x e^x \sin e^x\,dx \\ &= \tfrac{1}{2}e^{2x}\cos x - e^x\sin e^x - \cos e^x. \end{aligned}$$

Multiplying (53) by three, adding (54) to the resulting equation, and solving for v_1', we obtain

$$v_1' = \tfrac{1}{2}e^x\sin e^x.$$

Thus, $v_1 = -\frac{1}{2}\cos e^x$. Substituting v_1 and v_2 into (52) and collecting terms, we find

$$y_p = -e^{-2x}\sin e^x - e^{-3x}\cos e^x.$$

Of course, the general solution of the given equation is $y = y_c + y_p$.

EXAMPLE Verify that $y_1 = x$ and $y_2 = e^x$ are solutions of the homogeneous equation associated with

(55) $$(1 - x)y'' + xy' - y = 2(x - 1)^2e^{-x}$$

and find the general solution of this nonhomogeneous equation.

Differentiating yields $y_1' = 1$, $y_1'' = 0$, $y_2' = e^x$, and $y_2'' = e^x$. Substituting y_1 and y_2 into the associated homogeneous equation, we find, respectively,

$$(1 - x)\cdot 0 + x\cdot 1 - x = 0$$

and

$$(1 - x)e^x + xe^x - e^x = 0,$$

so

$$y_c = c_1 x + c_2 e^x.$$

Thus, we seek a particular solution of (55) of the form

$$y_p = v_1 x + v_2 e^x.$$

Dividing the differential equation by $(1 - x)$ and writing it in the standard form $y'' + py' + qy = g$, we find that $g(x) = -2(x - 1)e^{-x}$. Recalling equation (50), we note that v_1' and v_2' must satisfy

$$\begin{pmatrix} x & e^x \\ 1 & e^x \end{pmatrix}\begin{pmatrix} v_1' \\ v_2' \end{pmatrix} = \begin{pmatrix} 0 \\ -2(x - 1)e^{-x} \end{pmatrix}.$$

Hence,

$$v_1' = \frac{2(x - 1)}{(x - 1)e^x} = 2e^{-x}$$

and

$$v_2' = \frac{-2x(x - 1)e^{-x}}{(x - 1)e^x} = -2xe^{-2x}.$$

Integrating, we find

$$v_1 = -2e^{-x}$$

and

$$v_2 = -2\int^x xe^{-2x}\,dx = xe^{-2x} - \int^x e^{-2x}\,dx = (x + \tfrac{1}{2})e^{-2x},$$

so

$$y_p = (\tfrac{1}{2} - x)e^{-x}$$

and the general solution is

$$y = c_1 x + c_2 e^x + (\tfrac{1}{2} - x)e^{-x}.$$

The method of variation of parameters is easily extended to higher order equations. Assume that $y_1, y_2, \ldots, y_n$ are linearly independent solutions of the homogeneous equation associated with

$$y^{(n)} + a_1(x)y^{(n-1)} + \cdots + a_n(x)y = g(x), \tag{56}$$

where $a_1, \ldots, a_n$, and g are continuous on some interval I and $g(x) \not\equiv 0$ on I. Thus, the complementary function is

$$y_c = c_1 y_1 + c_2 y_2 + \cdots + c_n y_n,$$

where $c_1, c_2, \ldots, c_n$ are arbitrary constants. One assumes that there exists a particular solution of (56) of the form

$$y_p(x) = v_1(x)y_1(x) + v_2(x)y_2(x) + \cdots + v_n(x)y_n(x), \tag{57}$$

where $v_1, v_2, \ldots, v_n$ are unknown functions which are to be determined. Since there are n functions to be determined, we may specify n conditions which these functions are to satisfy. Obviously, one of the conditions is that y_p satisfy (56). Of course, we wish to specify the other $n - 1$ conditions in such a way that the computations required to determine v_i are as simple as possible. In order to accomplish this objective, we set the sum of all terms which involve v_i' to zero at each step in the differentiation process. Differentiating y_p, we find

$$y_p' = (v_1y_1' + v_2y_2' + \cdots + v_ny_n') + (v_1'y_1 + v_2'y_2 + \cdots + v_n'y_n).$$

Thus, the first condition that we impose is that

$$v_1'y_1 + v_2'y_2 + \cdots + v_n'y_n = 0, \tag{58}$$

so

$$y_p' = v_1y_1' + v_2y_2' + \cdots + v_ny_n'. \tag{59}$$

Repeating this procedure $n - 2$ times so that there will be a total of n conditions specified, we find the successive derivatives of y_p to be

$$y_p^{(m)} = v_1y_1^{(m)} + v_2y_2^{(m)} + \cdots + v_ny_n^{(m)}, \qquad m = 2, \ldots, n - 1 \tag{60}$$

and the accompanying conditions on the functions $v_1, v_2, \ldots, v_n$ to be

$$v_1'y_1^{(m-1)} + v_2'y_2^{(m-1)} + \cdots + v_n'y_n^{(m-1)} = 0, \qquad m = 2, \ldots, n - 1. \tag{61}$$

Differentiating $y^{(n-1)}$—the last equation represented by (60)—we get

$$y_p^{(n)} = v_1y_1^{(n)} + v_2y_2^{(n)} + \cdots + v_ny_n^{(n)} + v_1'y_1^{(n-1)} + v_2'y_2^{(n-1)} + \cdots + v_n'y_n^{(n-1)}. \tag{62}$$

Substituting (57), (59), (60), and (61) into (56), rearranging the terms, and using the fact that y_i satisfies the homogeneous equation associated with (56), the condition that y_p satisfy (56) reduces to

$$v_1'y_1^{(n-1)} + v_2'y_2^{(n-1)} + \cdots + v_n'y_n^{(n-1)} = g(x). \tag{63}$$

Thus, the n equations which $v_1', v_2', \ldots, v_n'$ must satisfy simultaneously for y_p to be a particular of (56) are (58), (61), and (63). Written in matrix-vector notation we see that $(v_1', v_2', \ldots, v_n')$ must satisfy

$$\begin{pmatrix} y_1 & y_2 & \cdots & y_n \\ y_1' & y_2' & & y_n' \\ \vdots & \vdots & & \vdots \\ y_1^{(n-1)} & y_2^{(n-1)} & \cdots & y_n^{(n-1)} \end{pmatrix} \begin{pmatrix} v_1' \\ v_2' \\ \vdots \\ v_n' \end{pmatrix} = \begin{pmatrix} 0 \\ 0 \\ \vdots \\ g(x) \end{pmatrix} \tag{64}$$

The determinant of the matrix in (64) is nonzero, since it is the Wronskian of $y_1, y_2, \ldots, y_n$ which are linearly independent solutions of the homogeneous equation associated with (56). Hence, (64) has a unique solution which is

$$v_i' = \frac{W_i}{W}, \qquad i = 1, 2, \ldots, n, \tag{65}$$

where W is the Wronskian of $y_1, y_2, \ldots, y_n$ and W_i is the determinant obtained from W by replacing the ith column by the column $(0, 0, \ldots, g(x))$. Thus, a particular solution of (56) is

$$y_p(x) = \sum_{i=1}^{n} y_i(x) \int^x v_i'(t)\,dt.$$

EXAMPLE Use the method of variation of parameters to find a particular solution of the equation $y''' - y' = x^2$.

One easily finds that the functions $y_1 = 1$, $y_2 = e^x$, and $y_3 = e^{-x}$ are linearly independent solutions of the associated homogeneous equation. Hence, if a particular solution y_p is to have the form

$$y_p(x) = v_1(x)y_1(x) + v_2(x)y_2(x) + v_3(x)y_3(x),$$

then v_1', v_2', v_3' should satisfy

$$\begin{pmatrix} 1 & e^x & e^{-x} \\ 0 & e^x & -e^{-x} \\ 0 & e^x & e^{-x} \end{pmatrix} \begin{pmatrix} v_1' \\ v_2' \\ v_3' \end{pmatrix} = \begin{pmatrix} 0 \\ 0 \\ x^2 \end{pmatrix}.$$

Therefore,

$$v_1' = \frac{W_1}{W} = \begin{vmatrix} 0 & e^x & e^{-x} \\ 0 & e^x & -e^{-x} \\ x^2 & e^x & e^{-x} \end{vmatrix} \Bigg/ \begin{vmatrix} 1 & e^x & e^{-x} \\ 0 & e^x & -e^{-x} \\ 0 & e^x & e^{-x} \end{vmatrix} = -\frac{2x^2}{2} = -x^2,$$

$$v_2' = \frac{W_2}{W} = \begin{vmatrix} 1 & 0 & e^{-x} \\ 0 & 0 & -e^{-x} \\ 0 & x^2 & e^{-x} \end{vmatrix} \Bigg/ 2 = \frac{x^2e^{-x}}{2},$$

and

$$v_3' = \frac{W_3}{W} = \begin{vmatrix} 1 & e^x & 0 \\ 0 & e^x & 0 \\ 0 & e^x & x^2 \end{vmatrix} \Bigg/ 2 = -\frac{x^2e^x}{2}.$$

So $v_1 = -x^3/3$, $v_2 = -\frac{1}{2}x^2e^{-x} - xe^{-x} - e^{-x}$, $v_3 = \frac{1}{2}x^2e^x - xe^x + e^x$, and therefore $y_p = -x^3/3 - 2x$.

EXERCISES

Use the method of variation of parameters to find a particular solution for the following equations.

1. $y'' + y = \sec x$
2. $y'' - 3y' + 2y = \cos e^{-x}$
3. $y'' - 2y' + y = e^{2x}(e^x + 1)^{-2}$
4. $y'' + 4y' + 4y = e^{-2x}/x^2$
5. $y''' - y' = x$
6. $y''' - 6y'' + 11y' - 6y = e^{4x}$
7. $y^{(4)} - y = \cos x$
8. $y^{(4)} + 2y^{(2)} + y = \sec^3 x$
9. $y^{(4)} - 4y^{(3)} + 6y^{(2)} - 4y^{(1)} + y = e^x$
10. Find the general solution of the differential equation
$$(x^2 + 1)y'' - 2xy' + 2y = (x^2 + 1)^2$$
given that $y_1 = x$ and $y_2 = x^2 - 1$ are solutions of the associated homogeneous equation.
11. Find the general solution of
$$x^2y'' - 2xy' + 2y = x^3$$
given that $y_1 = x$ is a solution of the associated homogeneous equation.
12. Solve
$$x^2y'' - (x + 4)xy' + 2(x + 3)y = 2x^4e^{2x}$$
by assuming that the associated homogeneous equation has a solution of the form x^n.
13. Given that $y_1 = 1/x$ is a solution of the associated homogeneous equation, solve the initial value problem
$$x(1 + 3x^2)y'' + 2y' - 6xy = (1 + 3x^2)^2; \qquad y(1) = 2, \quad y'(1) = 4$$

7

APPLICATIONS OF THE SECOND ORDER LINEAR DIFFERENTIAL EQUATION

In this chapter we shall consider some applications of linear differential equations to various physical and electrical systems. In particular we shall consider applications of the nonhomogeneous linear second order equation with constant coefficients

$$ay'' + 2by' + cy = f(t) \tag{1}$$

where a, b, and c are known constants and $f(t)$ is a known function. The independent variable t will usually represent time and the solution $y(t)$ of (1) will be the state of some parameter which describes the system under consideration.

7.1 THE HARMONIC OSCILLATOR

When there is no damping force (i.e., when $b = 0$) and there is no external force [i.e., when $f(t) = 0$ for all t] equation (1) reduces to

$$ay'' + cy = 0. \tag{2}$$

For $a \neq 0$ and $ac > 0$, equation (2) is a simple mathematical model of a vibrating system. This model, which is called a *harmonic oscillator*, is applicable to many systems, such as a mass on a spring, quantum-mechanical oscillators, a simple pendulum, and the oscillation of beams, to mention a few.

The auxiliary equation for (2),

$$ar^2 + c = 0, \tag{3}$$

has roots $\pm\sqrt{c/a}\,i$. Hence, the general solution of (2) is

$$y = A \sin \omega t + B \cos \omega t, \tag{4}$$

where $\omega^2 = c/a$ and A and B are arbitrary constants. At this point we wish to show that the general solution (4) may also be written in the form

$$y = C \sin (\omega t + \varphi) \tag{5}$$

where C and φ are arbitrary constants. In order to accomplish this goal we must be able to determine the constants C and φ in terms of the constants A and B. Applying the formula for the sine of the sum of two angles to (5), we find

$$y = C(\cos \varphi \sin \omega t + \sin \varphi \cos \omega t). \tag{6}$$

Equating the coefficients of $\sin \omega t$ and $\cos \omega t$ in equations (4) and (6), we find that the following relationships between the constants A and B of (4) and C and φ of (5) must hold in order for (5) to be a solution of (2):

$$A = C \cos \varphi \tag{7a}$$

$$B = C \sin \varphi. \tag{7b}$$

Squaring (7a) and (7b), adding the results, and solving for C, we get

$$C = \sqrt{A^2 + B^2}\,. \tag{8}$$

Substituting (8) into (7a) and (7b) we see that φ must simultaneously satisfy

$$\cos \varphi = \frac{A}{\sqrt{A^2 + B^2}} \tag{9a}$$

and

$$\sin \varphi = \frac{B}{\sqrt{A^2 + B^2}}\,. \tag{9b}$$

The reason for writing the general solution of (2) as (5) instead of (4) is because it is easier to perceive the physical significance of the arbitrary constants in (5). The constant C of (5) is the *amplitude of the oscillation* and $-C \leq y(t) \leq C$ for all t. The constant φ is called the *phase angle*. If φ is chosen so that $-\pi < \varphi \leq \pi$, then for $\varphi > 0$, φ is the angle by which (5) leads the function $y = C \sin \omega t$ and φ/ω is the time lead; while for $\varphi < 0$, φ is the angle by which (5) lags the function $y = C \sin \omega t$ and φ/ω is the time lag.

See Figure 7.1. The *period* of oscillation—the time interval between two successive maxima—is $P = 2\pi/\omega$, since this is the period of the sine function of equation (5). The maxima occur when $C = C \sin(\omega t + \varphi)$, which occurs when $\omega t + \varphi = (4n + 1)\pi/2$, or solving for t when $t = [(4n + 1)\pi/2 - \varphi]/\omega$, where $n = 0, 1, 2, \ldots$. The frequency of oscillation—the number of oscillations per unit time—is $F = 1/P = \omega/2\pi$.

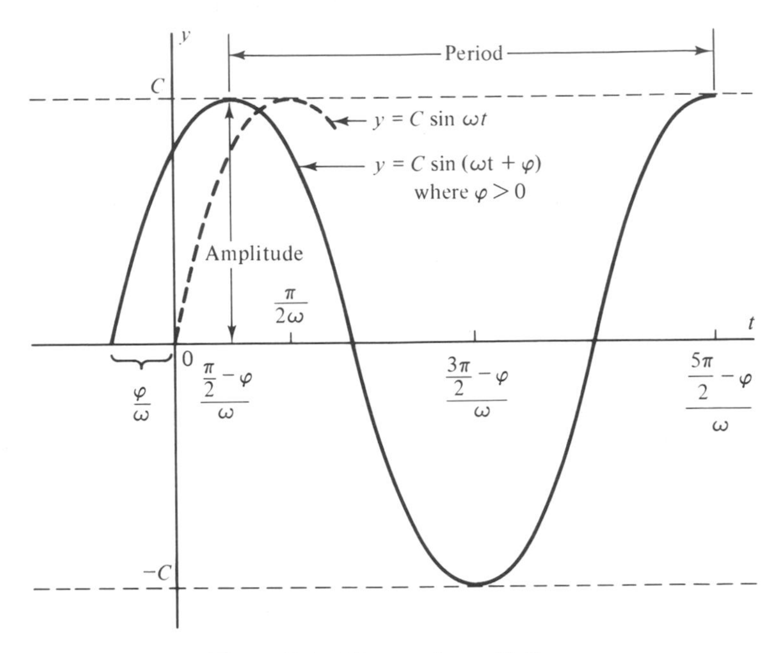

Figure 7.1 Harmonic oscillation.

Mass on a spring A spring of length L is suspended from a fixed support. A body of mass m is attached to the end of the spring and this system is allowed to come to rest at its equilibrium position. See Figure 7.2. We assume that the mass m is small compared to the mass of the spring to avoid exceeding the elastic limit of the spring. Suppose that in the equilibrium position the length of the elongated spring is $L + l$, where $l > 0$ is small compared to L. According to Hooke's law, the elongation ($l > 0$) and compression ($l < 0$) of the spring is directly proportional to the force that produces the elongation or compression. In the equilibrium position the only force acting on the system is the force of gravity: $F_g = mg$, where g is the acceleration due to gravity. Thus, by Hooke's law we must have

$$F_g = mg = kl,$$

where $k > 0$ is the constant of proportionality of the particular spring. One

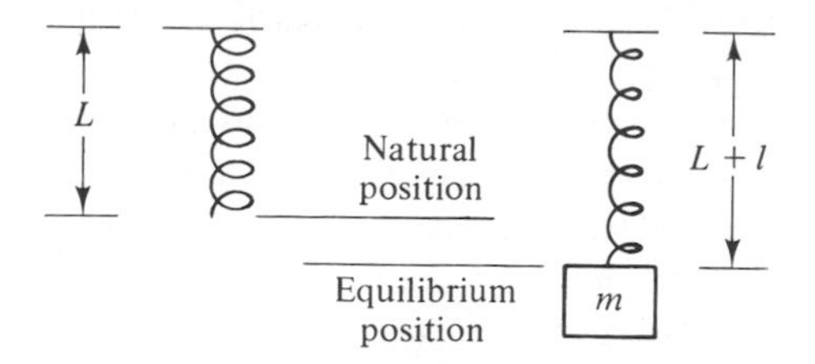

Figure 7.2 Mass on a spring.

can easily calculate k for a particular spring by attaching a body of known mass to the spring and accurately measuring the elongation produced.

The mechanical system consisting of the spring and the attached mass is set in motion (i) by pulling the mass downward from its equilibrium position and releasing it, (ii) by lifting the mass upward from its equilibrium position and releasing it, or (iii) by applying an instantaneous external force to the mass and thereby imparting a velocity to the mass and dislodging the system from its equilibrium position. Our problem is to derive the differential equation whose solution will describe the motion of the system. First we must choose a coordinate system. We shall choose the origin to be the point of equilibrium and choose the positive direction to be measured vertically downward. See Figure 7.3(a). In deriving the equation of motion, we employ Newton's second law, $F = ma$, and we shall assume that there is no resistance due to the medium (air in this case) in which the system is functioning. The forces acting on the system are the restoring force of the spring, F_s, which

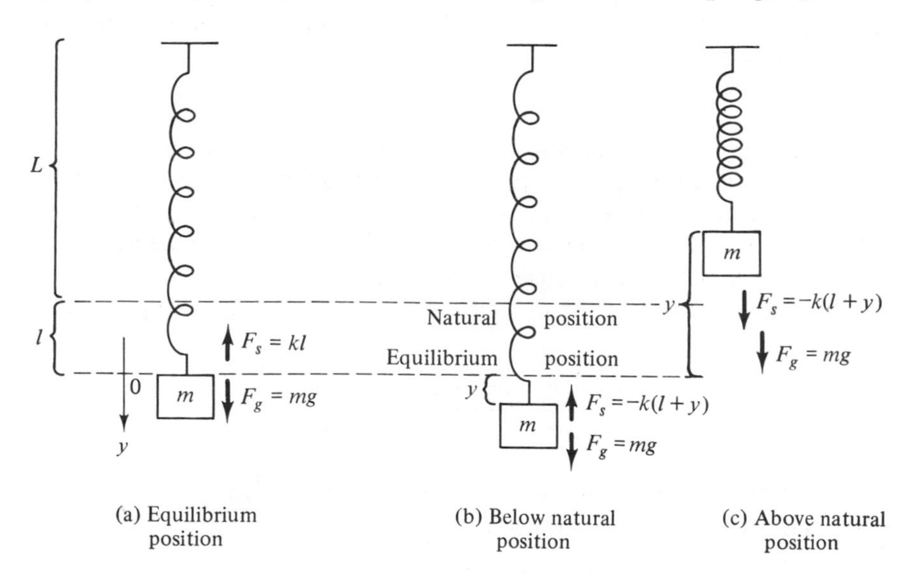

Figure 7.3 Forces acting on a mass on a spring.

always acts to restore the spring to its natural position ($y = -l$) and the force of gravity $F_g = mg$. When the mass is below the natural position, Figure 7.3(b), F_s acts in the negative direction and since $k > 0$ and $y + l > 0$, $F_s = -k(l + y) < 0$. When the mass is above the natural position, Figure 7.3(c), F_s acts in the positive direction and since $k > 0$ and $y + l < 0$, $F_s = -k(l + y) > 0$. Applying Newton's second law, we find, regardless of the position of the mass relative to the natural position,

$$F_s + F_g = ma$$

or

$$-k(l + y) + mg = m\frac{d^2y}{dt^2}, \tag{10}$$

since $mg = kl$ (10) reduces to

$$m\frac{d^2y}{dt^2} + ky = 0. \tag{11}$$

Comparing with (3) we see that the general solution to this equation is

$$y = A \sin \omega t + B \cos \omega t$$

or

$$y = C \sin (\omega t + \varphi),$$

where $\omega^2 = k/m = g/l$ and A and B or C and φ are constants which are determined by the conditions (initial conditions) under which the system is set in motion.

EXAMPLE Suppose that a 2-kg mass elongates a particular spring 30 cm. A 3-kg mass is attached to the spring and after the equilibrium position is reached, the mass is pulled downward 15 cm and released with a downward velocity of 50 cm/s. Find the equation of motion for this system and specify the amplitude, period, and frequency of the oscillation.

By Hooke's law $mg = kl$, so for this particular spring the spring constant is

$$k = \frac{mg}{l} = \frac{(2\text{ kg})(9.8\text{ m/s}^2)}{.3\text{ m}} = \frac{196}{3}\text{ kg/s}^2.$$

Since the mass attached to the spring is 3 kg, $\omega = \sqrt{\text{k/m}} = \sqrt{196/9\text{ s}^2} = 14/3$ s and the equation of motion is

$$y(t) = A \sin \tfrac{14}{3}t + B \cos \tfrac{14}{3}t.$$

The instant the system is set into motion ($t = 0$), $y(0) = .15$ m and $y'(0) = .5$ m/s. Imposing these initial conditions, we obtain the following equations

$$y(0) = .15 \text{ m} = B$$
$$y'(0) = .5 \text{ m/s} = \tfrac{14}{3}A/\text{s}.$$

So

$$B = .15 \text{ m} \quad \text{and} \quad A = \frac{1.5}{14} \text{ m}.$$

The amplitude of oscillation is

$$C = \sqrt{\left(\frac{1.5}{14}\right)^2 + (.15)^2} \text{ m} = .1845 \text{ m} = 18.45 \text{ cm}.$$

The period is

$$P = \frac{2\pi}{\omega} = \frac{3\pi}{7} \text{ s}.$$

The frequency of oscillation is

$$F = \frac{1}{P} = \frac{7}{3\pi} \text{ cycles/second}.$$

Simple pendulum A simple pendulum consists of a rigid straight wire of negligible mass and length L with a bob of mass m attached to one end. The other end of the wire is attached to a fixed support. The pendulum is free to move in a vertical plane. Let θ be the angle that the wire makes with the vertical—the equilibrium position of the system. We will choose θ to be positive if the wire to the right of the vertical and negative if the wire is to the left of the vertical. See Figure 7.4. If we neglect the resistance due to the medium in which the system is operating, then there are only two forces

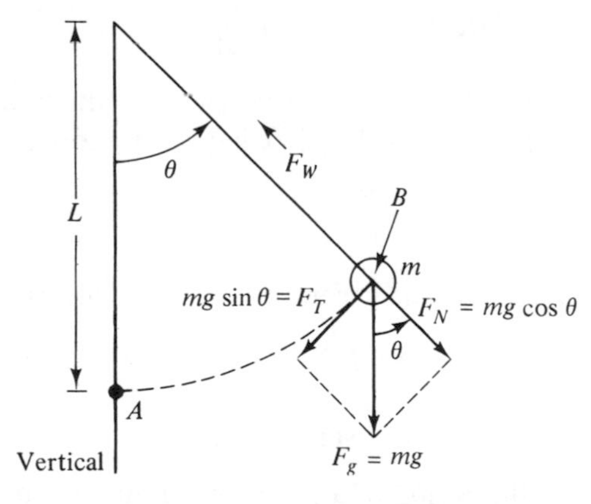

Figure 7.4 Simple pendulum.

acting on the mass m: F_W, the tension in the wire, which acts along the wire and toward the support; and $F_g = mg$, the force of gravity, which acts vertically downward. The force of gravity may be represented by two forces: one that acts parallel to F_W but in the opposite direction, F_N; and one that acts perpendicular to F_W, F_T. From Figure 7.4 we see that

$$F_N = mg \cos \theta = -F_W$$

and that

$$F_T = mg \sin \theta.$$

So the net force that tends to restore the system to equilibrium is F_T. If we let s represent the arc length AB, then applying Newton's second law ($F = ma$), we get

$$-mg \sin \theta = m \frac{d^2 s}{dt^2}, \tag{12}$$

since d^2s/dt^2 is the acceleration along the arc AB. Since $s = L\theta$,

$$\frac{d^2 s}{dt^2} = L \frac{d^2 \theta}{dt^2}$$

and equation (12) reduces to

$$-g \sin \theta = L \frac{d^2 \theta}{dt^2}$$

or

$$L \frac{d^2 \theta}{dt^2} + g \sin \theta = 0. \tag{13}$$

This equation is nonlinear and its solution involves elliptic integrals. The Maclaurin expansion for $\sin \theta$ is

$$\sin \theta = \theta - \frac{\theta^3}{3!} + \frac{\theta^5}{5!} - \cdots.$$

For θ small, $\sin \theta \doteq \theta$. Making this replacement in equation (13), we obtain the following linear equation, which approximates the motion of the pendulum:

$$L \frac{d^2 \theta}{dt^2} + g\theta = 0. \tag{14}$$

Comparing with (3) we find the general solution of (14) to be

$$\theta(t) = A \sin \omega t + B \cos \omega t$$

or

$$\theta(t) = C \sin (\omega t + \varphi),$$

where $\omega^2 = g/L$. So the period of a simple pendulum is approximately

$$P = 2\pi/\omega = 2\pi\sqrt{L/g}\,.$$

EXERCISES

1. A 50-g mass is attached to a spring whose natural length is 90 cm. The spring with the mass attached is suspended from the ceiling and allowed to come to rest in its equilibrium position, which is 110 cm from the ceiling. This system is set into motion by pulling the mass down an additional 5 cm from its equilibrium position and releasing it without imparting any initial velocity to the mass. Write the equation of motion for this system and determine the amplitude, period, and frequency of oscillation.

2. What is the equation of motion for the system described in Exercise 1, if the system is set into motion by lifting the mass upward 5 cm from its equilibrium position and releasing it without imparting any initial velocity? What is the amplitude, period, and frequency of oscillation?

3. A 100-g mass is attached to the end of a spring which is suspended from a fixed support and causes the spring to stretch 35 cm. The mass is struck from below with a hammer and thereby given an initial upward velocity of magnitude 14 cm/s. Write the equation of motion for this system. What is the amplitude, period, and frequency of oscillation?

4. A 250-g mass stretches a particular spring 28 cm. This system is set in motion by pulling the mass down 8 cm from the equilibrium position and releasing it with a downward velocity of magnitude 35 cm/s. Write the equation of motion and determine the amplitude, period, and frequency of oscillation and the phase angle.

5. What is the equation of motion, the amplitude, period, and frequency of oscillation, and the phase angle for the system described in Exercise 4, if the system is set in motion by pulling the mass down 28 cm from the equilibrium position and releasing it with an upward velocity of magnitude 35 cm/s.

6. Answer the following questions for any system that consists of a mass on a spring and which is executing simple harmonic motion:
(a) What is the velocity of the mass at the instant the displacement from the equilibrium position is a maximum?
(b) What is the position of the mass when the velocity is a maximum?

7. A mass m is attached to one end of a spring whose spring constant is $k = 3200$ g/s². Assuming that the system is executing simple harmonic motion and that the period is $\pi/2$ seconds, determine the mass m.

8. A 2-kg mass is attached to one end of a spring. The other end of the spring is attached to a fixed support and the system is allowed to come to rest at its equilibrium position. The mass is then pulled down X cm below its equilibrium position and released with an initial downward velocity of magnitude 24 cm/s. Find the distance X and the spring constant k, if the amplitude of the resultant harmonic oscillation is 5 cm and the period is $(\pi/4)$ s.

9. A 1-kg mass is attached to the end of a spring which is suspended from the ceiling and this system is set in motion. The period of the resulting oscillation is 2 seconds. Later the 1-kg mass is removed and replaced by an unknown mass m. The new system is set in motion and the new period of oscillation is found to be 4 s. What is the mass m?

10. Approximately how many centimeters long would a simple pendulum be whose period is 1 s?

11. A simple pendulum of length 19.6 cm is displaced $\theta(0) = \frac{1}{32}$ rad from its equilibrium position and released with an angular velocity which is directed toward the equilibrium position and whose magnitude is $|\theta'(0)| = (5\sqrt{2}/32)$ rad/s. Write the equation of motion, and find the amplitude, period, and frequency of oscillation, and the phase angle.

7.2 FREE, DAMPED MOTION

When there is a damping force acting on the system ($b \neq 0$) and no external force driving the system [$f(t) = 0$ for all t], equation (1) becomes

$$ay'' + 2by' + c = 0. \tag{15}$$

Since the damping force always acts to retard the system, $ab > 0$. The damping force may be due to the medium in which the system is operating or to some other force which tends to impede the motion of the system. The auxiliary equation

$$ar^2 + 2br + c = 0 \tag{16}$$

has roots

$$r = \frac{-2b \pm \sqrt{4b^2 - 4ac}}{2a} = -\frac{b}{a} \pm \frac{\sqrt{b^2 - ac}}{a}.$$

Hence, there are three cases to consider, depending upon the sign of the quantity $b^2 - ac$.

Damped, oscillatory motion When $b^2 - ac < 0$, we have the case of damped, oscillatory motion. The roots of the auxiliary equation are complex conjugates and the solution of (15) is

$$y = e^{-\alpha t}(A \sin \omega t + B \cos \omega t), \tag{17}$$

where $\alpha = b/a$, $\omega = \sqrt{ac - b^2}/|a|$, and A and B are arbitrary constants. As before, one could also write the solution in the form

$$y = Ce^{-\alpha t} \sin(\omega t + \varphi), \tag{18}$$

where $C = \sqrt{A^2 + B^2}$ and φ simultaneously satisfies

$$\cos \varphi = \frac{A}{\sqrt{A^2 + B^2}}$$

$$\sin \varphi = \frac{A}{\sqrt{A^2 + B^2}}.$$

The factor $Ce^{-\alpha t}$ is called the *damping factor*, and since $\alpha > 0$, the solution approaches zero as $t \to \infty$. The other factor of equation (18) is the solution of (2) and represents simple periodic, oscillatory motion. The product of these factors—the solution of (15)—represents oscillatory motion in which the amplitude of oscillation decreases with increasing time. See Figure 7.5. The time interval between two successive maxima is constant and is still called the period. As before, the period is $P = 2\pi/\omega$.

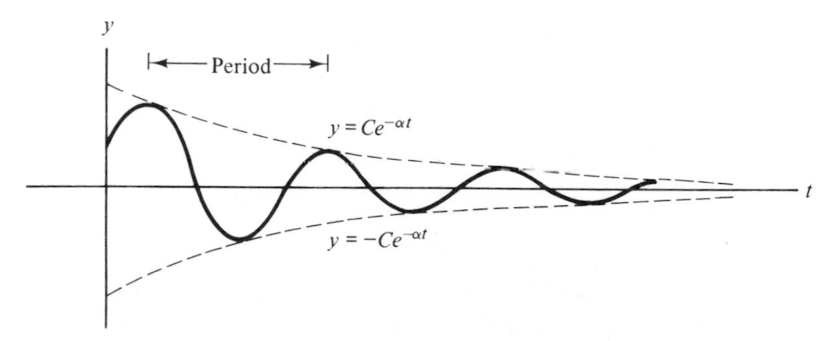

Figure 7.5 Damped, oscillatory motion.

Critically damped motion When $b^2 - ac = 0$, we have the case of critical damped motion. In this case the roots of the auxiliary equation (16) are real and equal, $r_1 = r_2 = -b/a$. So the general solution of (15) is

$$y = (A + Bt)e^{-\alpha t}, \tag{19}$$

where $\alpha = b/a > 0$ and A and B are arbitrary constants. The solution approaches zero (the equilibrium position) as $t \to \infty$, owing to the damping factor $e^{-\alpha t}$. The motion is not oscillatory, and there can be at most one passage through the equilibrium position since the factor $A + Bt$ is linear. Three typical graphs of the solution (19) are shown in Figure 7.6. The graph of the solution depends upon the constants A and B of (19), which in turn depend upon the initial conditions.

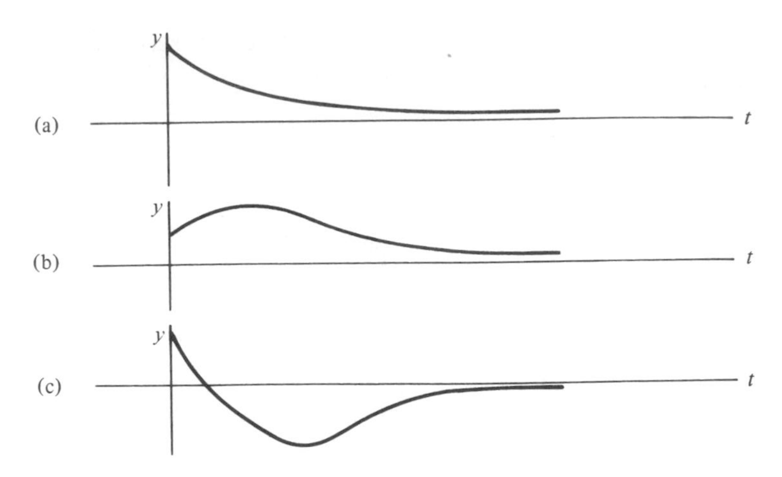

Figure 7.6 Typical critically damped and overdamped motion.

Overdamped motion When $b^2 - ac > 0$, the auxiliary equation (16) has two distinct, negative real roots:

$$r_1 = -\frac{b}{a} + \frac{\sqrt{b^2 - ac}}{a}$$

$$r_2 = -\frac{b}{a} - \frac{\sqrt{b^2 - ac}}{a}$$

and, therefore, in this case the general solution of (15) is

$$y = Ae^{r_1 t} + Be^{r_2 t}. \tag{19}$$

As $t \to \infty$, $y \to 0$ and no oscillation occurs since both $e^{r_1 t}$ and $e^{r_2 t}$ are monotone decreasing functions. The three graphs displayed in Figure 7.6 are also typical of the types of motion executed by overdamped systems.

Mass on a spring The motion of a mass suspended from one end of a spring, the other end of which is attached to a fixed support, is usually damped motion. The damping force is normally due to the resistance caused by the medium in which the system is operating or to the resistance caused by adding other components to the system, such as the dashpot shown in Figure 7.7. It has been shown experimentally that as long as the speed of the mass is not too large, the resistance is proportional to the speed, dy/dt.

Again we choose the origin to be the equilibrium point and choose the positive direction to be vertically downward. The damping force, F_d, always acts to retard the motion of the system. When the mass is moving downward, the velocity, dy/dt, is positive and so the damping force is directed upward and is negative. Thus, $F_d = -c(dy/dt) < 0$ where c is a positive constant. When the mass is moving upward the velocity, dy/dt, is negative and therefore

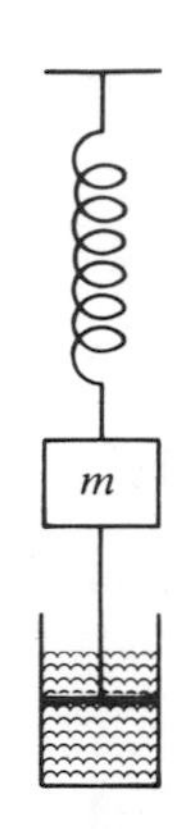

Figure 7.7 Damped mass on a spring system.

the damping force is directed downward, is positive, and again is $F_d = -c(dy/dt) > 0$. Applying Newton's second law of motion to the mass on a spring system with damping, we see that

$$F_d + F_s + F_g = ma,$$

where F_s is the force of the spring and F_g is the force of gravity. Referring to equation (10) and substituting for F_d, we find

$$-c\frac{dy}{dt} - k(l + y) + mg = m\frac{d^2y}{dt^2}. \tag{20}$$

Since $kl = mg$ we obtain, by rearrangement,

$$m\frac{d^2y}{dt^2} + c\frac{dy}{dt} + ky = 0. \tag{21}$$

EXAMPLE A 200-g mass is attached to one end of a spring. The other end of the spring is attached to the ceiling and the resulting system is allowed to come to rest at its equilibrium position. The length of the spring is measured and found to have increased 20 cm. The mass is pulled down 4 cm below its equilibrium position and released without imparting any initial velocity $[y'(0) = 0]$. The damping force equals $400(dy/dt)$ g-cm/s^2. Find the equation of motion for this system.

Since $k = mg/l = 200$ g $(980 \text{ cm/s}^2)/20$ cm $= 9800$ g/s^2, equation (21) becomes

$$200y'' + 400y' + 9800y = 0 \tag{22}$$

and the initial conditions are

$$y(0) = 4; \qquad y'(0) = 0. \tag{23}$$

The auxiliary equation for (22),

$$r^2 + 2r + 49 = 0,$$

has complex conjugate roots $-1 \pm 4\sqrt{3}\, i$. Hence, the general solution of (22) is

$$y = e^{-t}(A \sin 4\sqrt{3}\, t + B \cos 4\sqrt{3}\, t).$$

Differentiating and combining terms, we find

$$y' = e^{-t}[-(A + 4\sqrt{3}\, B) \sin 4\sqrt{3}\, t + (4\sqrt{3}\, A - B) \cos 4\sqrt{3}\, t].$$

In order to satisfy the initial conditions (23), A and B must simultaneously satisfy

$$y(0) = 4 = B$$
$$y'(0) = 0 = 4\sqrt{3}\, A - B.$$

Hence, $A = \sqrt{3}/3$, $B = 4$, and the equation of motion for this system is

$$y = e^{-t}[(\sqrt{3}/3) \sin 4\sqrt{3}\, t + 4 \cos 4\sqrt{3}\, t].$$

This is damped oscillatory motion. The damping factor is $\sqrt{(\sqrt{3}/3)^2 + 4^2}\, e^{-t} = (7\sqrt{3}/3)e^{-t}$ and the period is $(2\pi/4\sqrt{3})$ seconds or $(\sqrt{3}\,\pi/6)$ seconds.

EXAMPLE Find the equation of motion for the system described in the previous example if the resistance is increased to $2800(dy/dt)$ g-cm/s^2.

In this case equation (21) becomes

$$200y'' + 2800y' + 9800y = 0 \tag{24}$$

and the auxiliary equation

$$r^2 + 14r + 49 = 0$$

has two equal real roots $r_1 = r_2 = -7$. Thus, the general solution of (24) is

$$y = (A + Bt)e^{-7t}.$$

Differentiating and combining terms, we find

$$y' = (B - 7A - 7Bt)e^{-7t}. \tag{25}$$

In order to satisfy the initial conditions A and B must simultaneously satisfy

$$y(0) = 4 = A$$
$$y'(0) = 0 = B - 7A.$$

Consequently, $A = 4$, $B = 28$, and the equation of motion for this system is

$$y = (4 + 28t)e^{-7t}.$$

This is critically damped motion and the mass does not pass through the equilibrium position since $y > 0$ for $t > 0$. Substituting the values of A and B into (25), we find

$$y' = -196te^{-7t} < 0 \quad \text{for all } t > 0.$$

Consequently, the displacement, y, of the mass from the equilibrium position is a strictly monotone decreasing function of time, t, for all $t > 0$. Thus, the mass ascends monotonically toward the equilibrium position.

EXAMPLE Find the equation of motion for the system described previously if the resistance is increased to $3500(dy/dt)$ g-cm/s².

In this instance (21) becomes

$$200y'' + 3500y' + 9800y = 0. \tag{26}$$

The auxiliary equation

$$r^2 + 17.5r + 49 = 0$$

has roots $r_1 = -14$ and $r_2 = -7/2$. Hence, the general solution of (26) is

$$y = Ae^{-14t} + Be^{-7t/2}.$$

Differentiating, we find

$$y' = -14Ae^{-14t} - \tfrac{7}{2}Be^{-7t/2}.$$

In order to satisfy the initial conditions, A and B must simultaneously satisfy the equations:

$$y(0) = 4 = A + B$$
$$y'(0) = 0 = -14A - \tfrac{7}{2}B.$$

Solving this linear system of equations, we find that $A = -\frac{4}{3}$ and $B = \frac{16}{3}$. So the equation of motion in this case is

$$y = -\tfrac{4}{3}e^{-14t} + \tfrac{16}{3}e^{-7t/2}.$$

This is an example of overdamped motion. The mass ascends monotonically toward the equilibrium position and never passes through the equilibrium position as was the case of critically damped motion we just examined. However, the rate of ascension in the overdamped case is slower than the rate of ascension in the critically damped case, owing to the increased damping.

EXERCISES

1. A 100-g mass is attached to one end of a spring, the other end is attached to a fixed support, and this system is allowed to come to rest. In the equilibrium position the spring is stretched 5 cm. The mass is pulled downward 5 cm and released with a downward velocity of magnitude 7 cm/s. Find the equation of motion for this system for the following damping forces, F_d, and indicate if the motion is damped oscillatory motion, critically damped motion, or overdamped motion.

(a) $F_d = 2800(dy/dt)$ g-cm/s²
(b) $F_d = 400\sqrt{13}\,(dy/dt)$ g-cm/s²
(c) $F_d = 3500(dy/dt)$ g-cm/s²

2. What is the damping factor and period of the damped oscillatory motion in Exercise 1?

3. A 500-g mass is attached to the lower end of a spring which is suspended from the ceiling. The system is allowed to come to rest at the equilibrium position. In this position the spring is stretched 49 cm. The mass is pulled downward from the equilibrium position and released. The resistance of the system is $c\,dy/dt$. Under what condition is the motion oscillatory? Critically damped? Overdamped?

4. A 600-g mass is attached to the lower end of a spring which is suspended from a fixed support. The system is set in motion and the period is found to be 2 seconds when no damping force is present. When a damping force of $c\,dy/dt$ g-cm/s² is added to the system, the period is doubled. Calculate the spring constant k and the damping coefficient c.

7.3 FORCED MOTION

Let us now suppose that there is a periodic external force $f(t) = E \sin \bar{\omega} t$, where E and $\bar{\omega}$ are constants, acting on a system which is described by a parameter, $y(t)$, which is appropriately modeled mathematically by the equation

$$ay'' + 2by' + cy = f(t). \tag{1}$$

In the case of the mass on a spring system, the periodic external force $f(t)$ might be due to a motor which vibrates the support (i.e., the support is no longer fixed) or to a magnetic field which acts upon a suspended iron mass.

Undamped, forced motion When there is no damping force ($b = 0$) and there is a periodic external force, equation (1) becomes

$$ay'' + cy = E \sin \bar{\omega} t. \tag{27}$$

Recall from Section 7.1 that the solution of the associated homogeneous equation $ay'' + cy = 0$ is

$$y_c = A \sin \omega t + B \cos \omega t,$$

where $\omega^2 = c/a$.

If $\bar{\omega} \neq \omega$, then we seek a particular solution of (27) of the form

$$y_p = C \sin \bar{\omega} t + D \cos \bar{\omega} t. \tag{28}$$

Differentiating, we obtain

$$y_p' = C\bar{\omega} \cos \bar{\omega} t - D\bar{\omega} \sin \bar{\omega} t \tag{29}$$

and

$$y_p'' = -C\bar{\omega}^2 \sin \bar{\omega} t - D\bar{\omega}^2 \cos \bar{\omega} t. \tag{30}$$

Substituting $y = y_p$ in (27) and rearranging, we find

$$C(c - a\bar{\omega}^2) \sin \bar{\omega} t + D(c - a\bar{\omega}^2) \cos \bar{\omega} t = E \sin \bar{\omega} t.$$

Equating coefficients and solving for C and D, we get

$$C = \frac{E}{c - a\bar{\omega}^2} \quad \text{and} \quad D = 0.$$

Hence, when $\bar{\omega} \neq \omega$, the general solution of (27) is

$$y = y_c + y_p = A \sin \omega t + B \cos \omega t + \frac{E}{c - a\bar{\omega}^2} \sin \bar{\omega} t. \tag{31}$$

This solution remains bounded for all t and the constants A and B are determined by the initial conditions.

If $\bar{\omega} = \omega$, then the frequency of the forcing function $f(t)$ is identical to the frequency of the system—a phenomenon known as *resonance*. In this case we cannot find a particular solution of (27) of the form (28) but must seek a particular solution of the form

$$y_p = Ct \sin \bar{\omega} t + Dt \cos \bar{\omega} t. \tag{32}$$

Differentiating twice, substituting into (27), and solving for C and D, one finds

$$C = 0 \quad \text{and} \quad D = \frac{-E}{2a\bar{\omega}}.$$

Thus, when $\bar{\omega} = \omega$, the general solution of (27) is

$$y = y_c + y_p = A \sin \omega t + B \cos \omega t - \frac{E}{2a\bar{\omega}} t \cos \bar{\omega} t. \tag{31}$$

Regardless of the initial conditions which determine the constants A and B, this solution becomes unbounded as $t \to \infty$ due to the term $t \cos \bar{\omega} t$. Any real system will, of course, be destroyed by the large amplitudes induced by a forcing function which has the same frequency as the system.

Damped, forced motion When there is a damping force ($b \neq 0$)—which is the case for all realizable systems—and when the external force is periodic, $f(t) = E \sin \bar{\omega} t$, equation (1) becomes

$$ay'' + 2by' + cy = E \sin \bar{\omega} t. \tag{32}$$

We found in Section 7.2 that the solution of the associated homogeneous equation—equation (15)—depended upon the sign of the quantity $b^2 - ac$. Consequently, the complementary solution is

$$y_c = \begin{cases} e^{-\alpha t}(A \sin \omega t + B \cos \omega t), & \text{if } b^2 - ac < 0 \\ e^{-\alpha t}(A + Bt), & \text{if } b^2 - ac = 0, \\ Ae^{r_1 t} + Be^{r_2 t}, & \text{if } b^2 - ac > 0 \end{cases}$$

where A and B are arbitrary constants,

$$\alpha = \frac{b}{a} > 0, \quad \omega = \frac{\sqrt{ac - b^2}}{|a|},$$

$$r_1 = -\frac{b}{a} + \frac{\sqrt{b^2 - ac}}{a} < 0, \quad \text{and} \quad r_2 = -\frac{b}{a} - \frac{\sqrt{b^2 - ac}}{a} < 0.$$

Notice that regardless of the form of y_c, $y_c \to 0$ as $t \to \infty$.

As a result of the forms of the complementary solution, y_c, and the assumed form of the forcing function, $f(t)$, we seek a particular solution of (32) of the form

$$y_p = C \sin \bar{\omega} t + D \cos \bar{\omega} t. \tag{28}$$

Substituting y_p, y_p', and y_p'' from (28), (29), and (30), respectively, into (32) and rearranging, we obtain

$$[(c - a\bar{\omega}^2)C - 2b\bar{\omega}D] \sin \bar{\omega} t + [2b\bar{\omega}C + (c - a\bar{\omega}^2)D] \cos \bar{\omega} t = E \sin \bar{\omega} t. \tag{33}$$

Equating coefficients, we see that C and D must simultaneously satisfy

$$(c - a\bar{\omega}^2)C - 2b\bar{\omega}D = 0$$
$$2b\bar{\omega}C + (c - a\bar{\omega}^2)D = E.$$

Solving this linear system, we find

$$C = \frac{(c - a\bar{\omega}^2)E}{H(\bar{\omega})}$$

and

$$D = \frac{-2b\bar{\omega}E}{H(\bar{\omega})},$$

where

$$H(\bar{\omega}) = (c - a\bar{\omega}^2)^2 + 4b^2\bar{\omega}^2. \tag{34}$$

So a particular solution of (32) is

$$y_p = \frac{E}{H(\bar{\omega})}\left[(c - a\bar{\omega}^2)\sin \bar{\omega}t - 2b\bar{\omega}\cos \bar{\omega}t\right], \tag{35}$$

or, equivalently,

$$y_p = \frac{E}{\sqrt{H(\bar{\omega})}}\sin(\bar{\omega}t + \varphi), \tag{36}$$

where φ simultaneously satisfies

$$\cos\varphi = \frac{c - a\bar{\omega}^2}{\sqrt{H(\bar{\omega})}}$$

$$\sin\varphi = \frac{-2b\bar{\omega}}{\sqrt{H(\bar{\omega})}}.$$

The general solution of (32) is $y = y_c + y_p$. Since $y_c \to 0$ as $t \to \infty$, this term of the general solution is known as the *transient solution.* The transient solution contains the arbitrary constants A and B which are determined once the initial conditions are specified. Since the initial conditions influence only the transient solution and since the transient solution becomes negligible after a sufficiently long period of time due to damping, the initial conditions affect the general solution for only a small period of time. After a sufficiently long period of time, the general solution is for all practical purposes the particular solution, y_p, which is, therefore, also called the *steady state solution.* The steady state solution is oscillatory with the same frequency as the frequency of the forcing function and with amplitude

$$A(\bar{\omega}) = \frac{E}{\sqrt{H(\bar{\omega})}} = \frac{E}{\sqrt{(c - a\bar{\omega}^2)^2 + 4b^2\bar{\omega}^2}}, \tag{37}$$

which depends on the forcing function. Thus, after a long period of time the solution depends upon the external force which is driving the system but not upon the initial conditions under which the system was started.

Let us consider the amplitude function $A(\bar{\omega})$ for $0 < \bar{\omega} < \infty$. As $\bar{\omega} \to 0$, $A(\bar{\omega}) \to E/|c|$ and as $\bar{\omega} \to \infty$, $A(\bar{\omega}) \to 0$. Also notice that $A(\bar{\omega})$ is finite for all positive real $\bar{\omega}$ since $b \neq 0$. Consequently, when there is damping present in a system, the amplitude of oscillation remains finite, whereas when there is no damping present and when $\bar{\omega} = \omega$, the amplitude of the solution increases without bound as time increases until the system is destroyed. The amplitude $A(\bar{\omega})$ will be a maximum when $H(\bar{\omega})$ is a minimum. Differentiating $H(\bar{\omega})$ and equating to the result to zero, we get

$$\bar{\omega}[a^2\bar{\omega}^2 - ac + 2b^2] = 0.$$

Solving for $\bar{\omega}$, we find that the maximum amplitude occurs when

$$\bar{\omega} = \frac{\sqrt{ac - 2b^2}}{|a|}. \tag{38}$$

The forcing function is said to be in *resonance* with the system when $\bar{\omega}$ has the value given by (38). Notice that this can only occur when $ac - 2b^2 > 0$, which implies that $ac > 2b^2 > b^2$, which in turn implies that $b^2 - ac < 0$. That is, the corresponding free system $[f(t) = 0]$ must exhibit damped oscillatory motion and not critically damped motion or overdamped motion for resonance to occur. The resonance frequency is

$$F_R = \frac{\sqrt{ac - 2b^2}}{2\pi|a|}$$

and is less than the frequency of the corresponding free system, which is

$$F = \frac{\omega}{2\pi} = \frac{\sqrt{ac - b^2}}{2\pi|a|}.$$

EXERCISES

1. A 200-g mass is attached to one end of a spring. The other end of the spring is attached to a movable support. The support is held fixed and the system is allowed to come to rest in its equilibrium position. In the equilibrium position the spring is stretched 20 cm. The mass is pulled down 4 cm below its equilibrium position and released without imparting any velocity and at the same instant a motor starts to drive the support with a force $f(t) = 9600 \sin \bar{\omega}t$ g-m/s^2.

   (a) What value of $\bar{\omega}$ causes resonance?
   (b) Assuming that $\bar{\omega} = 5$ cycles/second, what is the general solution for this system?

2. Find the general solution of the following differential equation, which represents forced, undamped motion: $ay'' + cy = E \sin \bar{\omega}t$, where $a \neq 0$, $ac > 0$, and $\bar{\omega} \neq \omega = \sqrt{c/a}$ for the general initial conditions $y(0) = c_0$; $y'(0) = c_1$.

3. Find the resonance frequency, F_R, and the general solution to the system which results from adding a damping force equal to $400y'$ g-cm/s^2 to the system described in Exercise 1.

4. Assuming that $b^2 - ac < 0$, find the general solution of the following differential, which represents forced, damped oscillatory motion:

$$ay'' + 2by' + cy = E \sin \bar{\omega}t$$

   for the general initial conditions $y(0) = c_0$; $y'(0) = c_1$.

5. Suppose that the equation of motion of a certain system is

$$y'' + Ky' + 12y = 3 \sin \bar{\omega} t.$$

(a) What value of K will produce resonance?
(b) If $K = 4$, what value of $\bar{\omega}$ will produce resonance?

7.4 ELECTRICAL CIRCUITS

In this section we shall consider the flow of electric current in simple series circuits. In a later chapter we shall consider the flow of electric current in electric networks. Table 7.1 contains a list of some common electric circuit components and quantities, the alphabetic symbol which is used to denote the numeric value of the component or quantity, the graphic symbol which is used to represent the component in schematic drawings, and the unit of the component or quantity. The units of the various components and quantities are named in honor of the following physicists: André Marie Ampère (1775–1836, French), Charles Augustin De Coulomb (1736–1806, French), Michael Faraday (1791–1867, English), Joseph Henry (1797–1878, American), Georg Simon Ohm (1789–1854, German), and Allessandro Volta (1745–1827, Italian).

Table 7.1

	Symbol		
Circuit Component or Quantity	*Alphabetic*	*Graphic*	*Unit*
Capacitor	C		Farad (F)
Electric charge	q		Coulomb (C)
Electric current	i		Ampere (A)
Electromotive Force (emf)			
Battery	E		Volt (V)
Generator	$E \sin \omega t$		Volt (V)
Inductor	L		Henry (H)
Resistor	R		Ohm (Ω)
Time	t		Second (s)

The electric charge, q, and current, i, will be functions of time. The capacitance, C, inductance, L, and the resistance, R, are all positive quantities and in general are functions of time. However, in a great many instances the quantities C, L, and R may be assumed to be constants. In all of our discussions we shall make this assumption. The electromotive force of a battery, E, shall be assumed to be a constant and the electromotive force of a generator shall be assumed to have the form $E \cos \bar{\omega} t$, where E and $\bar{\omega}$ are constants. A resistor uses electrical energy and converts that energy into heat. An electric light bulb is an example of a resistor. A capacitor stores electrical energy and an inductor opposes a change in electric current. The electromotive force (battery or generator) causes the flow of electric current.

The behavior of an electrical network (system) is determined by the following two laws, which were stated by the German physicist Gustav Kirchhoff (1824–1887):

Kirchhoff's first law (current law) At any junction in a network, the sum of the current flowing into the junction is equal to the sum of the current flowing out of the junction.

Kirchhoff's second law (voltage law) The algebraic sum of the voltage drops around any loop of a network is equal to the algebraic sum of the impressed electromotive forces around the loop.

According to the fundamental laws of electricity, the voltage drop across a resistor is iR, the voltage drop across a capacitor is q/C, and the voltage drop across an inductor is $L\, di/dt$. The following relation is known to exist between the charge on a capacitor, q, and the current flowing through the capacitor, i:

$$i = \frac{dq}{dt}. \tag{39}$$

The RLC series circuit Let us now consider the circuit depicted schematically by Figure 7.8. The arrow in the figure provides an orientation for the loop. The current i is understood to be positive when it flows in the direction of the arrow and negative when it flows in the other direction. This circuit consists of a resistor of resistance R, an inductor with inductance L, a capacitor with capacitance C, an electromotive force $E(t)$, and a switch s connected in series.

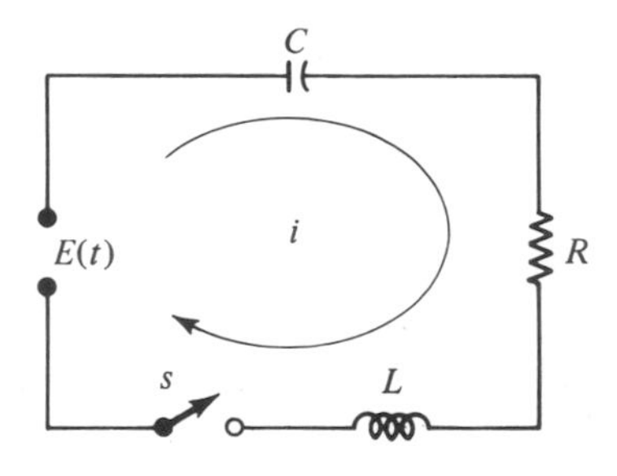

Figure 7.8 *RLC* series circuit.

Since there are no junctions in this network, we need only apply Kirchhoff's second law. Doing so, we find

$$L\frac{di}{dt} + Ri + \frac{q}{C} = E(t). \tag{40}$$

Substituting (39) into (40), we see that the charge on the capacitor must satisfy the second order linear differential equation

$$L\frac{d^2q}{dt^2} + R\frac{dq}{dt} + \frac{q}{C} = E(t). \tag{41}$$

Appropriate initial conditions at some time $t = 0$ are

$$q(0) = q_0, \qquad \frac{dq(0)}{dt} = i(0) = i_0. \tag{42}$$

Thus, the appropriate initial conditions to specify for this equation are the initial charge on the capacitor q_0 and the initial current in the circuit i_0. Differentiating (40) with respect to t and substituting $i = dq/dt$, we find that the current must satisfy the second order linear differential equation

$$L\frac{d^2i}{dt^2} + R\frac{di}{dt} + \frac{i}{C} = \frac{dE}{dt}. \tag{43}$$

Appropriate initial conditions for this equation at $t = 0$ are

$$i(0) = i_0, \qquad \frac{di(0)}{dt} = i_0'. \tag{44}$$

That is, the appropriate conditions to specify for this equation at $t = 0$ are the initial current in the circuit i_0 and the initial rate of change of current in the system i_0'. Notice that the initial charge on the capacitor, q_0, and the initial current in the circuit, i_0, are physically measurable quantities while the initial rate of change of current, i_0', is not. However, if we measure q_0, i_0, and $E(0) = E_0$—the electromotive force at $t = 0$—we can determine i_0' from (40). Substituting $t = 0$ into (40) and solving for i_0', we find

$$i_0' = \frac{1}{L}\left[E_0 - Ri_0 + \frac{q_0}{C}\right]. \tag{45}$$

Comparing equation (41) with the following equation for the forced, damped motion of a mass on a spring,

$$m\frac{d^2y}{dt^2} + c\frac{dy}{dt} + ky = f(t). \tag{46}$$

we discover the correspondence between mechanical and electrical systems shown in Table 7.2. This correspondence enables one to simulate mechanical systems which might be very expensive to construct—an airplane, for example

—by electrical systems which are relatively inexpensive. By measuring the response of the actual electrical system, one determines the response of the hypothetical mechanical system.

Table 7.2

Mechanical systems		*Electrical systems*	
m	mass	L	inductance
y	displacement	q	charge
$\frac{dy}{dt} = v$	velocity	$\frac{dq}{dt} = i$	current
c	damping	R	resistance
k	spring constant	$1/C$	reciprocal of capacitance
$f(t)$	driving force	$E(t)$	electromotive force

EXAMPLE Find the general solution for the charge on the capacitor, $q(t)$, and the electric current, $i(t)$, for the RLC series circuit shown in Figure 7.8, if $E(t) = E$, a constant.

Since the electromotive force is a constant, equation (41) becomes

$$Lq'' + Rq' + \frac{1}{C}q = E. \tag{47}$$

Comparing the associated homogeneous equation

$$Lq'' + Rq' + \frac{1}{C}q = 0 \tag{48}$$

with equation (15), we find the transient solution (complementary solution) to be

$$q_c(t) = \begin{cases} e^{-\alpha t}(A \sin \omega t + B \cos \omega t), & \text{if } R^2 - 4L/C < 0 \\ e^{-\alpha t}(A + Bt), & \text{if } R^2 - 4L/C = 0, \\ Ae^{r_1 t} + Be^{r_2 t}, & \text{if } R^2 - 4L/C > 0 \end{cases} \tag{49}$$

where

$$\alpha = \frac{R}{2L}, \qquad \omega = \frac{\sqrt{L/C - R^2/4}}{L},$$

$$r_1 = -\frac{R}{2L} + \frac{\sqrt{R^2/4 - L/C}}{L} < 0, \quad \text{and} \quad r_2 = -\frac{R}{2L} - \frac{\sqrt{R^2/4 - L/C}}{L} < 0.$$

In order to find the steady state solution (particular solution), we assume that

$$q_p(t) = K.$$

Differentiating twice, substituting into (47), and solving for K, we find that

$$K = EC.$$

Hence, the general solution of (47) is

$$q(t) = q_c(t) + EC. \tag{50}$$

Since $q_c(t) \to 0$ as $t \to \infty$, the charge on the capacitor $q(t)$ approaches the constant value EC regardless of whether the quantity $R^2 - 4L/C$ is negative, zero, or positive. The constants A and B of equation (49) are determined from the initial conditions $q(0) = q_0$ and $q'(0) = i(0) = i_0$.

Differentiating (50) with respect to t, we find the following expression for the current:

$$i(t) = \frac{dq}{dt} = \frac{dq_c}{dt}.$$

Thus, we could find an expression for the current by differentiating equation (49). However, since the electromotive force is assumed to be a constant, we shall find an expression for the current by solving the differential equation which results from equation (43) by setting $dE/dt = 0$, namely

$$Li'' + Ri' + \frac{i}{C} = 0. \tag{51}$$

This homogenous linear equation is the same as equation (48) when q is replaced by i. Thus, the right-hand side of equation (49) is also an expression for the current as a function of time. The constants A and B in this case would be determined from the initial conditions $i(0) = i_0$ and $i'(0) = i_0'$ and normally would not have the same value as the corresponding constant in the expression for q. What we discover from this expression for the current is that $i(t) \to 0$ as $t \to \infty$, regardless of whether the quantity $R^2 - 4L/C$ is negative, zero, or positive.

EXERCISES

1. An initial charge q_0 has been placed on the capacitor shown in Figure 7.9. At $t = 0$ the switch is closed and the capacitor starts discharging.

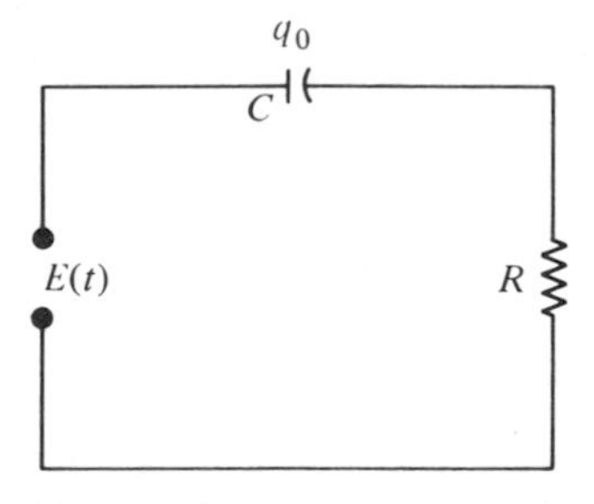

Figure 7.9 *RC* series circuit.

(a) Write an expression for $i_0 = i(0)$.
(b) Find an expression for the charge, $q(t)$, the current, $i(t)$, and the voltage across the capacitor, $V(t)$, if
  (i) $E(t) = E$, a constant.
  (ii) $E(t) = E \cos \bar{\omega} t$.

2. Consider the LC series circuit shown in Figure 7.10. Assume that the capacitor has initial charge q_0 and the initial current is i_0.

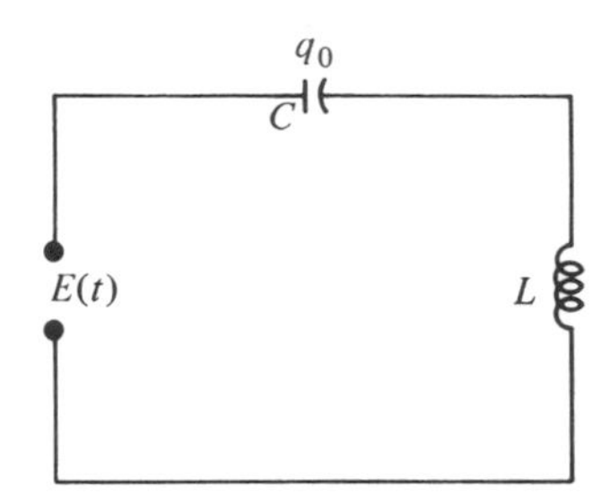

Figure 7.10 *LC* series circuit.

(a) Write an expression for i_0'.
(b) Find the current, $i(t)$, if
  (i) $E(t) = E$, a constant. (Notice that the solution is oscillatory with frequency $F = \omega/2\pi = 1/2\pi\sqrt{LC}$.)
  (ii) $E(t) = E \cos \bar{\omega} t$, $\bar{\omega} \neq 1/\sqrt{LC}$.
  (iii) $E(t) = E \cos \bar{\omega} t$, $\bar{\omega} = 1/\sqrt{LC}$. (Notice that this causes resonance and that $i(t) \to \infty$ as $t \to \infty$.)

3. If the initial current of a RL series circuit shown in Figure 7.11 is i_0, find the current, $i(t)$, for
(a) $E(t) = E$, a constant.
(b) $E(t) = E \cos \bar{\omega} t$.

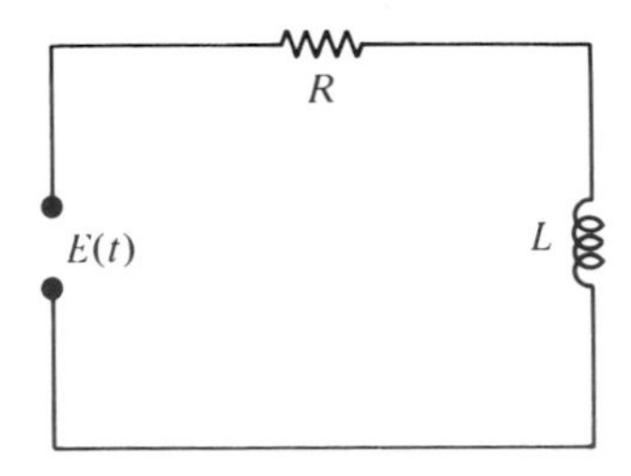

Figure 7.11 *RL* series circuit.

4. Find the steady state current for the RLC series circuit in Figure 7.8 if $E(t) = E \sin \bar{\omega} t$. For what value of $\bar{\omega}$ will the amplitude of the steady state current be a maximum?

8

POWER SERIES METHODS

Most ordinary differential equations cannot be solved in closed form. That is, the solution of most differential equations cannot be written in terms of elementary functions. Consequently, one must seek alternative forms by which to express the solution. As early as 1676 Newton considered the possibility of representing the solution of ordinary differential equations by infinite series. There are many types of infinite series. The most familiar are power series and Fourier series. In this chapter we shall consider the power series method for solving differential equations. This method is applicable to linear differential equations with constant and nonconstant coefficients, and most important of all, this method is applicable to nonlinear differential equations. The power series method can also be used to obtain numerical solutions to both linear and nonlinear differential equations. However, this method is not always effective, since difficulties sometimes arise in connection with the generation of the coefficients of the series.

8.1 BASIC PROPERTIES OF POWER SERIES

We shall state without proof several basic properties of power series. These and other properties are stated and proven in most introductory calculus texts.

A *power series* is a series of the form

$$a_0 + a_1(x - x_0) + a_2(x - x_0)^2 + \cdots + a_n(x - x_0)^n + \cdots = \sum_{n=0}^{\infty} a_n(x - x_0)^n. \tag{1}$$

The numbers a_n are called the *coefficients* of the series and equation (1) is often referred to as a power series expansion about x_0. If

$$\lim_{N \to \infty} \sum_{n=0}^{N} a_n(x_1 - x_0)^n$$

exists, then the power series (1) is said to *converge* (or *be convergent*) at x_1. If the limit does not exist, then the series (1) is said to *diverge* (or *be divergent*) at x_1. The power series (1) *converges absolutely* at x_1 if the series $\sum_{n=0}^{\infty} |a_n(x_1 - x_0)^n|$ converges at x_1.

Property 1 If the power series (1) converges absolutely at x_1, then (1) converges at x_1.

The converse of this theorem is false, since the series $\sum_{n=0}^{\infty} (-1)^n/(n + 1)$ is convergent but not absolutely convergent.

Property 2 The power series (1) either

(i) converges only at x_0, or

(ii) converges for all real x, or

(iii) there exists a positive real number R, called the *radius of convergence*, such that (1) is absolutely convergent for $|x - x_0| < R$ and divergent for $|x - x_0| > R$.

Property 3 The radius of convergence of (1), R, is given by

(i) $$R = \lim_{n \to \infty} \left| \frac{a_n}{a_{n+1}} \right|, \quad \text{if the limit exists}$$

or

(ii) $$R = \frac{1}{\lim_{n \to \infty} \sqrt[n]{|a_n|}}, \quad \text{if the limit exists.}$$

The interval $(x_0 - R, x_0 + R)$ is called the *interval of convergence* of the power series (1). The series (1) may or may not converge at the endpoints of the interval of convergence, $x = x_0 \pm R$.

Suppose that the series $\sum_{n=0}^{\infty} a_n(x - x_0)^n$ converges for $|x - x_0| < R_1$ and the series $\sum_{n=0}^{\infty} b_n(x - x_0)^n$ converges for $|x - x_0| < R_2$. Then we may define functions f and g by

$$f(x) = \sum_{n=0}^{\infty} a_n(x - x_0)^n \quad \text{for } |x - x_0| < R_1$$

and

$$g(x) = \sum_{n=0}^{\infty} b_n(x - x_0)^n \quad \text{for } |x - x_0| < R_2.$$

Let $R = \min(R_1, R_2)$. The following algebraic properties hold for $|x - x_0| < R$:

Property 4 If $f(x) = \sum_{n=0}^{\infty} a_n(x - x_0)^n = \sum_{n=0}^{\infty} b_n(x - x_0)^n = g(x)$ for all x, then $a_n = b_n$ for all n.

Property 5 $cf(x) = \sum_{n=0}^{\infty} ca_n(x - x_0)^n$ for any real constant c.

Property 6 $f(x) \pm g(x) = \sum_{n=0}^{\infty} (a_n \pm b_n)(x - x_0)^n$.

Property 7

$$f(x)g(x) = \sum_{n=0}^{\infty} \left(\sum_{m=0}^{n} a_{n-m}b_m\right)(x - x_0)^n = \sum_{n=0}^{\infty} \left(\sum_{m=0}^{n} a_m b_{n-m}\right)(x - x_0)^n.$$

(Notice that

$$\sum_{m=0}^{n} a_{n-m}b_m = \sum_{m=0}^{n} a_m b_{n-m} = a_0 b_n + a_1 b_{n-1} + \cdots + a_{n-1}b_1 + a_n b_0.)$$

Property 8 If $g(x_0) \neq 0$ and $\rho > 0$ is the minimum distance in the complex plane from x_0 to a zero of $g(x)$, then

$$\frac{f(x)}{g(x)} = \sum_{n=0}^{\infty} c_n(x - x_0)^n \quad \text{for } |x - x_0| < R',$$

where $R' = \min(R_1, R_2, \rho)$ and $c_n = (a_n - \sum_{m=0}^{n-1} c_m b_{n-m})/b_0$. [Notice that $g(x_0) \neq 0$ implies that $b_0 \neq 0$ and that c_n is computed using the values of $a_n, b_0, b_1, \ldots, b_n, c_0, c_1, \ldots, c_{n-1}$.]

A function f which is defined on an interval containing x_0 is said to be *analytic at* x_0 if f can be represented by a power series expansion about x_0 which has a positive radius of convergence. That is, f is analytic at x_0 if

$$f(x) = \sum_{n=0}^{\infty} a_n(x - x_0)^n \tag{2}$$

and the series converges for $|x - x_0| < R$, where $R > 0$. By definition any power series is analytic on its interval of convergence. All polynomials and the functions e^{ax}, sin bx, and cos bx, where a and b are any real constants, are analytic on the entire real line and therefore are called *entire functions*. It follows from properties 5–8 that if c is a constant and f and g are analytic functions at x_0, then the functions cf, $f \pm g$, fg, and f/g provided $g(x_0) \neq 0$ are all analytic at x_0.

Property 9 If f is analytic at x_0, then f has derivatives of all orders at x_0. That is, if

$$f(x) = \sum_{n=0}^{\infty} a_n(x - x_0)^n \quad \text{for } |x - x_0| < R$$

then

$$(3) \qquad f'(x) = \sum_{n=1}^{\infty} na_n(x - x_0)^{n-1}$$

$$= \sum_{n=0}^{\infty} (n + 1)a_{n+1}(x - x_0)^n \quad \text{for } |x - x_0| < R,$$

$$(4) \qquad f''(x) = \sum_{n=2}^{\infty} n(n - 1)a_n(x - x_0)^{n-2}$$

$$= \sum_{n=0}^{\infty} (n + 2)(n + 1)a_{n+2}(x - x_0)^n \quad \text{for } |x - x_0| < R,$$

and in general

$$(5) \qquad f^{(k)}(x) = \sum_{n=k}^{\infty} n(n - 1)\cdots(n - k + 1)a_n(x - x_0)^{n-k}$$

$$= \sum_{n=0}^{\infty} \frac{(n + k)!}{n!} a_{n+k}(x - x_0)^n \quad \text{for } |x - x_0| < R.$$

The second series in equations (3), (4), and (5) are obtained from the first series in each equation by "shifting" the index of summation by 1, 2, and k, respectively.

Setting $x = x_0$ in equation (5), we find that $f^{(k)}(x_0) = k!\, a_k$ for $k = 0, 1, \ldots$. Thus,

$$(6) \qquad a_k = \frac{f^{(k)}(x_0)}{k!} \quad \text{for } k = 0, 1, \ldots.$$

Substituting (6) into (2), we obtain the *Taylor series expansion of f about x_0*, namely

$$(7) \qquad f(x) = \sum_{n=0}^{\infty} \frac{f^{(n)}(x_0)}{n!}(x - x_0)^n \quad \text{for } |x - x_0| < R.$$

It should be noted that a real-valued function of a real variable which has derivatives of all orders at x_0 need not be analytic at x_0. That is, the converse of property 9 is false. Consider, for instance, the function

$$f(x) = \begin{cases} e^{-1/x^2}, & x \neq 0 \\ 0, & x = 0. \end{cases}$$

Calculating successive derivatives of f and evaluating at $x_0 = 0$, we find that $f(0) = f'(0) = f''(0) = \cdots = f^{(k)}(0) = \cdots 0$. Hence, we see from equation (7) that the series expansion of f about 0 would necessarily have coefficients which were all zero, and therefore the series would reduce to the zero function. Thus, f has derivatives of all orders, but its series expansion about 0 converges to the zero function on any interval, no matter how small, with center 0.

EXAMPLE Find the interval of convergence of the series $\sum_{n=0}^{\infty} n^2(x-2)^n$.

For this series $a_n = n^2$. Calculating the radius of convergence, we find

$$R = \lim_{n\to\infty}\left|\frac{a_n}{a_{n+1}}\right| = \lim_{n\to\infty}\left|\frac{n^2}{(n+1)^2}\right| = 1.$$

Since the series expansion is about $x_0 = 2$, the interval of convergence is $(1, 3)$.

EXAMPLE Calculate the Taylor series expansion for $f(x) = \sin x$ about $x_0 = 0$ and determine the radius of convergence.

Differentiating, we find $f^{(1)}(x) = \cos x$, $f^{(2)}(x) = -\sin x$, $f^{(3)}(x) = -\cos x$, $f^{(4)}(x) = \sin x$, $f^{(5)}(x) = \cos x, \ldots$. Evaluating f and its derivatives at 0, we obtain $f(0) = 0$, $f^{(1)}(0) = 1$, $f^{(2)}(0) = 0$, $f^{(3)}(0) = -1$, $f^{(4)}(0) = 0$, $f^{(5)}(0) = 1, \ldots$. From equation (6) we see that the series coefficients are $a_{2n} = 0$ and $a_{2n+1} = (-1)^n/(2n+1)!$ for $n = 0, 1, 2, \ldots$. Hence, the desired expansion is

$$\sin x = \sum_{n=0}^{\infty} \frac{(-1)^n x^{2n+1}}{(2n+1)!} \tag{8}$$

Since consecutive terms of the series (8) are not consecutive powers of x, we cannot determine the radius of convergence of (8) directly from property 3. However, we can use the ratio test. The *ratio test* states that the series $\sum_{n=0}^{\infty} u_n$ whose terms are all nonzero

(i) converges, if $\lim_{n\to\infty}\left|\dfrac{u_{n+1}}{u_n}\right| = L < 1$.

(ii) diverges, if $\lim_{n\to\infty}\left|\dfrac{u_{n+1}}{u_n}\right| = L > 1$.

(iii) Otherwise, the test fails.

(See any good calculus text for a proof of the ratio test.) Applying the ratio test to the series (8) with $u_n = (-1)^n x^{2n+1}/(2n+1)!$, we find that the series converges for all real x, since

$$\lim_{n\to\infty}\left|\frac{u_{n+1}}{u_n}\right| = \lim_{n\to\infty}\left|\frac{(-1)^{n+1}x^{2n+3}/(2n+3)!}{(-1)^n x^{2n+1}/(2n+1)!}\right|$$

$$= \lim_{n\to\infty}\left|\frac{x^2}{(2n+3)(2n+2)}\right| = 0 < 1$$

for all real x.

EXERCISES

1. Determine the interval of convergence for each of the following power series.

(a) $\sum_{n=0}^{\infty} x^n$ (b) $\sum_{n=0}^{\infty} nx^n$

(c) $\sum_{n=0}^{\infty} 2^n x^n$ (d) $\sum_{n=0}^{\infty} n!\, x^n$

(e) $\sum_{n=0}^{\infty} 3^n (x+2)^n$ (f) $\sum_{n=0}^{\infty} \frac{(-1)^n x^{2n}}{3^n}$

(g) $\sum_{n=0}^{\infty} \frac{(x-1)^n}{n!}$ (h) $\sum_{n=0}^{\infty} \frac{(2x-1)^n}{n}$

2. Calculate the power series expansion for the following functions about the point x_0 indicated, and determine the interval of convergence.

(a) e^x, $x_0 = 0$ (b) $\cos x$, $x_0 = 0$
(c) x^2, $x_0 = 1$ (d) $\ln x$, $x_0 = 1$
(e) $1/x^2$, $x_0 = 1$ (f) $1/(1+x)$, $x_0 = 0$
(g) $1/(1+x)$, $x_0 = -3$ (h) $1/(1+x^2)$, $x_0 = 0$

3. Consider $f(x) = x/(x^2 - 1)$.

(a) For what real values is f analytic?
(b) If f is expanded in a power series about $x_0 = \frac{1}{2}$, what will be the interval of convergence?
(c) If f is expanded in a power series about $x_0 = 3$, what will be the interval of convergence?

8.2 DETERMINATION OF SOLUTION SERIES COEFFICIENTS

The power series method is based upon the assumption that there is a solution to a given differential equation of the form

$$y(x) = \sum_{n=0}^{\infty} a_n (x - x_0)^n. \tag{9}$$

In the event that the problem is an initial value problem, x_0 is normally taken to be the point at which the initial conditions are specified. There are two different ways in which one may determine the coefficients a_n of equation (9). The first method is called the *Taylor series method* and the second method is called the *method of undetermined coefficients.* The Taylor series method, which was employed in Section 4.3.1.1 to produce numerical solutions to first order equations, consists of calculating successive derivatives from the given differential equation, evaluating the derivatives at the initial point, x_0, and substituting the results into the Taylor series—equation (7). The difficulty of this method lies with the recursive calculation of higher order derivatives. The method of undetermined coefficients consists of assuming that there is a solution of the form (9), formally differentiating (9) as many times as necessary to obtain series expressions for all the derivatives appearing in the differential equation, substituting these expressions into the differential equation and solving for the coefficients a_n. The difficulty of this method lies with the series manipulations which may be required and with solving for the series coefficients. In either case one must determine by some means—either from the series coefficients or from other considerations—the interval of convergence of the assumed solution. For instance, Theorem 8.1 provides a lower bound for the radius of convergence of a series solution to a second order linear differential equation.

Let us use both of these methods to solve the first order linear initial value problem:

$$y' = e^{-x^2}; \qquad y(0) = 1. \tag{10}$$

This is an initial value problem which we considered in Chapter 2. You should recall that we could write the solution as $y(x) = 1 + \int_0^x e^{-x^2}\,dx$. However, there is no elementary function $g(x)$ for which $g'(x) = e^{-x^2}$. So we could not write the solution in closed form and therefore had to be satisfied with some numerical approximation.

Calculating successive derivatives and then evaluating at $x = 0$, we find

$$\begin{aligned}
y^{(2)}(x) &= -2xe^{-x^2}; & y^{(2)}(0) &= 0\\
y^{(3)}(x) &= (4x^2 - 2)e^{-x^2}; & y^{(3)}(0) &= -2\\
y^{(4)}(x) &= (-8x^3 + 12x)e^{-x^2}; & y^{(4)}(0) &= 0\\
y^{(5)}(x) &= (16x^4 - 48x^2 + 12)e^{-x^2}; & y^{(5)}(0) &= 12.
\end{aligned}$$

Noting that $y(0) = 1$ and $y'(0) = 1$ and substituting into equation (7), we obtain the following solution by the Taylor series method:

$$y(x) = 1 + x - \frac{x^3}{3} + \frac{x^5}{10} - \cdots. \tag{11}$$

Now let us assume that equation (10) has a solution of the form

$$y(x) = \sum_{n=0}^{\infty} a_n x^n. \tag{12}$$

Letting $x = 0$ in equation (12) and imposing the initial condition yields $y(0) = 1 = a_0$. Differentiating equation (12), we obtain

$$y'(x) = \sum_{n=1}^{\infty} n a_n x^{n-1} = \sum_{n=0}^{\infty} (n+1) a_{n+1} x^n. \tag{13}$$

Since $e^x = \sum_{n=0}^{\infty} x^n/n!$,

$$e^{-x^2} = \sum_{n=0}^{\infty} \frac{(-x^2)^n}{n!} = \sum_{n=0}^{\infty} \frac{(-1)^n x^{2n}}{n!}. \tag{14}$$

Substituting (13) and (14) into (10), we find

$$\sum_{n=0}^{\infty} (n+1) a_{n+1} x^n = \sum_{n=0}^{\infty} \frac{(-1)^n x^{2n}}{n!}.$$

Or, equivalently,

$$a_1 + 2a_2 x + 3a_3 x^2 + 4a_4 x^3 + 5a_5 x^4 + \cdots = 1 - x^2 + \frac{x^4}{2} + \cdots.$$

Using property 4—that is, equating coefficients, we find that $a_1 = 1$, $a_2 = 0$, $a_3 = -\frac{1}{3}$, $a_4 = 0$, $a_5 = \frac{1}{10}, \ldots$. In general, we have $a_{2n} = 0$ and $a_{2n-1} = (-1)^{n-1}/(2n-1)(n-1)!$ for $n = 1, 2, \ldots$. Thus, by using the method of undetermined coefficients we again obtain the series solution (11) of the initial value problem (10). Applying the ratio test to the series (11) whose terms after the first term are $u_n = (-1)^n x^{2n+1}/(2n+1)n!$ for $n = 0, 1, \ldots$, we find that the series converges for all real x, since

$$\lim_{n\to\infty} \left| \frac{u_{n+1}}{u_n} \right| = \lim_{n\to\infty} \left| \frac{(-1)^{n+1} x^{2n+3}/(2n+3)(n+1)!}{(-1)^n x^{2n+1}/(2n+1)n!} \right|$$

$$= \lim_{n\to\infty} \left| \frac{(2n+1)x^2}{(2n+3)(n+1)} \right| = 0 < 1$$

for all real x. It should be obvious that it is easier to obtain additional values for the series coefficients by using the method of undetermined coefficients than by using the Taylor series method. Consequently, we shall usually employ the method of undetermined coefficients.

EXAMPLE Find the power series solution for the initial value problem:

$$y' = 1 + y^2; \qquad y(0) = 0. \tag{15}$$

The differential equation in (15) is nonlinear; however, the solution of the initial value problem is easily found by the method of separation of variables to be $y(x) = \tan x$ where $x \in (-\pi/2, \pi/2)$. Employing the method of undetermined coefficients, we assume that (15) has a solution of the form

$$y(x) = \sum_{n=0}^{\infty} a_n x^n. \tag{16}$$

Differentiating, we see that

$$y'(x) = \sum_{n=1}^{\infty} n a_n x^{n-1} = \sum_{n=0}^{\infty} (n+1) a_{n+1} x^n. \tag{17}$$

Evaluating (16) at $x = 0$ and imposing the initial condition, we find that $y(0) = 0 = a_0$. Substituting (16) and (17) into (15), we see that the series coefficients, a_n, must satisfy

$$\sum_{n=0}^{\infty} (n+1) a_{n+1} x^n = 1 + \left(\sum_{n=0}^{\infty} a_n x^n\right)^2 = 1 + \sum_{n=0}^{\infty} \left(\sum_{k=0}^{n} a_k a_{n-k}\right) x^n.$$

Equating coefficients, we obtain

for $n = 0$: $\quad a_1 = 1 + a_0^2 = 1$

for $n = 1$: $\quad 2a_2 = 2a_0 a_1 = 0, \quad$ so $a_2 = 0$

for $n = 2$: $\quad 3a_2 = 2a_0 a_2 + a_1^2 = 1, \quad$ so $a_3 = \frac{1}{3}$

for $n = 3$: $\quad 4a_4 = 2a_0 a_3 + 2a_1 a_2 = 0, \quad$ so $a_4 = 0$

for $n = 4$: $\quad 5a_5 = 2a_0 a_4 + 2a_1 a_3 + a_2^2 = \frac{2}{3}, \quad$ so $a_5 = \frac{2}{15}$.

In general,

$$a_n = \begin{cases} 0, & \text{if } n \text{ is even} \\ 1 + a_0^2, & \text{if } n = 1 \\ \dfrac{1}{n} \displaystyle\sum_{k=0}^{n-1} a_k a_{n-1-k}, & \text{if } n \text{ is odd and } n > 1. \end{cases}$$

Thus, we are able to recursively calculate the series coefficients but we are unable to easily express a_n explicitly as a function of n. Therefore, we can not calculate the radius of convergence directly. However, we do know that the radius of convergence is $\pi/2$, since we know that the solution written in closed form is $y(x) = \tan x$ for $x \in (-\pi/2, \pi/2)$.

EXAMPLE Find the power series solution to the initial value problem

$$y' = x + y^2; \qquad y(0) = 0. \tag{18}$$

The differential equation in (18) is nonlinear and in some respects is similar to the differential equation in (15). However, in this instance we cannot obtain a closed form solution by any of the techniques discussed previously in this text. So a numerical method is necessary and the power series method is a very appropriate method for solving this initial value problem. Again we assume that (18) has a power series solution of the form (16). By differentiating (16), we obtained (17). Evaluating (16) at $x = 0$ and imposing the initial condition $y(0) = 0$, we find that $a_0 = 0$. Substituting the assumed series solution (16) and its derivative (17) into (18), we see that the series coefficients must satisfy

$$\sum_{n=0}^{\infty} (n+1)a_{n+1}x^n = x + \left(\sum_{n=0}^{\infty} a_n x^n\right)^2 = x + \sum_{n=0}^{\infty}\left(\sum_{k=0}^{n} a_k a_{n-k}\right)x^n.$$

Equating coefficients we find

$$\begin{aligned}
&\text{for } n = 0: && a_1 = a_0^2 = 0\\
&\text{for } n = 1: && 2a_2 = 1 + 2a_0a_1 = 1, \quad \text{so } a_2 = \tfrac{1}{2}\\
&\text{for } n = 2: && 3a_3 = 2a_0a_2 + a_1^2 = 0, \quad \text{so } a_3 = 0\\
&\text{for } n = 3: && 4a_4 = 2a_0a_3 + 2a_1a_2 = 0, \quad \text{so } a_4 = 0\\
&\text{for } n = 4: && 5a_5 = 2a_0a_4 + 2a_1a_3 + a_2^2 = \tfrac{1}{4}, \quad \text{so } a_5 = \tfrac{1}{20}.
\end{aligned}$$

So the series solution begins

$$y(x) = \frac{x^2}{2} + \frac{x^5}{20} + \cdots.$$

Again we can recursively calculate as many series coefficients as necessary to produce a solution with a desired accuracy for all real x sufficiently close to $x = 0$. However, we cannot express the general series coefficient a_n as an explicit function of n and, therefore, cannot directly calculate the radius of convergence for this solution. Unfortunately, this is the situation which one usually encounters when generating power series solutions to nonlinear equations.

The method of undetermined coefficients does not always readily produce results, as the following example shows. Let us again consider the nonlinear differential equation which we derived in Chapter 7 for describing the motion of a simple pendulum, namely,

$$Ly'' + g \sin y = 0. \tag{19}$$

Here we have replaced $\theta(t)$ of equation (13) of Chapter 7 by $y(x)$ so that the power series notation for the solution will look more familiar. Let us assume that equation (19) has a power series solution of the form

$$y(x) = \sum_{n=0}^{\infty} a_n x^n. \tag{20}$$

Differentiating twice, we find

(21) $$y'(x) = \sum_{n=1}^{\infty} na_n x^{n-1} = \sum_{n=0}^{\infty} (n+1)a_{n+1}x^n$$

and

(22) $$y''(x) = \sum_{n=2}^{\infty} n(n-1)a_n x^{n-2} = \sum_{n=0}^{\infty} (n+2)(n+1)a_{n+2}x^n.$$

From equation (8) we know that

(23) $$\sin y = \sum_{n=0}^{\infty} \frac{(-1)^n y^{2n+1}}{(2n+1)!}.$$

Substituting (20), (22), and (23) into (19), we see that the series coefficients must satisfy

(24)

$$L\sum_{n=0}^{\infty} (n+2)(n+1)a_{n+2}x^n + g\sum_{n=0}^{\infty} (-1)^n\left(\sum_{m=0}^{\infty} a_m x^m\right)^{2n+1}/(2n+1)! = 0.$$

The coefficients $a_0 = y(0)$ and $a_1 = y'(0)$ are determined by the initial conditions. But it is impossible to determine the remaining coefficients from a_0, a_1, and equation (24). A power series expansion can be obtained using the method of undetermined coefficients if $\sin y$ is expanded about $y(0) = a_0$ instead of about 0 as in equation (23). However, the power series solution to (19) is obtained more easily using the Taylor series method. Other methods available for solving equation (19) include various numerical techniques and approximations obtained by altering equation (19) in some manner, such as replacing $\sin y$ by y or $y - y^3/3!$.

EXERCISES

Use the method of undetermined coefficients to find the power series solution to each of the following linear initial value problems.

1. $y' - y = 0;\ y(0) = 1$

2. $y' - xy = 0;\ y(0) = 1$

3. $y'' + y = 0;\ y(0) = 0,\ y'(0) = 1$

4. $y' - y = e^x;\ y(0) = 1$

5. $y'' + y = \sin x;\ y(0) = 0,\ y'(0) = 1$

Find the first five nonzero terms of the power series solution to each of the following nonlinear initial value problems.

6. $y' = xy^2 + 1;\ y(0) = 1$

7. $y' = x + e^y$; $y(0) = 1$ (*Hint:* Use the Taylor series method.)

8. $y' = y^2 + xy + 1$; $y(0) = 1$

9. $y'' = x^2 - y^2$; $y(0) = 1$, $y'(0) = 0$

10. $yy'' + 3(y')^2 = 0$; $y(0) = 1$, $y(0) = \frac{1}{2}$

11. $y''' - yy' - xy = 0$; $y(0) = 0$, $y'(0) = 1$, $y''(0) = 2$

12. $Ly'' + g \sin y = 0$; $y(0) = \pi/60$, $y'(0) = \pi/2$ (*Hint:* Use the Taylor series method.)

8.3 LINEAR DIFFERENTIAL EQUATIONS WITH VARIABLE COEFFICIENTS

Throughout this section we shall consider only linear differential equations. At least one coefficient function will be variable, since we have already discussed in detail how to solve linear equations with constant coefficients. We will state definitions and theorems for second order linear equations, since most of the equations that we shall examine will be second order. These definitions and theorems can usually be extended in an obvious fashion to higher order equations. There are several reasons for considering second order linear equations. First, since the equations are linear, the superposition principle applies. That is, a linear combination of any two solutions is again a solution. It is the lack of a superposition principle for nonlinear equations which prohibits a systematic study of such equations. Second, we can adequately illustrate the power series method with second order linear equations. And third, a large number of problems from various diverse disciplines require the solution of the same second order linear equations. Consequently, we shall examine some of the more famous second order linear equations in this section.

Consider the second order linear homogeneous equation of the form

$$y'' + p(x)y' + q(x)y = 0. \tag{25}$$

The point x_0 is called an *ordinary point* of equation (25) if both $p(x)$ and $q(x)$ are analytic at x_0—that is, if both $p(x)$ and $q(x)$ have power series expansions about x_0 which converge to $p(x)$ and $q(x)$, respectively, in some interval about x_0. If either $p(x)$ or $q(x)$ is not analytic at x_0, the point x_0 is called a *singular point* of equation (25).

The differential equation

$$x(1 - x)y'' + xy' + (1 - x^2)y = 0 \tag{26}$$

written in the form of equation (25) is

$$y'' + \frac{x}{1 - x}y' + \frac{1 + x}{x}y = 0.$$

The function $p(x) = x/(1 - x)$ is analytic except at $x = 1$ and the function $q(x) = (1 + x)/x$ is analytic except at $x = 0$. So $x = 0$ and $x = 1$ are singular points of equation (26) and all other finite points are ordinary points of equation (26). We shall concern ourselves only with ordinary and singular points which are finite. Consequently, we shall not discuss ordinary and singular points at infinity. The equation

$$y'' + xy' + 2xy = 0$$

has no finite singular points, while the equation

$$x^2y'' + 2y' - y = 0$$

has only one finite singular point—the point $x = 0$. The equation

$$(x^2 + 1)y'' + xy' - y = 0$$

has two singular points $x = \pm i$.

8.3.1 Solutions Near an Ordinary Point

The following theorem, which we state without proof, guarantees the existence and uniqueness of a power series solution near an ordinary point, x_0, for linear differential equations which have coefficient functions which are analytic at x_0 and provides a lower bound for the interval of convergence. Thus, the necessity of calculating the radius of convergence of the series solution is virtually eliminated for linear equations.

Theorem 8.1 Consider the initial value problem

$$y'' + p(x)y' + q(x)y = 0; \qquad y(x_0) = a, \quad y'(x_0) = b, \tag{27}$$

where a and b are arbitrary constants. If x_0 is an ordinary point of the differential equation of (27), if $p(x)$ has a power series expansion about x_0 which converge to $p(x)$ for $|x - x_0| < R_1$ where $R_1 > 0$, and if $q(x)$ has a power series expansion about x_0 which converges to $q(x)$ for $|x - x_0| < R_2$ where $R_2 > 0$, then there exists a unique power series solution of the initial value problem (27) of the form

$$y(x) = \sum_{n=0}^{\infty} a_n(x - x_0)^n, \tag{28}$$

which converges at least for $|x - x_0| < R = \min(R_1, R_2)$.

This theorem states that the series solution converges inside the circle of radius R about x_0. But it does not state that the solution diverges outside the circle. In some instances, the solution may converge inside a circle—the circle of convergence—with radius greater than the minimum radius of convergence, R, specified by the theorem. In any event the circle of convergence of the solution passes through a singular point of the differential equation—but perhaps not the nearest singular point.

EXAMPLE Determine a lower bound for the radius of convergence of a series solution of the differential equation

$$(x^2 + 4)y'' - x^2y' + 2xy = 0 \tag{29}$$

about the point $x_0 = 0$ and about the point $x_0 = 2$.

For this differential equation $p(x) = -x^2/(x^2 + 4)$ and $q(x) = 2x/(x^2 + 4)$. Both of these functions are analytic except at $x = \pm 2i$. The Euclidean distance from $x_0 = 0$ to $2i$ or $-2i$ is 2 and the distance from $x_0 = 2$ to $2i$ or $-2i$ is $2\sqrt{2}$. So a power series solution of (29) expanded about $x_0 = 0$ will have a radius of convergence of at least 2—that is, it will converge at least in the interval $(-2, 2)$. While a series solution of (29) expanded about $x_0 = 2$ will have a radius of convergence of at least $2\sqrt{2}$ and, therefore, will converge at least in the interval $(2 - 2\sqrt{2}, 2 + 2\sqrt{2})$.

Sir George Airy (1801–1892), English astronomer and mathematician, studied the following equation which bears his name in connection with the diffraction of light:

$$y'' + xy = 0 \quad \text{(Airy's equation)}. \tag{30}$$

Airy was interested in this equation because the solutions are oscillatory for x positive—that is, they behave like $\sin x$ and $\cos x$—and the solutions are exponential in nature for x negative—that is, they behave like e^x and e^{-x}. Although this equation appears to be very simple, it cannot be solved by any technique presented previously. In equation (30) $p(x) = 0$ and $q(x) = x$. Both of these functions are analytic for all x, so all points are ordinary points and we know from Theorem 8.1 that a power series expansion about any real point converges for all real x.

EXAMPLE Find two linearly independent series solutions for Airy's equation in powers of x.

We know that equation (30) has a solution of the form

$$y(x) = \sum_{n=0}^{\infty} a_n x^n \tag{31}$$

and that this solution is valid for all real x and all choices of the initial conditions $y(0) = a = a_0$ and $y'(0) = b = a_1$. Differentiating (31) twice, we obtain

$$y'(x) = \sum_{n=1}^{\infty} a_n x^{n-1} \tag{32}$$

and

$$y''(x) = \sum_{n=2}^{\infty} n(n-1)a_n x^{n-2}. \tag{33}$$

Substituting (31) and (33) into (30), we get

$$\sum_{n=2}^{\infty} n(n-1)a_n x^{n-2} + x\sum_{n=0}^{\infty} a_n x^n = 0.$$

or

$$\sum_{n=2}^{\infty} n(n-1)a_n x^{n-2} + \sum_{n=0}^{\infty} a_n x^{n+1} = 0.$$

Shifting the index of summation of the first series by 2 and the second series by -1 and writing the first term of the first series separately, we get

$$2a_2 + \sum_{n=1}^{\infty} (n+2)(n+1)a_{n+2}x^n + \sum_{n=1}^{\infty} a_{n-1}x^n = 0,$$

so $2a_2 = 0$ and $(n+2)(n+1)a_{n+2} + a_{n-1} = 0$ for $n \geq 1$. Thus, $a_2 = 0$ and

$$a_{n+2} = \frac{-a_{n-1}}{(n+2)(n+1)} \quad \text{for } n \geq 1. \tag{34}$$

We see from equation (34) that each coefficient except the first three coefficients—a_0, a_1, and a_2—is recursively determined from the coefficient in the series whose subscript precedes its subscript by three. Thus, a_0 determines a_3, which determines a_6, etc.; a_1 determines a_4, which determines a_7, etc.; and a_2 determines a_5, which determines a_8, etc. Consequently, it is appropriate to arrange the calculation of the series coefficients into three columns. Doing so and using equation (34), we find the values for the coefficients shown on page 242.

Since a_0 and a_1 are arbitrary, two linearly independent solutions of Airy's equation (30) are

$$y_1(x) = \sum_{n=1}^{\infty} \frac{(-1)^n x^{3n}}{(3n)(3n-1)(3n-3)(3n-4)\cdots 3\cdot 2} + 1$$

and

$$y_2(x) = \sum_{n=1}^{\infty} \frac{(-1)^n x^{3n+1}}{(3n+1)(3n)(3n-2)(3n-3)\cdots 4\cdot 3} + 1.$$

Series Coefficients for Airy's Equation (30)

$$a_3 = \frac{-a_0}{3\cdot 2} \quad (n = 1) \qquad a_4 = \frac{-a_1}{4\cdot 3} \quad (n = 2) \qquad a_5 = \frac{-a_2}{5\cdot 4} = 0 \quad (n = 3)$$

$$a_6 = \frac{-a_3}{6\cdot 5} = \frac{a_0}{6\cdot 5\cdot 3\cdot 2} \qquad a_7 = \frac{-a_4}{7\cdot 6} = \frac{a_1}{7\cdot 6\cdot 4\cdot 3} \qquad a_8 = \frac{-a_5}{8\cdot 7} = 0$$

$$a_9 = \frac{-a_6}{9\cdot 8} = \frac{-a_0}{9\cdot 8\cdot 6\cdot 5\cdot 3\cdot 2} \qquad a_{10} = \frac{-a_7}{10\cdot 9} = \frac{-a_1}{10\cdot 9\cdot 7\cdot 6\cdot 4\cdot 3} \qquad a_{11} = \frac{-a_8}{11\cdot 10} = 0$$

$$\vdots \qquad\qquad \vdots \qquad\qquad \vdots$$

$$a_{3n} = \frac{(-1)^n a_0}{(3n)(3n-1)(3n-3)(3n-4)\cdots 3\cdot 2} \qquad a_{3n+1} = \frac{(-1)^n a_1}{(3n+1)(3n)(3n-2)(3n-3)\cdots 4\cdot 3} \qquad a_{3n+2} = \frac{-a_{3n-1}}{(3n+2)(3n+1)} = 0$$

The function y_1 was obtained by choosing $a_0 = 1$ and $a_1 = 0$ and the function y_2 was obtained by choosing $a_0 = 0$ and $a_1 = 1$. The general solution of Airy's equation is $y(x) = a_0 y_1(x) + a_1 y_1(x)$ where $a_0 = y(0)$ and $a_1 = y'(0)$ are arbitrary.

So far, we have only calculated power series solutions which are expansions about $x_0 = 0$. This is sufficient. If one wishes a power series solution which is an expansion about some other point $x_0 \neq 0$, the problem can be reduced to finding an expansion about zero by performing the linear transformation $v = x - x_0$. In order to illustrate this important concept, we present the following example.

EXAMPLE Find the series solution of Airy's equation which satisfies the initial conditions $y(-2) = 1$ and $y'(-2) = \frac{1}{2}$.

In this case we seek a series solution about the point $x_0 = -2$. We set $v = x + 2$. Then $dv = dx$ and $x = v - 2$. Making this change of variable the initial value problem

$$\frac{d^2y}{dx^2} + xy = 0; \qquad y(-2) = 1, \quad y'(-2) = \tfrac{1}{2}$$

becomes

$$\frac{d^2y}{dv^2} + (v - 2)y = 0; \qquad y(0) = 1, \quad y'(0) = \tfrac{1}{2}. \tag{35}$$

We now seek a series solution of (35) in powers of v. After finding this solution, we make the substitution $v = x + 2$ to obtain the solution in powers of $x + 2$. Assuming that (35) has a solution of the form

$$y(v) = \sum_{n=0}^{\infty} a_n v^n, \tag{36}$$

differentiating (36) twice and substituting into (35), we see that the coefficients a_n must satisfy

$$\sum_{n=2}^{\infty} n(n-1)a_n v^{n-2} + (v-2)\sum_{n=0}^{\infty} a_n v^n = 0,$$

or

$$\sum_{n=2}^{\infty} n(n-1)a_n v^{n-2} + \sum_{n=0}^{\infty} a_n v^{n+1} + \sum_{n=0}^{\infty} -2a_n v^n = 0.$$

Shifting the index of summation of the first series by 2 and the second series by -1 yields

$$\sum_{n=0}^{\infty}(n+2)(n+1)a_{n+2}v^n + \sum_{n=1}^{\infty}a_{n-1}v^n + \sum_{n=0}^{\infty}-2a_nv^n = 0.$$

For $n = 0$, we get

$$2a_2 - 2a_0 = 0,$$

since the second series does not contribute anything unless $n \geq 1$. For $n \geq 1$, we find

$$(n+2)(n+1)a_{n+2} + a_{n-1} - 2a_n = 0.$$

Hence, $a_0 = y(0) = 1$, $a_1 = y'(0) = \frac{1}{2}$, $a_2 = a_0 = 1$, and

$$a_{n+2} = \frac{2a_n - a_{n-1}}{(n+2)(n+1)} \quad \text{for } n \geq 1.$$

Solving for a_3, a_4, and a_5, we find

$$a_3 = \frac{2a_1 - a_0}{3\cdot 2} = 0, \quad a_4 = \frac{2a_2 - a_1}{4\cdot 3} = \frac{3}{24}, \quad \text{and} \quad a_5 = \frac{2a_3 - a_2}{5\cdot 4} = \frac{-1}{20}.$$

In this case we cannot easily derive a general expression for a_n in terms of n. However, we know the desired series solution begins

$$y(v) = 1 + \frac{v}{2} + v^2 + \frac{3v^3}{24} - \frac{v^4}{20} + \cdots$$

or equivalently

$$y(x) = 1 + \frac{x+2}{2} + (x+2)^2 + \frac{3(x+2)^3}{24} - \frac{(x+2)^4}{20} + \cdots.$$

EXERCISES

1. For each of the following differential equations, determine the singular points and a lower bound for the radius of convergence of a series solution about each given point x_0.
   (a) $y'' + xy' - x^2y = 0$; $x_0 = 0$, $x_0 = 3$
   (b) $(x^2 + x - 2)y'' - 2y' + xy = 0$; $x_0 = -3$, $x_0 = 0$, $x_0 = 3$
   (c) $(x^2 + 4x + 5)y'' + 2xy' - y = 0$; $x_0 = -2$, $x_0 = 0$, $x_0 = 2$
   (d) $(\cos x)y'' + xy' + x^2y = 0$; $x_0 = 0$
   (e) $xy'' + (\sin x)y' + x^2y = 0$; $x_0 = 0$
   (f) $(1 + e^x)y'' - 2xy' + y = 0$; $x_0 = 0$, $x_0 = 1$ [*Hint:* Solve $1 + e^{\alpha + i\beta} = 1 + e^{\alpha}(\cos\beta + i\sin\beta) = 0$.]

2. (a) What is a lower bound of the radius of convergence of a series solution in powers of x of the equation $(1 + x^2)y'' - xy' - 3y = 0$?
(b) Find two linearly independent solutions in powers of x of this equation.
(c) On what interval is each of these solutions convergent?

3. (a) What is a lower bound of the radius of convergence of a series solution in powers of x of the equation $(1 + x^2)y'' - 4xy' + 6y = 0$?
(b) Find two linearly independent solutions in powers of x of this equation.
(c) On what interval is each of these solutions convergent?

4. Find power series solutions of the following initial value problems.
(a) $y'' + xy' - x^2y = 0$; $y(0) = 1$, $y'(0) = -1$
(b) $xy'' + 2y' - y = 0$; $y(1) = 2$, $y'(1) = 3$
(c) $x^2y'' + xy' + y = 0$; $y(-1) = -1$, $y'(-1) = 1$.

5. Hermite's equation is

$$y'' - 2xy' + 2\lambda y = 0,$$

where λ is a constant.
(a) Find two linearly independent solutions in powers of x.
(b) If λ is a nonnegative integer, Hermite's equation has a polynomial solution. Find polynomial solutions for $\lambda = 0$, 1, 2, and 3. After being normalized—that is, after being multiplied by a constant chosen so that the polynomial has a certain value at some specified point or chosen so that a particular definite integral has a specified value—the polynomial solutions to Hermite's equation are known as *Hermite polynomials* $H_n(x)$.

6. Legendre's equation is

$$(1 - x^2)y'' - 2xy' + \lambda(\lambda + 1)y = 0,$$

where λ is a constant.
(a) Find two linearly independent solutions in powers of x.
(b) If λ is a nonnegative constant, Legendre's equation has a polynomial solution. Find polynomial solutions for $\lambda = 0$, 1, 2, and 3.
(c) The Legendre polynomial $P_n(x)$ is defined to be the polynomial solution of Legendre's equation with $\lambda = n$, a nonnegative integer, which satisfies the normalization condition $P_n(1) = 1$. Find $P_0(x)$, $P_1(x)$, $P_2(x)$, and $P_3(x)$.

7. Chebyshev's equation is

$$(1 - x^2)y'' - xy' + \lambda^2 y = 0,$$

where λ is a constant.

(a) Find two linearly independent solutions in powers of x.
(b) Find polynomial solutions for $\lambda = 0$, 1, 2, and 3. After being properly normalized, the polynomial solutions of Chebyshev's equation for $\lambda = 0, 1, 2, \ldots$ are called *Chebyshev's polynomials.*

8.3.2 Solutions Near a Regular Singular Point

Recall that the point x_0 is an ordinary point of the second order linear homogeneous differential equation

$$y'' + p(x)y' + q(x)y = 0, \tag{37}$$

if both $p(x)$ and $q(x)$ are analytic at x_0. When x_0 is an ordinary point of equation (37), the solution may be expressed in terms of a power series. Moreover, this solution is valid and analytic in some interval containing x_0. If either $p(x)$ or $q(x)$ is not analytic at x_0, then x_0 is a singular point of equation (37). When x_0 is a singular point of equation (37), the differential equation itself, at least as written, is undefined at x_0. Therefore, one should not expect to be able to express the solution as a power series about x_0, since such a series will always be defined and analytic at x_0. However, for a large class of differential equations with the form of equation (37) but in which either $p(x)$ or $q(x)$ is not analytic at x_0, the power series method can be modified so that a solution may be obtained which is valid in a deleted interval about a singular point x_0. One wants to study a differential equation near its singular points because the behavior of the solution near these points tells one a lot about the solution in general and because the behavior of any physical system which is described by the differential equation is usually most interesting near a singular point.

The following examples illustrate that there exist differential equations of the form (37) with singularities at x_0 for which there are two linearly independent solutions (i) both of which are defined and finite at x_0, (ii) one of which is defined and finite at x_0 while the other becomes unbounded as x approaches x_0, and (iii) both of which become unbounded as x approaches x_0.

EXAMPLE The differential equation

$$y'' - \frac{2y'}{x} + \frac{2y}{x^2} = 0$$

has a singular point $x_0 = 0$. Two linearly independent solutions of this equation are $y_1 = x$ and $y_2 = x^2$. Both of these functions are defined and finite at $x_0 = 0$.

EXAMPLE The differential equation

$$y'' + \frac{2y'}{x} - \frac{2y}{x^2} = 0$$

has a singular point $x_0 = 0$. Two linearly independent solutions of this equation are $y_1 = x$ and $y_2 = x^{-2}$. The function $y_1 = x$ is defined and finite at the singular point while the function $y_2 = x^{-2} \to \infty$ as $x \to 0$.

EXAMPLE The differential equation

$$y'' + \frac{4y'}{x} + \frac{2y}{x^2} = 0$$

has a singular point $x_0 = 0$. Two linearly independent solutions of this equation are $y_1 = x^{-1}$ and $y_2 = x^{-2}$. Both of these functions become unbounded as x approaches zero.

If x_0 is a singular point of equation (37) and if

$$P(x) = (x - x_0)p(x) \quad \text{and} \quad Q(x) = (x - x_0)^2 q(x)$$

are analytic at x_0, then x_0 is called a *regular singular point* of (37). If either $P(x)$ or $Q(x)$ is not analytic at x_0, then x_0 is called an *irregular singular point* of (37). The mathematical theory which we shall develop presently will apply only at regular singular points.

EXAMPLE Classify all finite singular points of the differential equation

$$x(x - 1)^2 y'' + 2xy' + y = 0. \tag{38}$$

Writing this equation in the form of equation (37), we have

$$y'' + \frac{2y'}{(x - 1)^2} + \frac{y}{x(x - 1)^2} = 0. \tag{39}$$

In this case $p(x) = 2/(x - 1)^2$ and $q(x) = 1/x(x - 1)^2$. The function $p(x)$ is analytic except at $x = 1$ and the function $q(x)$ is analytic except at $x = 0$ and $x = 1$. Thus, $x = 0$ and $x = 1$ are singular points of (39)—and the equivalent equation (38). We consider the functions $P(x) = xp(x) = 2x/(x - 1)^2$ and $Q(x) = x^2 q(x) = x/(x - 1)^2$ at $x = 0$. Since both $P(x)$ and $Q(x)$ are analytic at $x = 0$, the point $x = 0$ is a regular singular point of (39). Next we consider the functions $P(x) = (x - 1)p(x) = 2/(x - 1)$ and $Q(x) = (x - 1)^2 q(x) = 1/x$ at $x = 1$. Since the function $P(x)$ is not analytic at $x = 1$, the point $x = 1$ is an irregular singular point of equation (39).

EXAMPLE Classify all finite singular points of the differential equation

$$(1 + x^2)y'' + 2xy' - x^2 y = 0.$$

In this case $p(x) = 2x/(1 + x^2)$ and $q(x) = -x^2/(1 + x^2)$. Both of these functions are analytic except for $x = \pm i$. So the finite singular points of this equation are $\pm i$. Let us consider the singular point at $x = i$. Multiplying $p(x)$ by the factor $(x - i)$ and $q(x)$ by the factor $(x - i)^2$, we find $P(x) = 2x/(x + i)$ and $Q(x) = -x^2(x - i)/(x + i)$. Since both $P(x)$ and $Q(x)$ are analytic at $x = i$, the point $x = i$ is a regular singular point of the differential equation. Likewise, $x = -i$ is found to be a regular singular point.

Suppose that the differential equation

$$y'' + p(x)y' + q(x)y = 0 \tag{37}$$

has a regular singular point at x_0. Multiplying this equation by $(x - x_0)^2$, we obtain the equivalent differential equation

$$(x - x_0)^2y'' + (x - x_0)p^*(x)y' + q^*(x)y = 0, \tag{40}$$

where $p^*(x) = (x - x_0)p(x)$ and $q^*(x) = (x - x_0)^2q(x)$ are analytic at x_0 and, therefore, are representable by power series expansions about x_0 which are valid for $|x - x_0| < R$ where $R > 0$. The German mathematician Ferdinand Georg Frobenius (1849–1917) first proposed seeking a series solution of (40) of the form

$$y(x) = |x - x_0|^r \sum_{n=0}^{\infty} a_n(x - x_0)^n, \tag{41}$$

where r is a constant and $a_0 \neq 0$. Notice that if $a_0 = 0$, then some factor $(x - x_0)^m$, where m is a positive integer, could be factored out of the series portion of (41) and combined with $|x - x_0|^r$. The factor $|x - x_0|^r$ appears in equation (41) instead of the factor $(x - x_0)^r$ because the function $|x - x_0|^r$ is defined for all real constants r and all real $x \neq x_0$, whereas the function $(x - x_0)^r$ is not defined for all real constants r unless $x > x_0$. The modified power series of equation (41) is known as a *Frobenius series*. Obviously, a Frobenius series is a generalization of a power series, since it reduces to a power series when r is a nonnegative integer. For the time being, let us denote the power series portion of (41) by $S(x)$. Thus, we assume that equation (40) has a solution of the form

$$y(x) = |x - x_0|^rS(x). \tag{42}$$

Differentiating twice, we obtain

$$y'(x) = r|x - x_0|^{r-1}\frac{|x - x_0|}{x - x_0}S(x) + |x - x_0|^rS'(x) \quad \text{for } x \neq x_0. \tag{43}$$

and

$$y''(x) = r(r - 1)|x - x_0|^{r-2}S(x) + \frac{2r|x - x_0|^r}{x - x_0}S'(x) + |x - x_0|^rS''(x) \quad \text{for } x \neq x_0. \tag{44}$$

Substituting (42), (43), and (44) into (40) and rearranging, we get

$$|x - x_0|^r\{(x - x_0)^2S''(x) + (x - x_0)[2r + p^*(x)]S'(x) + [r(r - 1) + rp^*(x) + q^*(x)]S(x)\} = 0.$$

Since the solution is to be valid for $x \neq x_0$, we must have

$$(x - x_0)^2S''(x) + (x - x_0)[2r + p^*(x)]S'(x) + [r(r - 1) + rp^*(x) + q^*(x)]S(x) = 0. \tag{45}$$

The series coefficients a_n and the constant r must be chosen so that equation (45) is satisfied. The power series expansions for S, S', and S'' begin with the constant terms a_0, a_1, and $2a_2$, respectively. In view of Taylor's theorem the power series expansions for $p^*(x)$ and $q^*(x)$ about x_0 begins with the constant terms $p^*(x_0)$ and $q^*(x_0)$. When we equate the coefficient of $(x - x_0)^n$ in equation (45) to zero, we see that the first term in equation (45) makes no contribution unless $n \geq 2$ and the second term in equation (45) makes no contribution unless $n \geq 1$. Thus, the constant term of equation (45) is

$$[r(r - 1) + rp^*(x_0) + q^*(x_0)]a_0 = 0.$$

Since $a_0 \neq 0$, r must be chosen so that it satisfies the *indicial equation*

$$r(r - 1) + rp^*(x_0) + q^*(x_0) = r^2 + [p^*(x_0) - 1]r + q^*(x_0) = 0. \tag{46}$$

The indicial equation is a quadratic equation in the unknown r and the roots of this equation determine the form of the two linearly independent solutions of (40) just as the roots of the auxiliary equation determine the form of the linearly independent solutions of linear differential equations with constant coefficients.

The following theorem which we present without proof is due to the German mathematician L. Fuchs (1833–1902).

Theorem 8.2 Consider the differential equation

$$(x - x_0)^2y'' + (x - x_0)p^*(x)y' + q^*(x)y = 0, \tag{40}$$

where p^* and q^* are analytic at x_0 and are representable by power series expansions about x_0 which are valid for $|x - x_0| < R$, where $R > 0$. Let r_1 and r_2 be the roots of the associated indicial equation

$$r^2 + [p^*(x_0) - 1]r + q^*(x_0) = 0 \tag{46}$$

and let $Re(r_2) \leq Re(r_1)$. Then there exist two linearly independent solutions y_1 and y_2 of (40) which are valid at least for $0 < |x - x_0| < R$ and the solutions have the following forms:

1. If $r_1 - r_2$ is not an integer,

$$y_1(x) = |x - x_0|^{r_1} \sum_{n=0}^{\infty} a_n(x - x_0)^n$$

$$y_2(x) = |x - x_0|^{r_2} \sum_{n=0}^{\infty} b_n(x - x_0)^n.$$

2. If $r_1 = r_2$,

$$y_1(x) = |x - x_0|^{r_1} \sum_{n=0}^{\infty} a_n(x - x_0)^n$$

$$y_2(x) = y_1(x) \ln |x - x_0| + |x - x_0|^{r_1} \sum_{n=0}^{\infty} b_n(x - x_0)^n.$$

3. If $r_1 - r_2$ is a positive integer,

$$y_1(x) = |x - x_0|^{r_1} \sum_{n=0}^{\infty} a_n(x - x_0)^n$$

$$y_2(x) = cy_1(x) \ln |x - x_0| + |x - x_0|^{r_2} \sum_{n=0}^{\infty} b_n(x - x_0)^n.$$

The coefficients a_n and b_n and the constant c are determined by substituting the proper form for the solution and its derivatives into the specific differential equation to be solved. The roots r_1 and r_2 are often called the *exponents of the singularity* at the regular singular point x_0.

The differential equation

$$x^2y'' + xy' + (x^2 - \lambda^2)y = 0, \tag{47}$$

where λ is a nonnegative constant is known as *Bessel's equation of order* λ. Any solution of Bessel's equation of order λ is called a *Bessel function of order* λ. Many physical systems give rise to Bessel's equation and the associated Bessel functions. Clearly, $x = 0$ is a regular singular point of Bessel's equation and $p^*(x) = 1$ and $q^*(x) = x^2 - \lambda^2$. So $p^*(0) = 1$ and $q^*(0) = -\lambda^2$. Consequently, the indicial equation for Bessel's equation of order λ is

$$r^2 - \lambda^2 = 0.$$

The roots of this equation are $r_1 = \lambda$ and $r_2 = -\lambda$. From Theorem 8.2 we know that there is always one solution of the form

$$y(x) = |x|^\lambda \sum_{n=0}^{\infty} a_n x^n. \tag{48}$$

Differentiating twice, we obtain

$$y'(x) = \lambda |x|^{\lambda-1} \frac{|x|}{x} \sum_{n=0}^{\infty} a_n x^n + |x|^\lambda \sum_{n=0}^{\infty} n a_n x^{n-1}$$

$$y''(x) = \lambda(\lambda - 1)|x|^{\lambda-2} \sum_{n=0}^{\infty} a_n x^n + 2\lambda |x|^{\lambda-1} \frac{|x|}{x} \sum_{n=0}^{\infty} n a_n x^{n-1}$$
$$+ |x|^\lambda \sum_{n=0}^{\infty} n(n-1) a_n x^{n-2}.$$

Substituting into (47), rearranging, and eliminating the factor $|x|^\lambda$, we see that the coefficients a_n must satisfy

$$\sum_{n=0}^{\infty} n(n + 2\lambda) a_n x^n + \sum_{n=0}^{\infty} a_n x^{n+2} = 0.$$

Shifting the index on the second series by 2, we have

$$\sum_{n=0}^{\infty} n(n + 2\lambda) a_n x^n + \sum_{n=2}^{\infty} a_{n-2} x^n = 0.$$

We now equate the coefficients of x^n to zero. For $n = 0$, we see that $0 \cdot a_0 = 0$, so a_0 is arbitrary. We choose $a_0 = 1$. For $n = 1$, we must have $(1 + 2\lambda)a_1 = 0$, so $a_1 = 0$ since $\lambda \geq 0$. For $n \geq 2$,

$$a_n = \frac{-a_{n-2}}{n(n + 2\lambda)}. \tag{49}$$

Calculating a few coefficients, we find that

$$a_2 = \frac{-1}{2(2 + 2\lambda)}, \qquad a_3 = 0$$

$$a_4 = \frac{1}{4(4 + 2\lambda)2(2 + 2\lambda)}, \qquad a_5 = 0$$

$$a_6 = \frac{-1}{6(6 + 2\lambda)4(4 + 2\lambda)2(2 + 2\lambda)}, \qquad a_7 = 0.$$

So, in general,

$$a_{2n} = \frac{(-1)^n}{2^{2n} n! \, (\lambda + 1)(\lambda + 2) \cdots (\lambda + n)} \quad \text{and} \quad a_{2n+1} = 0 \ \text{ for } n = 1, 2, \ldots.$$

Therefore, one solution of Bessel's equation of order λ is

$$(50)\qquad y(x) = |x|^{\lambda}\left[1 + \sum_{n=1}^{\infty} \frac{(-1)^n x^{2n}}{2^{2n} n!\,(\lambda + 1)(\lambda + 2)\cdots(\lambda + n)}\right].$$

Since $p^*(x) = 1$ and $q^*(x) = x^2 - \lambda^2$ are analytic for all real x, this solution is valid at least for $0 < |x| < \infty$. That is, the solution is valid for all real x except possibly $x = 0$.

EXAMPLE Find the general solution of Bessel's equation of order $\frac{1}{3}$ which is valid on a deleted interval about the regular singular point $x = 0$.

In this case $r_1 = \frac{1}{3}$ and $r_2 = -\frac{1}{3}$. From equation (50), we see that the solution corresponding to the larger root $r_1 = \frac{1}{3}$ is

$$y_1(x) = |x|^{1/3}\left[1 + \sum_{n=1}^{\infty} \frac{(-1)^n x^{2n}}{2^{2n} n! \left(\frac{4}{3}\right)\left(\frac{7}{3}\right)\cdots\left(\frac{3n+1}{3}\right)}\right].$$

Since $r_1 - r_2 = \frac{2}{3}$ is not an integer, according to Theorem 8.2 there is a second linearly independent solution corresponding to the smaller root $r_2 = -\frac{1}{3}$ of the form

$$y_2(x) = |x|^{-1/3} \sum_{n=0}^{\infty} b_n x^n.$$

This is the same form as equation (48). In deriving an expression for the coefficients a_n of equation (48), we had assumed that λ was nonnegative. However, reviewing the derivation we see that the result—equation (50)—holds unless $\lambda = -\frac{1}{2}$, in which case a_1 could be chosen arbitrarily, and unless λ is a negative integer, in which case one could not write the general equation for the coefficients a_n in the form of equation (49) but must write it in the form

$$(51)\qquad n(n + 2\lambda)a_n = -a_{n-2}, \qquad n \geq 2,$$

since for some value of n the factor $n + 2\lambda$ would be zero. Since $r_2 = -\frac{1}{3}$ and not $-\frac{1}{2}$ or some negative integer, we obtain the following linearly independent solution of Bessel's equation of order $\frac{1}{3}$ by substituting $\lambda = -\frac{1}{3}$ into equation (50):

$$y_2(x) = |x|^{-1/3}\left[1 + \sum_{n=1}^{\infty} \frac{(-1)^n x^{2n}}{2^{2n} n! \left(\frac{2}{3}\right)\left(\frac{5}{3}\right)\cdots\left(\frac{3n-1}{3}\right)}\right].$$

The general solution for Bessel's equation of order $\frac{1}{3}$ is $y(x) = c_0 y_1(x) + c_1 y_2(x)$, where c_0 and c_1 are arbitrary constants. The general solution is valid for all real x except $x = 0$.

Euler's equation,

$$x^2y'' + axy' + by = 0,$$

where a and b are real constants, has a regular singular point at $x = 0$. This equation is of interest primarily because the behavior of the solution of the equation

$$x^2y'' + xp(x)y' + q(x)y = 0,$$

where $p(x)$ and $q(x)$ are analytic at $x = 0$, is similar to the behavior of the associated Euler equation,

$$x^2y'' + p(0)xy' + q(0)y = 0$$

in a "small" interval about $x = 0$. The indicial equation for Euler's equation is

$$r^2 + (a - 1)r + b = 0.$$

EXAMPLE Find two linearly independent real solution of the Euler equation

$$x^2y'' - 3xy' + 5 = 0 \tag{52}$$

which are valid on a deleted interval about $x = 0$.

The indicial equation $r^2 - 4r + 5 = 0$ has complex roots $r_1 = 2 + i$ and $r_2 = 2 - i$. At first glance it might appear that Theorem 8.2 does not apply when the indicial equation has complex roots.

Let us consider the general case of an indicial equation with complex roots $r_1 = \alpha + \beta i$ and $r_2 = \alpha - \beta i$ where $\beta > 0$ and then return to this specific example. Since $r_1 - r_2 = 2\beta i \neq 0$, the difference between the roots is not an integer and we have case 1 of Theorem 8.2. Consequently, there are two linearly independent solutions of the differential equation of the form

$$y_1^*(x) = |x|^{r_1} \sum_{n=0}^{\infty} a_n x^n$$

and

$$y_2^*(x) = |x|^{r_2} \sum_{n=0}^{\infty} b_n x^n,$$

assuming that the regular singular point is $x_0 = 0$. If $x_0 \neq 0$, the singularity may be translated to the origin by the change of variable $v = x - x_0$. The solutions y_1^* and y_2^* are complex valued, since r_1 and r_2 are complex. Normally, the series coefficients a_n and b_n will also be complex valued. As before, we can produce two linearly independent real solutions by finding the real and imaginary parts of y_1^*. Suppose that $a_n = \gamma_n + \delta_n i$ for $n = 0, 1, 2, \ldots$. Then

$$\sum_{n=0}^{\infty} a_n x^n = \sum_{n=0}^{\infty} \gamma_n x^n + i \sum_{n=0}^{\infty} \delta_n x^n = S_1(x) + iS_2(x).$$

The following calculations show how to write $|x|^{r_1}$ in the form $f_1(x) + if_2(x)$ where $f_1(x)$ and $f_2(x)$ are real-valued functions.

$$\begin{aligned} |x|^{r_1} &= e^{r_1 \ln |x|} = e^{(\alpha+\beta i)\ln|x|} = e^{\alpha \ln |x|} e^{i\beta \ln |x|} \\ &= |x|^{\alpha}[\cos(\beta \ln |x|) + i \sin(\beta \ln |x|)] \\ &= |x|^{\alpha} \cos(\beta \ln |x|) + i|x|^{\alpha} \sin(\beta \ln |x|). \end{aligned}$$

Since

$$\begin{aligned} y_1^*(x) &= [f_1(x) + if_2(x)][S_1(x) + iS_2(x)] \\ &= [f_1(x)S_1(x) - f_2(x)S_2(x)] + i[f_1(x)S_2(x) + f_2(x)S_1(x)]. \end{aligned}$$

The two linearly independent real solutions are

$$\begin{aligned} (53) \qquad y_1(x) &= \operatorname{Re}(y_1^*(x)) \\ &= |x|^{\alpha}\left[\cos(\beta \ln |x|) \sum_{n=0}^{\infty} \gamma_n x^n - \sin(\beta \ln |x|) \sum_{n=0}^{\infty} \delta_n x^n\right] \end{aligned}$$

and

$$\begin{aligned} (54) \qquad y_2(x) &= \operatorname{Im}(y_1^*(x)) \\ &= |x|^{\alpha}\left[\cos(\beta \ln |x|) \sum_{n=0}^{\infty} \delta_n x^n + \sin(\beta \ln |x|) \sum_{n=0}^{\infty} \gamma_n x^n\right]. \end{aligned}$$

Returning to the problem at hand, we seek a solution of equation (52) of the form

$$y(x) = |x|^r \sum_{n=0}^{\infty} a_n x^n.$$

Differentiating twice, substituting into equation (52), rearranging, and canceling the factor $|x|^r$, we see that the coefficients a_n must satisfy

$$\sum_{n=0}^{\infty} (n^2 - 4n + 2rn + r^2 - 4r + 5)a_n x^n = 0.$$

Equating the coefficients of x^n to zero and substituting $r = 2 + i$ (noting that since r satisfies the indicial equation, $r^2 - 4r + 5 = 0$), we find

$$n(n + 2i)a_n = 0 \quad \text{for } n \geq 0.$$

Consequently, a_0 is arbitrary and $a_n = 0$ for $n \geq 1$. We had anticipated that a_n would be complex valued. Fortunately, in this case all the coefficients are easily determined and real, if we choose $a_0 = 1$, for instance. Thus, $a_0 = 1 = \gamma_0 + \delta_0 i$ implies that $\gamma_0 = 1$ and $\delta_0 = 0$ and $a_n = 0 = \gamma_n + \delta_n i$ for $n \geq 1$ implies that $\gamma_n = \delta_n = 0$ for $n \geq 1$. Since $r = 2 + i = \alpha + \beta i$, $\alpha = 2$ and $\beta = 1$. Substituting the values for α, β, γ_n, and δ_n into equations (53) and (54), we obtain the following two real linearly independent solutions to the Euler equation (52):

$$y_1(x) = x^2 \cos(\ln|x|), \qquad x \neq 0$$

and

$$y_2(x) = x^2 \sin(\ln|x|), \qquad x \neq 0.$$

It can be shown that if Euler's equation has complex conjugate roots $r = \alpha + \beta i$, then two real linearly independent solutions are

$$y_1(x) = |x|^\alpha \cos(\beta \ln|x|), \qquad x \neq 0$$

and

$$y_2(x) = |x|^\alpha \sin(\beta \ln|x|), \qquad x \neq 0.$$

EXAMPLE Find two linearly independent solutions of Bessel's equation of order zero which are valid on a deleted interval about the regular singular point $x = 0$.

Setting $\lambda = 0$ in equation (47), we find Bessel's equation of order zero to be

$$x^2y'' + xy' + x^2y = 0. \tag{55}$$

In this case the roots of the indicial equation $r^2 - \lambda^2 = 0$ are $r_1 = r_2 = 0$. So we have case 2 of Theorem 8.2. Substituting $\lambda = 0$ into equation (50), we find one solution of Bessel's equation of order zero to be

$$y_1(x) = \sum_{n=0}^{\infty} \frac{(-1)^n x^{2n}}{2^{2n}(n!)^2}. \tag{56}$$

This series converges for all real x, including $x = 0$. So $y_1(x)$ is a solution of equation (55) for all real x. The function $y_1(x)$ is called *Bessel's function of the first kind of order zero* and often denoted by $J_0(x)$.

We now seek a second solution of equation (55) of thc form

$$y_2(x) = y_1(x) \ln|x| + \sum_{n=0}^{\infty} b_n x^n \qquad (x \neq 0) \tag{57}$$

as indicated by Theorem 8.2. Differentiating, we get

$$y_2'(x) = y_1'(x) \ln|x| + \frac{y_1(x)}{x} + \sum_{n=1}^{\infty} nb_n x^{n-1} \qquad (x \neq 0)$$

and

$$y_2''(x) = y_1''(x) \ln|x| + \frac{2y_1'(x)}{x} - \frac{y_1(x)}{x^2} + \sum_{n=2}^{\infty} n(n-1)b_n x^{n-2} \qquad (x \neq 0).$$

Substituting into equation (55) and collecting like terms, we obtain

$$\ln|x|(x^2y_1'' + xy_1' + x^2y_1) + 2xy_1'$$

$$+ \sum_{n=2}^{\infty} n(n-1)b_nx^n + \sum_{n=1}^{\infty} nb_nx^n + \sum_{n=0}^{\infty} b_nx^{n+2} = 0 \qquad (x \neq 0).$$

Since y_1 satisfies equation (55), the first term is zero. Calculating y_1' from equation (56), shifting the index on the last series by 2, and combining series, we obtain

$$\sum_{n=0}^{\infty} \frac{4n(-1)^n x^{2n}}{2^{2n}(n!)^2} + \sum_{n=2}^{\infty} (n^2b_n + b_{n-2})x^n + b_1x = 0.$$

Equating coefficients, we see that

$$b_1 = 0, \tag{58}$$

$$\frac{4n(-1)^n}{2^{2n}(n!)^2} + 4n^2b_{2n} + b_{2n-2} = 0 \quad \text{for } n = 1, 2, \ldots, \tag{59}$$

and

$$(2n+1)^2b_{2n+1} + b_{2n-1} = 0 \qquad \text{for } n = 1, 2, \ldots, \tag{60}$$

since the first series makes a contribution only to the coefficients of even powers of x. Equations (58) and (60) yield $b_1 = b_3 = b_5 = \cdots = b_{2n+1} = \cdots = 0$. So the coefficients of the odd powers of x of the assumed solution are all zero. Solving equation (59) for b_{2n}, we find

$$b_{2n} = \frac{(-1)^{n+1}}{n2^{2n}(n!)^2} - \frac{b_{2n-2}}{4n^2} \quad \text{for } n = 1, 2, \ldots. \tag{61}$$

The coefficient b_0 is arbitrary as usual. Choosing $b_0 = 0$ and using the recursion (61), we find

$$b_2 = \tfrac{1}{4},$$

$$b_4 = \frac{-1}{2\cdot 2^4(2!)^2} - \frac{\frac{1}{4}}{4\cdot 2^2} = \frac{-1}{2^4(2!)^2}\left[1 + \frac{1}{2}\right].$$

$$b_6 = \frac{1}{3\cdot 2^6(3!)^2} + \frac{1+\frac{1}{2}}{4\cdot 3^2\cdot 2^4(2!)^2} = \frac{1}{2^6(3!)^2}\left[1 + \frac{1}{2} + \frac{1}{3}\right],$$

and, in general,

$$b_{2n} = \frac{(-1)^{n+1}}{2^{2n}(n!)^2}\left[1 + \frac{1}{2} + \cdots + \frac{1}{n}\right].$$

Substituting the values of b_n which we have generated into equation (57), we obtain the following linearly independent solution of Bessel's equation of order zero:

$$y_2(x) = y_1(x) \ln |x| + \sum_{n=1}^{\infty} \frac{(-1)^{n+1}\left(1 + \frac{1}{2} + \cdots + \frac{1}{n}\right)x^{2n}}{2^{2n}(n!)^2} \qquad (x \neq 0).$$

The function $y_2(x)$ satisfies the differential equation (55) for all real x except $x = 0$.

EXAMPLE Find two linearly independent solutions of Bessel's equation of order $\frac{1}{2}$ which are valid on a deleted interval about $x = 0$.

Putting $\lambda = \frac{1}{2}$ in equation (47), we see that Bessel's equation of order $\frac{1}{2}$ is

$$x^2y'' + xy' + (x^2 - \tfrac{1}{4})y = 0. \tag{62}$$

The roots of the indicial equation $r^2 - \frac{1}{4} = 0$ are $r_1 = \frac{1}{2}$ and $r_2 = -\frac{1}{2}$. The roots themselves are not integers; however, the difference of the roots $r_1 - r_2 = 1$ is an integer, so we have case 3 of Theorem 8.2. Setting $\lambda = \frac{1}{2}$ in equation (50), we obtain the following solution of Bessel's equation of order $\frac{1}{2}$:

$$y_1(x) = |x|^{1/2}\left[1 + \sum_{n=1}^{\infty} \frac{(-1)^n x^{2n}}{2^{2n} n! \left(\frac{3}{2}\right)\left(\frac{5}{2}\right)\cdots\left(\frac{2n+1}{2}\right)}\right]$$

$$= |x|^{1/2}\left[1 + \sum_{n=1}^{\infty} \frac{(-1)^n x^{2n}}{(2n+1)!}\right] = |x|^{-1/2}\left[x + \sum_{n=1}^{\infty} \frac{(-1)^n x^{2n+1}}{(2n+1)!}\right]$$

$$= |x|^{-1/2} \sin x \qquad (x \neq 0).$$

The *Bessel function of the first kind of order* $\frac{1}{2}$ is defined by

$$J_{1/2}(x) = \left(\frac{2}{\pi}\right)^{1/2} y_1(x) = \left(\frac{2}{\pi|x|}\right)^{1/2} \sin x.$$

This function is a solution of equation (62) for all real x except $x = 0$. We seek a second solution of equation (62) of the form

$$y_2(x) = cy_1(x) \ln |x| + |x|^{-1/2} \sum_{n=0}^{\infty} b_n x^n \qquad (x \neq 0). \tag{63}$$

Differentiating twice, we find

$$y_2'(x) = cy_1'(x) \ln |x| + \frac{cy_1(x)}{x} - \frac{1}{2} |x|^{-3/2} \frac{|x|}{x} \sum_{n=0}^{\infty} b_n x^n + |x|^{-1/2} \sum_{n=1}^{\infty} nb_n x^{n-1}$$

and

$$y_2''(x) = cy_1''(x) \ln |x| + \frac{2cy_1'(x)}{x} - \frac{cy_1(x)}{x^2} + \frac{3}{4} |x|^{-5/2} \sum_{n=0}^{\infty} b_n x^n$$

$$- |x|^{-3/2} \frac{|x|}{x} \sum_{n=1}^{\infty} nb_n x^{n-1} + |x|^{-1/2} \sum_{n=2}^{\infty} n(n-1) b_n x^{n-2}.$$

Substituting into equation (62) and rearranging, we obtain

$$c \ln |x| [x^2 y_1'' + xy_1' + (x^2 - \tfrac{1}{4}) y_1] + 2cxy_1' - cy_1$$

$$+ |x|^{-1/2} \left[\sum_{n=2}^{\infty} n(n-1) b_n x^n + \sum_{n=0}^{\infty} b_n x^{n+2} \right] = 0.$$

The first term in this equation is zero, since y_1 satisfies equation (62). Calculating y_1' and substituting the series expansions for y_1 and y_1' into the equation above, we see that the constant c and the coefficients b_n must satisfy

$$4c|x|^{1/2} \sum_{n=1}^{\infty} \frac{n(-1)^n x^{2n}}{(2n+1)!} + |x|^{-1/2} \left(\sum_{n=2}^{\infty} n(n-1) b_n x^n + \sum_{n=0}^{\infty} b_n x^{n+2} \right) = 0.$$

Multiplying this equation by $|x|^{1/2}$ and shifting the index on the third series by 2, we obtain

$$4c|x| \sum_{n=1}^{\infty} \frac{n(-1)^n x^{2n}}{(2n+1)!} + \sum_{n=2}^{\infty} [n(n-1) b_n + b_{n-2}] x^n = 0.$$

Let us assume that $x > 0$. Then the first series begins with x^3 and includes only higher odd powers of x while the second series begins with x^2 and includes all higher integer powers of x. Equating the coefficients of x^2, x^3, x^4, x^5, and x^6 to zero, we get

$$\begin{aligned} 2b_2 + b_0 &= 0 \\ -\tfrac{2}{3}c + 6b_3 + b_1 &= 0 \\ 12b_4 + b_2 &= 0 \\ \tfrac{1}{15}c + 20b_5 + b_3 &= 0 \\ 30b_6 + b_4 &= 0. \end{aligned}$$

So b_0, b_1, and c are arbitrary; $b_2, b_4, b_6, \ldots$ are determined recursively from b_0 and c; and $b_3, b_5, b_7, \ldots$ are determined recursively from b_1. To make the computations simpler we choose $b_0 = b_1 = 1$ and $c = 0$. Then

$$b_2 = -\frac{1}{2}; \qquad b_3 = -\frac{1}{6}$$

$$b_4 = -\frac{b_2}{12} = \frac{1}{24}; \qquad b_5 = -\frac{b_3}{20} = \frac{1}{120}$$

and, in general,

$$b_{2n} = \frac{-b_{2n-2}}{2n(2n-1)}; \qquad b_{2n+1} = \frac{-b_{2n-1}}{2n(2n+1)} \quad \text{for } n = 1, 2, \ldots.$$

By induction we find

$$b_{2n} = \frac{(-1)^n}{(2n)!} \quad \text{and} \quad b_{2n+1} = \frac{(-1)^n}{(2n+1)!} \quad \text{for } n = 0, 1, \ldots.$$

Substituting these values into equation (63), we find a second linearly independent solution of Bessel's equation of order $\frac{1}{2}$ to be

$$\begin{aligned} y_2(x) &= |x|^{-1/2}\left[\sum_{n=0}^{\infty} \frac{(-1)^n x^{2n}}{(2n)!} + \sum_{n=0}^{\infty} \frac{(-1)^n x^{2n+1}}{(2n+1)!}\right] \\ &= |x|^{-1/2}(\cos x + \sin x) \qquad (x \neq 0). \end{aligned}$$

The function

$$y_3(x) = y_2(x) - y_1(x) = |x|^{-1/2} \cos x$$

is usually chosen as the second linearly solution of Bessel's equation of order 1/2. When multiplied by the normalizing factor $(2/\pi)^{1/2}$, $y_3(x)$ is denoted by $J_{-1/2}(x)$. Thus,

$$J_{-1/2}(x) = \left(\frac{2}{\pi|x|}\right)^{1/2} \cos x.$$

This function could be obtained directly from the recursion formulas by choosing $b_0 = (2/\pi)^{1/2}$, $b_1 = 0$, and $c = 0$. The reader should notice that choosing $b_0 = c = 0$ and leaving b_1 arbitrary yields $b_1 y_1(x)$ and does not produce a second linearly independent solution. This example illustrates that even though the roots of the indicial equation differ by an integer, it is sometimes possible (when c is arbitrary) to obtain two linearly independent solutions of the form $|x|^r \sum_{n=0}^{\infty} a_n x^n$. We shall encounter other examples of this nature in the exercises.

EXAMPLE Find two linearly independent solutions of Bessel's equation of order 1,

$$x^2y'' + xy' + (x^2 - 1)y = 0, \tag{64}$$

which are valid on a deleted interval about $x = 0$.

The roots of the indicial equation are $r_1 = 1$ and $r_2 = -1$. The solution obtained by substituting $\lambda = 1$ into equation (50) is

$$y_1(x) = |x|\left[1 + \sum_{n=1}^{\infty} \frac{(-1)^n x^{2n}}{2^{2n} n!\,(n+1)!}\right] \qquad (x \neq 0).$$

Bessel's function of the first kind of order 1,

$$J_1(x) = \frac{x}{2}\left[\sum_{n=0}^{\infty} \frac{(-1)^n x^{2n}}{2^{2n} n!\,(n+1)!}\right],$$

converges for all real x and satisfies equation (64) for all x. Since $r_1 - r_2 = 2$ is an integer, we again have case 3 of Theorem 8.2 and seek a second solution of equation (64) of the form

$$y_2(x) = cy_1(x)\ln|x| + |x|^{-1}\sum_{n=0}^{\infty} b_n x^n.$$

Differentiating twice, substituting into equation (64), and rearranging, we get

$$c\ln|x|[x^2y_1'' + xy_1' + (x^2 - 1)y_1] + 2cxy_1'$$
$$+ |x|^{-1}\left[\sum_{n=1}^{\infty} n(n-2)b_n x^n + \sum_{n=0}^{\infty} b_n x^{n+2}\right] = 0.$$

Since y_1 satisfies equation (64), the first term is zero. Calculating y_1', substituting into the equation above, multiplying the resulting equation by $|x|$, and shifting the index on the last series by 2, we see that the constant c and the coefficients b_n must satisfy

$$2c\sum_{n=0}^{\infty} \frac{(-1)^n(1+2n)x^{2n+2}}{2^{2n} n!\,(n+1)!} + \sum_{n=2}^{\infty} [n(n-2)b_n + b_{n-2}]x^n - b_1 x = 0.$$

Since the first series in this equation contains only even powers of x, we obtain the following formulas by equating the coefficients of powers of x to zero:

$$b_1 = 0 \tag{65}$$

$$2c + b_0 = 0 \tag{66}$$

$$(2n+1)(2n-1)b_{2n+1} + b_{2n-1} = 0, \qquad n = 1, 2, \ldots \tag{67}$$

$$\frac{2c(-1)^n(1+2n)}{2^{2n} n!\,(n+1)!} + (2n+2)(2n)b_{2n+2} + b_{2n} = 0, \qquad n = 1, 2, \ldots. \tag{68}$$

It follows from equations (65) and (67) that $b_1 = b_3 = b_5 = \cdots = b_{2n+1} = \cdots = 0$. From equations (66) and (68) we see that b_0 and b_2 are arbitrary, $c = -b_0/2$, and that $b_4, b_6, \ldots$ are determined recursively from b_2. Choosing $b_0 = 1$, requires $c = -\frac{1}{2}$. Substituting this value of c into equation (68) and solving for b_{2n+2}, we obtain the recursion

$$b_{2n+2} = \frac{(-1)^n(1 + 2n)}{2^{2n}n!\,(n + 1)!\,4n(n + 1)} - \frac{b_{2n}}{4n(n + 1)}, \qquad n = 1, 2, \ldots.$$

Choosing $b_2 = \frac{1}{4}$, we get

$$b_4 = -\frac{5}{2^6},$$

$$b_6 = \frac{5}{2^7 3^2},$$

and in general it can be shown by mathematical induction that

$$b_{2n} = \frac{(-1)^{2n+1}}{2^{2n}n!\,(n - 1)!}\left[2\left(1 + \frac{1}{2} + \cdots + \frac{1}{n}\right) - \frac{1}{n}\right], \qquad n = 1, 2, \ldots.$$

So a second linearly independent solution of Bessel's equation of order 1 is

$$y_2(x) = \frac{-y_1(x)\ln|x|}{2} + |x|^{-1}\left[1 + \sum_{n=1}^{\infty}\frac{(-1)^{n+1}\left[2\left(1 + \frac{1}{2} + \cdots + \frac{1}{n}\right) - \frac{1}{n}\right]x^{2n}}{2^{2n}n!\,(n - 1)!}\right] \qquad (x \neq 0).$$

The solution is valid for all real x except $x = 0$.

EXERCISES

1. Locate and classify the singular points of the following differential equations.

(a) $x^2(x - 2)^2y'' - 2x(x + 1)y' + (x^2 + 1)y = 0$
(b) $x^2(x - 2)y'' - 2(x + 1)y' + (x^2 + 1)y = 0$
(c) $xy'' + (\sin x)y' + 2xy = 0$
(d) $x^2y'' + (\sin x)y' + 2xy = 0$
(e) $x^3y'' + (\sin x)y' + 2xy = 0$
(f) $(x^2 + 4)^2y'' + x(x^2 + 4)y' - y = 0$

2. (a) Find two linearly independent solutions of the following Euler equations.

(i) $x^2y'' - \frac{1}{2}xy' + \frac{1}{2}y = 0$
(ii) $x^2y'' - 4xy' + 6y = 0$

(b) Show that if the indicial equation of Euler's equation has real, unequal roots r_1 and r_2, then two linearly independent solutions are

$$y_1(x) = |x|^{r_1} \quad\text{and}\quad y_2(x) = |x|^{r_2}.$$

3. (a) Find two linearly independent solutions of the following Euler equations.
   (i) $x^2y'' + xy' = 0$
   (ii) $x^2y'' + 5xy' + 4y = 0$
   (b) Show that if the indicial equation of Euler's equation has two equal roots r, then two linearly independent solutions are $y_1(x) = |x|^r$ and $y_2(x) = |x|^r \ln |x|$.

4. Find two linearly independent solutions of the following differential equations which are valid on a deleted interval about the regular singular point $x = 0$. For what values of x does each solution converge?
   (a) $x^2y'' + 3xy' + x(1 + x)y = 0$
   (b) $2x^2(x + 1)y'' + x(3x - 1)y' + y = 0$
   (c) $xy'' + (2x + 3)y' + 4y = 0$
   (d) $xy'' + y = 0$
   (e) $2xy'' + (3 - x)y' - y = 0$
   (f) $x^2y'' - xy' + (x^2 + 1)y = 0$
   (g) $x^2y'' + x^3y' - 2y = 0$
   (h) $xy'' + 2y' + y = 0$

5. (a) Show that $x_0 = 0$ is a regular singular point of Laguerre's equation

$$xy'' + (1 - x)y' + \lambda y = 0.$$

   (b) What are the roots of the indicial equation?
   (c) Find the Frobenius series solution.
   (d) When λ is a nonnegative integer, the solution of part (c) reduces to a polynomial. After being appropriately normalized these polynomials are known as *Laguerre polynomials*. Find polynomial solutions for $\lambda = 0, 1, 2, 3$.
   (e) What is the form of the second linearly independent solution of Laguerre's equation?

6. (a) Show that Legendre's equation

$$(1 - x^2)y'' - 2xy' + \lambda y = 0$$

   has a regular singular point at $x_0 = 1$ and at $x_0 = -1$.
   (b) What are the roots of the indicial equation associated with the point $x_0 = 1$?
   (c) Find a Frobenius series solution which is valid in a deleted interval about $x_0 = 1$.
   (d) For what values of x does the series converge?

7. Show that $x = 0$ is an irregular singular point of the differential equation

$$x^2y'' + (3x - 1)y' + y = 0$$

Suppose that this equation has a Frobenius series solution. Show that the "indicial equation"—the coefficient of the lowest power of x upon substituting the assumed series solution into the differential equation—is a linear equation whose solution is $r = 0$. Show that the formal Frobenius series corresponding to this root is

$$\sum_{n=0}^{\infty} n!\, x^n.$$

Notice that this series converges only for $x = 0$. Thus, the Frobenius series formally satisfies the given differential equation but is not a solution. This example shows that if x_0 is an irregular singular point of a differential equation and if the differential equation has a "solution" in the form of a Frobenius series expansion about x_0, the "solution" may not be valid on any interval about x_0.

8.4 NUMERICAL POWER SERIES SOLUTIONS

Most methods for producing numerical solutions to differential equation are derived in such a way that the results are equivalent to the results that would be obtained by using a Taylor series solution of a particular order. The difficulty with using the Taylor series itself to solve an initial value problem such as

$$(69)\qquad \begin{aligned} y^{(n)} &= f(x, y, y^{(1)}, \ \ldots, \ y^{(n-1)}); \\ y(x_0) &= c_0, \ \ldots, \ y^{(n-1)}(x_0) = c_{n-1} \end{aligned}$$

lies in calculating the successive derivatives of f. This usually becomes exceedingly complex for complicated functions as the order of differentiation required increases. Therefore, most numerical methods have been derived in such a way that they do not require the calculation of any derivatives of f. However, as we have seen, it is often possible to obtain recursion formulas for the power series (Taylor series) coefficients. These formulas can be used to generate numerical values for the coefficients. If we assume that the differential equation (69) has a solution of the form

$$(70)\qquad y(x) = \sum_{n=0}^{\infty} a_n(x - x_0)^n$$

and we have been able to obtain an expression for the coefficients a_n, then we also know that $y^{(n)}(x_0) = n!\, a_n$. A recursive formula for calculating the coefficients a_n is often more desirable from the standpoint of efficiency and sometimes even accuracy than a set of explicit formulas for calculating the derivatives $y^{(n)}(x_0)$—one formula for each derivative. Modern high-speed computers have made the fast and accurate evaluation of recursion formulas possible. Therefore, it is now feasible to use power series expansions to produce numerical solutions to complex initial value problems. As a matter of fact, this is the method commonly used to produce tables for functions such as $\sin x$, $\cos x$, $\tan x$, and $\ln x$. When $\sin x$ or some other such function is evaluated for a particular value of x by a computer, the computer performs this evaluation to the accuracy employed within the computer by calculating as many series coefficients as necessary and evaluating the series at x. Usually the series expansion used is not the power series expansion but an expansion in some other set of orthogonal functions. The reason for using a series expansion other than the power series expansion is because the other series converges more rapidly than the power series—that is, fewer terms are required in the series to achieve the same accuracy.

Of course, one literally cannot compute all the coefficients of an infinite series but must be satisfied with determining and using a finite number of coefficients, say $N + 1$. Let

$$y_N(x) = \sum_{n=0}^{N} a_n(x - x_0)^n$$

and

$$\epsilon_N(x) = \sum_{n=N+1}^{\infty} a_n(x - x_0)^n.$$

Thus, $y_N(x)$ is the approximation at x to the solution $y(x)$ which is obtained by using $N + 1$ terms of the power series expansion and $\epsilon_N(x) = y(x) - y_N(x)$ is the error associated with this approximation. There are at least three different ways in which to use the power series expansion to produce numerical solutions to a given initial value problem. The first method consists of simply choosing N, evaluating $y_N(x)$ at those points x where a value of the solution is desired, and estimating the error $\epsilon_N(x)$ at these points. Notice that we said estimating $\epsilon_N(x)$. Obviously, if we could calculate $\epsilon_N(x)$ exactly, we could calculate $y(x) = y_N(x) + \epsilon_N(x)$. The second method consists of specifying the point x farthest away from x_0—the point at which the initial conditions are given—for which a value of the solution is required and specifying the maximum error, E, desired on the interval I with endpoints x_0 and x. The problem in this case is to determine N so that $|\epsilon_N(x)| \le E$ for $x \in I$. The third method of employing the power series expansion to generate a numerical solution to an initial value problem consists of choosing N, specifying a maximum desirable error E, and determining an interval I about x_0 on which $|\epsilon_N(x)| \le E$ for $x \in I$. In this case the use of the series solution to produce solution values for the original initial value problem is limited to the interval I. If a value for the solution is desired at some point outside the interval I, then the series expansion is reinitialized. That is, a new point x_0 is chosen near one endpoint of the interval I, the power series, and as many of its derivatives as necessary are evaluated at x_0 to produce new initial conditions. Then $N + 1$ coefficients of the power series solution of the new initial value problem consisting of the given differential equation and the new initial conditions are generated. A new interval I about the new x_0 is determined in such a way that $|\epsilon_N(x)| \leq E$ for $x \in I$. The new series is evaluated at the desired points which lie within the new interval I to produce numerical values for the solution. We shall now illustrate how to use the power series technique to generate numerical solutions by each of the three methods just described.

EXAMPLE Use the power series expansion to produce numerical solutions to the initial value problem

$$y' = 1 + y^2; \qquad y(0) = 0 \tag{71}$$

at the points .05, .10, .15, . . ., 1.55 and compare the results with the actual solution.

The exact solution of the initial value problem (71) is $y(x) = \tan x$ for $x \in (-\pi/2, \pi/2)$. Earlier in this chapter—see equation (15)—we found that there was a power series solution of (71) of the form $y(x) = \sum_{n=0}^{\infty} a_n x^n$ and that the coefficients a_n satisfied

$$a_n = \begin{cases} 1 + a_0^2, & n = 1 \\ \dfrac{1}{n} \displaystyle\sum_{k=0}^{n-1} a_k a_{n-1-k}, & n \geq 2. \end{cases} \tag{72}$$

Actually the even coefficients a_{2m} ($m = 0, 1, \ldots$) are all zero, since $a_0 = 0$. However, since we want to write computer programs which will also be valid when $a_0 \neq 0$, we will program the recursion (72) the way it stands.

First, we shall choose $N = 19$ and let the computer recursively calculate and store the 20 coefficients $a_0, a_1, a_2, \ldots, a_{19}$. We shall store the coefficients a_n in the computer in locations named A(1), A(2), A(3), . . ., A(20). Then for $x - .05, .10, \ldots, 1.55$ we shall calculate the approximate solution

$$y_{19}(x) = \sum_{n=0}^{19} a_n x^n = \sum_{I=1}^{20} \mathrm{A(I)} x^{I-1}. \tag{73}$$

The error at x, $\epsilon_{19}(x)$, can be calculated from $\epsilon_{19}(x) = \tan x - y_{19}(x)$, since we know the exact solution for this example. Owing to the fact that we will not normally know the exact solution, we would like to produce a function that will provide an estimate of the error and compare this estimate with the true error. Taylor's Formula with a Remainder Theorem says that if $y^{(n)}(x)$ exists at every point in an interval I containing x_0 and if $x \in I$, then there exists a point ξ between x_0 and x such that

$$y(x) = y(x_0) + y'(x_0)(x - x_0) + \frac{y''(x_0)}{2}(x - x_0)^2 + \cdots \tag{74}$$
$$+ \frac{y^{(n-1)}(x_0)}{(n-1)!}(x - x_0)^{n-1} + \frac{y^{(n)}(\xi)}{n!}(x - x_0)^n.$$

For the problem under consideration $x_0 = 0$ and $y^{(n)}(0)/n! = a_n$. Assuming that we have calculated a_n for $n = 0, 1, \ldots, 19$, the error—the last term of equation (74)—is

$$\epsilon_{19}(x) = \frac{y^{(20)}(\xi)}{20!}x^{20},$$

where ξ is between 0 and x. Usually, there are two problems which one encounters in calculating $\epsilon_N(x)$: the first is the calculation of the twentieth derivative of y and the second is the determination of a value of ξ. All that one really knows is that there is at least one value of ξ between 0 and x which satisfies equation (74). Since $a_{19} = y^{(19)}(0)/19!$ is available and since we can hope that $y^{(20)}(\xi)x^{20}/20!$ is not too different from $y^{(19)}(0)x^{19}/19!$, we shall use the following function to estimate the error

$$e_{19}(x) = a_{19}x^{19} = A(20)x^{19}. \tag{75}$$

That is, we shall use the last term which we include in our approximate power series solution as our estimation of the error. This is a common method for estimating the error and the remarks above provide a rational for using this estimate.

The computer program for producing a 20-term numerical power series solution to the given initial value problem is displayed in Figure 8.1. The series coefficient a_1 is calculated by the sixth statement in the program. The remaining coefficients $a_2, a_3, \ldots, a_{19}$ are recursively calculated using equation (72). This calculation is accomplished in the computer program by statements 7–13. The student should verify that these statements do indeed produce the coefficients a_n of equation (72). Statements 14–26 perform the following functions:

1. Calculate the actual solution at $x = .05, .10, \ldots 1.55$. Within the computer program the actual solution is denoted by YA.
2. Calculate the error estimate at x using equation (75). The error estimate is denoted by ERR within the computer program.
3. Calculate the power series solution at x using equation (73). In the computer program the power series solution is denoted by SUM.
4. Calculate the difference between the actual solution and the power series solution at x. The difference between the solutions is denoted by DIF.
5. Print out the results for each x.

```
      C--TWENTY TERM POWER SERIES SOLUTION TO THE IVP Y'=1+Y*Y; Y(0) = 0 ON (0,1.55)
0001        DIMENSION A(20)
0002     22 FORMAT(1H1,12X,'X',9X,'TAN(X)',9X,'APPROX',7X,'DIFFERENCE',5X,'ERR
           1OR EST')
0003     40 FORMAT(11X,F4.2,4(4X,F11.7))
0004        WRITE(3,22)
0005        A(1) = 0.
0006        A(2) = 1. + A(1)*A(1)
      C---RECURSIVELY CALCULATE THE SERIES COEFFICIENTS
0007        DO 20 I = 3,20
0008        N = I - 1
0009        SUM = 0.
0010        DO 15 K = 1,N
0011        L = I - K
0012     15 SUM = SUM + A(K)*A(L)
0013     20 A(I) = SUM/N
0014        X = 0.
0015        DO 30 I = 1,31
0016        X = X + .05
      C---EVALUATE THE ACTUAL SOLUTION AT X
0017        YA = TAN(X)
      C---EVALUATE THE POWER SERIES EXPANSION AT X AND CALCULATE THE
      C---VALUE OF THE LAST TERM OF THE POWER SERIES AT X
0018        ERR = A(20)
0019        SUM = 0.
0020        DO 25 J = 1,19
0021        ERR = ERR*X
0022        L = 21 - J
0023     25 SUM = (SUM + A(L))*X
0024        SUM = SUM + A(1)
      C--FIND THE DIFFERENCE BETWEEN THE ACTUAL SOLUTION AND THE POWER SERIES SOLUTION
0025        DIF = YA - SUM
0026     30 WRITE(3,40) X, YA, SUM, DIF, ERR
0027        CALL EXIT
0028        END
```

Figure 8.1 Computer program for the first method of using a power series expansion to produce numerical solutions to the IVP: $y' = 1 + y^2$; $y(0) = 0$.

Notice that in the computer program the error estimate—equation (75)—is calculated as

$$e_{19}(x) = (\cdots(((a_{19})x)x)\cdots x) \tag{76}$$

and the power series solution at x—equation (73)—is calculated as

$$y_{19}(x) = (\cdots(((a_{19}x) + a_{18})x + a_{17})x + \cdots + a_1)x + a_0. \tag{77}$$

The results of the calculations are shown in Figure 8.2. The power series solution is accurate to six decimal places for $.05 \le x \le .75$ and the error estimate is nearly the same order of magnitude as the difference between the actual solution and the power series solution for $.05 \le x \le 1.30$. Thus, in this case the error estimate produced by the last term used in the series is "near" the true error as long as x is not "too far" from x_0—the point about which the power series expansion is made.

Figure 8.3 is a graph of the actual solution to the given initial value problem on the interval [0, 1.55] and the 20-, 50-, 100-, and 150-term power series solutions obtained by using the first method. The error estimate at $x = 1.55$ when using the 150-term power series expansion, $e_{149}(1.55)$ is .175. So when we use the second method and specify the error we wish to achieve, we should not specify an error of less than .175 unless we intend to calculate and store more than 150 series coefficients.

The second method of using the power series expansion to produce numerical solutions to initial value problems consists of specifying the interval, I, on which the solution is to be calculated, specifying the maximum error, E,

X	TAN(X)	APPROX	DIFFERENCE	ERROR EST
0.05	0.0500417	0.0500417	0.0000000	0.0000000
0.10	0.1003346	0.1003346	0.0	0.0000000
0.15	0.1511351	0.1511350	0.0000001	0.0000000
0.20	0.2027100	0.2027098	0.0000002	0.0000000
0.25	0.2553417	0.2553415	0.0000002	0.0000000
0.30	0.3093361	0.3093358	0.0000003	0.0000000
0.35	0.3650283	0.3650281	0.0000002	0.0000000
0.40	0.4227928	0.4227924	0.0000004	0.0000000
0.45	0.4830545	0.4830543	0.0000002	0.0000000
0.50	0.5463018	0.5463018	0.0	0.0000000
0.55	0.6131045	0.6131040	0.0000005	0.0000000
0.60	0.6841359	0.6841358	0.0000001	0.0000000
0.65	0.7602034	0.7602031	0.0000004	0.0000001
0.70	0.8422873	0.8422869	0.0000004	0.0000003
0.75	0.9315951	0.9315947	0.0000004	0.0000010
0.80	1.0296364	1.0296354	0.0000010	0.0000034
0.85	1.1383305	1.1383257	0.0000048	0.0000109
0.90	1.2601557	1.2601395	0.0000162	0.0000323
0.95	1.3983793	1.3983269	0.0000525	0.0000902
1.00	1.5574045	1.5572405	0.0001640	0.0002391
1.05	1.7433081	1.7428188	0.0004892	0.0006042
1.10	1.9647474	1.9633389	0.0014086	0.0014624
1.15	2.2344770	2.2305450	0.0039320	0.0034030
1.20	2.5721207	2.5614119	0.0107088	0.0076392
1.25	3.0095215	2.9808702	0.0286512	0.0165916
1.30	3.6020241	3.5260305	0.0759935	0.0349558
1.35	4.4550886	4.2527351	0.2023535	0.0716032
1.40	5.7976360	5.2456684	0.5519676	0.1428961
1.45	8.2375507	6.6338921	1.6036587	0.2783408
1.50	14.0997047	8.6145945	5.4851103	0.5300491
1.55	48.0567932	11.4891815	36.5676117	0.9882853

Figure 8.2 Printout of the computer program of Figure 8.1.

desired on this interval, and then determining the number of terms, $N + 1$, that one must calculate in order for the error, $\epsilon_N(x)$, to satisfy $|\epsilon_N(x)| \leq E$ for all $x \in I$. For the given initial value problem $I = [0, 1.55]$ and we arbitrarily choose $E = .5$. Since we do not normally know $\epsilon_N(x)$, we shall again approximatc it by

$$e_N(x) = a_N x^N.$$

Since $|e_N(x)|$ is an increasing function on the interval $[0, 1.55]$, we want to choose N so that

$$|e_N(1.55)| = |a_N|(1.55)^N < .5 = E, \tag{78}$$

if possible. The initial condition $y(0) = 0 = a_0$ together with the recursive equation (72) imply $a_0 = a_2 = \cdots = a_{2m} = \cdots 0$, so equation (78) is satisfied for all even N. We, therefore, want to determine, if possible, the smallest odd integer N for which equation (78) is satisfied and then use the $(N + 1)$-term power series expansion to generate our numerical solution. In order to determine N, we calculate a_1 and see if equation (78) is satisfied. If not, we calculate a_3 and see if equation (78) is satisfied. We continue this process until we find the smallest odd integer N for which a_N satisfies equation (78) or until N becomes as large as some prescribed maximum. We shall require that $N \leq 99$.

A flowchart for a computer program to solve the given initial value problem by the second power series method on the interval $[0, 1.55]$ with $E = .5$ and N the smallest odd integer less than or equal to 99 which satisfies equation (78) is shown in Figure 8.4. The number of the corresponding program statement or statements which accomplishes each of the operations indicated by the flowchart is also shown in Figure 8.4. The computer program

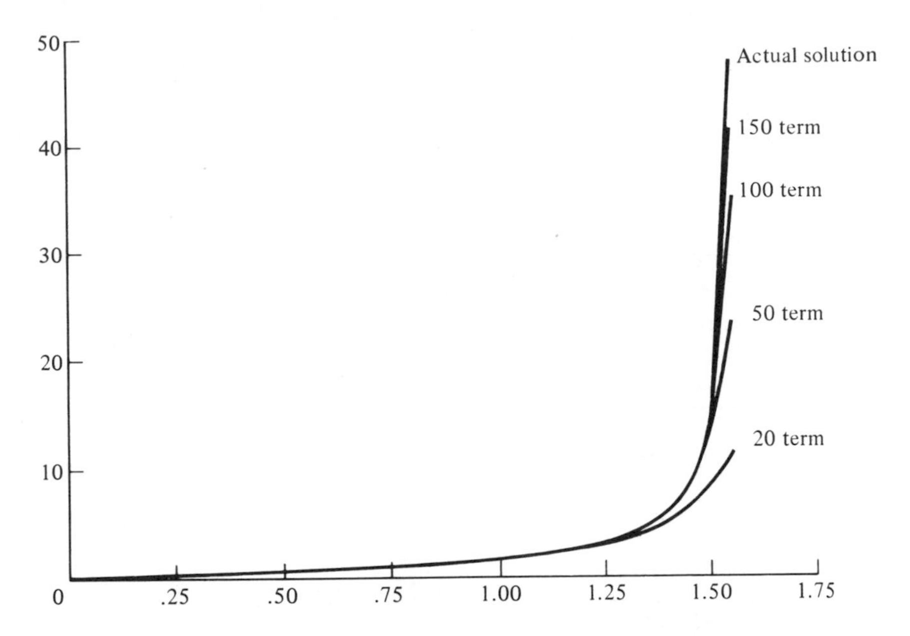

Figure 8.3 Power series solutions to the IVP $y' = 1 + y^2$; $y(0) = 0$.

itself is displayed in Figure 8.5. This program is similar to the program for the first method of solution—see Figure 8.1—except in this program the number of terms to be used in the series solution is not fixed at 20 but is variable and yet must not exceed 100. Notice that program statements 16–20 have been added in order to determine the number of series coefficients N that must be to calculate in order to satisfy equation (78).

The printout of this computer program is shown in Figure 8.6. Observe that $|e_{71}(x)|$, which is shown in the column labeled ERROR EST, is less than .5 for all $x \in [0, 1.55]$, but the true error $\epsilon_{71}(x)$, which is shown in the column labeled DIFFERENCE, is as large as 18.28. Thus, to obtain a true error whose absolute value is less than .5 for $x \in [0, 1.55]$, the maximum error desired E would have to be chosen much smaller than .5 and as a consequence additional terms would have to be calculated and used in the series expansion. As a general rule it is not usually wise or efficient to use the power series expansion $y_N(x)$ to approximate the true solution $y(x)$ when the distance from x to x_0—the point about which the expansion is made—is much greater than 1. This is because for $|x - x_0| > 1$ all the factors $(x - x_0)^n$ are greater than 1 in absolute value and therefore tend to hinder the convergence of

$$y_N(x) = \sum_{n=0}^{N} a_n(x - x_0)^n$$

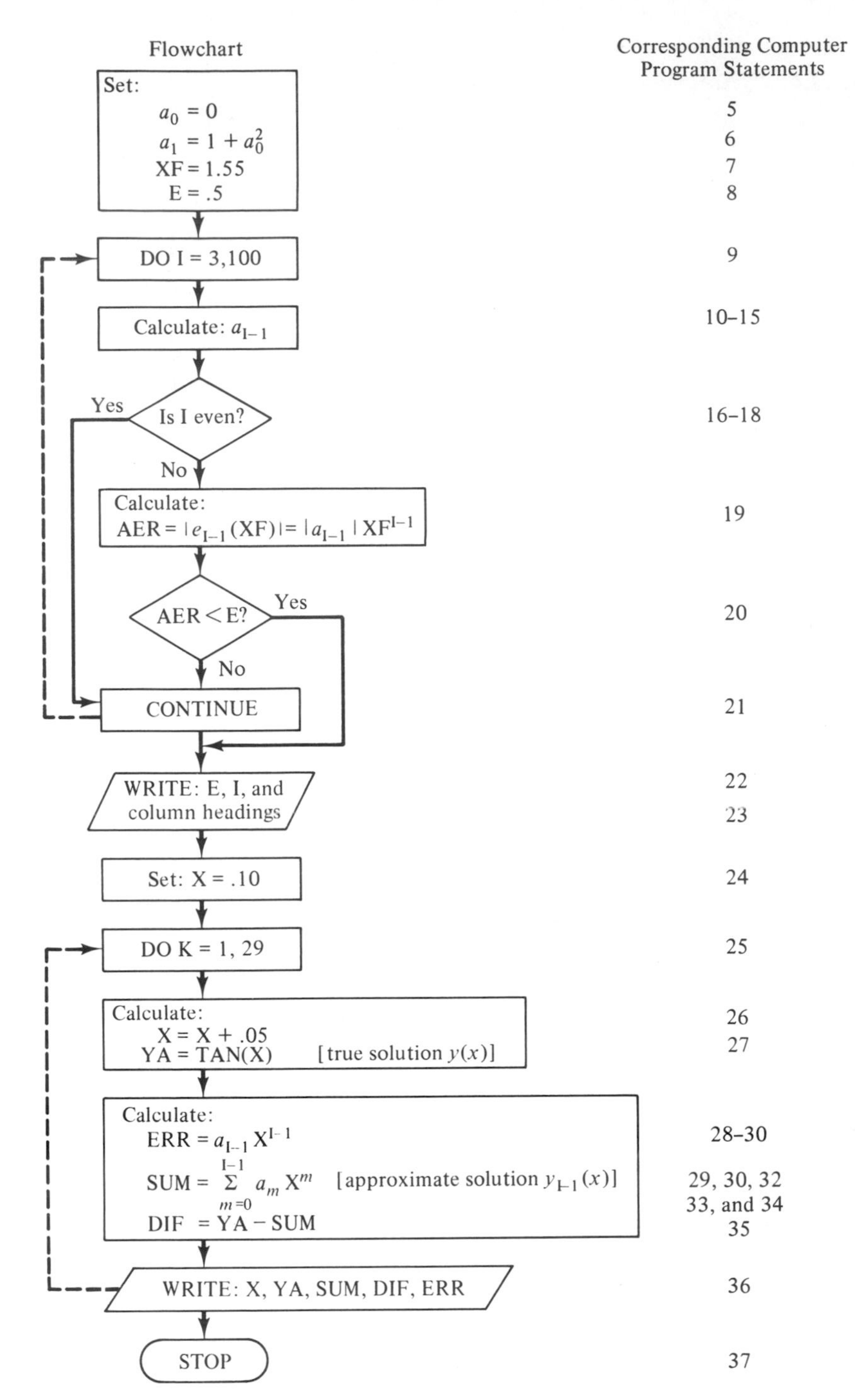

Figure 8.4 Flowchart of the second method for using power series expansions to produce numerical solutions to the IVP: $y' = 1 + y^2$; $y(0) = 0$.

```
      C--POWER SERIES SOLUTION TO THE IVP Y'=1+Y*Y; Y(0) = 0 ON (0,1.55)
0001        DIMENSION A(100)
0002     10 FORMAT(1H1,5X,'THE NUMBER OF TERMS REQUIRED TO ACHIEVE AN ACCURACY
           1 OF ',E9.2, ' ON THE INTERVAL (0,1.55) IS ', I3)
0003     22 FORMAT(1H0,12X,'X',9X,'TAN(X)',9X,'APPROX',7X,'DIFFERENCE',5X,'ERR
           1OR EST')
0004     40 FORMAT(11X,F4.2,4(4X,F11.7))
0005        A(1) = 0.
0006        A(2) = 1. + A(1)*A(1)
0007        XF = 1.55
0008        E = .5
      C---RECURSIVELY CALCULATE THE SERIES COEFFICIENTS
0009        DO 20 I = 3,100
0010        N = I - 1
0011        SUM = 0.
0012        DO 15 K = 1,N
0013        L = I - K
0014     15 SUM = SUM + A(K)*A(L)
0015        A(I) = SUM/N
0016        M = I/2
0017        J = M*2
0018        IF(I.NE.J) GO TO 20
0019        AER = ABS(A(I))*XF**N
0020        IF(AER.LT.E) GO TO 23
0021     20 CONTINUE
0022     23 WRITE(3,10) E, I
0023        WRITE(3,22)
0024        X = .10
0025        DO 30 K = 1,29
0026        X = X + .05
      C---EVALUATE THE ACTUAL SOLUTION AT X
0027        YA = TAN(X)
      C---EVALUATE THE POWER SERIES EXPANSION AT X AND CALCULATE THE
      C---VALUE OF THE LAST TERM OF THE POWER SERIES AT X
0028        ERR = A(I)
0029        SUM = 0.
0030        DO 25 J = 1,N
0031        ERR = ERR*X
0032        L = I + 1 - J
0033     25 SUM = (SUM + A(L))*X
0034        SUM = SUM + A(1)
      C--FIND THE DIFFERENCE BETWEEN THE ACTUAL SOLUTION AND THE POWER SERIES SOLUTION
0035        DIF = YA - SUM
0036     30 WRITE(3,40) X, YA, SUM, DIF, ERR
0037        CALL EXIT
0038        END
```

Figure 8.5 Computer program for the second method of using a power series expansion to produce numerical solutions to the IVP: $y' = 1 + y^2$; $y(0) = 0$.

as N increases instead of enhancing the convergence as it does when $|x - x_0| < 1$.

The third method of employing the power series expansion to produce a numerical solution to an initial value problem consists of specifying the number of terms, $N + 1$, to be used in the power series expansion, specifying the maximum desirable error E, and determining an interval I about x_0—the point about which the power series expansion made—so that for $x \in I$, $|\epsilon_N(x)| \leq E$. A value for the solution is calculated for each point at which the solution is desired and which lies within the interval I. If a value for the solution is desired at one or more points which lie outside the interval I, then the problem is reinitialized. That is, a new point x_0 is chosen near an endpoint of the interval I, new initial conditions are produced, the power series solution to the new initial value problem consisting of the given differential equation and the new initial conditions is generated, a new interval I is found for the new point x_0, and a value of the solution is calculated at each point at which a solution is desired and which lies within the new interval I. This process is continued until a value of the solution is determined for all the points at which the solution is desired or until some other specified condition causes the process to terminate.

For the given initial value problem we shall choose $N + 1 = 20$ and $E = 5 \times 10^{-7}$. Suppose that we have obtained the 20-term power series

THE NUMBER OF TERMS REQUIRED TO ACHIEVE AN ACCURACY OF 0.50E 00 ON THE INTERVAL (0,1.55) IS 72

X	TAN(X)	APPROX	DIFFERENCE	ERROR EST
0.15	0.1511351	0.1511350	0.0000001	0.0000000
0.20	0.2027100	0.2027098	0.0000002	0.0000000
0.25	0.2553417	0.2553415	0.0000002	0.0000000
0.30	0.3093361	0.3093358	0.0000003	0.0000000
0.35	0.3650283	0.3650281	0.0000002	0.0000000
0.40	0.4227928	0.4227924	0.0000004	0.0000000
0.45	0.4830545	0.4830543	0.0000002	0.0000000
0.50	0.5463018	0.5463018	0.0	0.0000000
0.55	0.6131045	0.6131040	0.0000005	0.0000000
0.60	0.6841359	0.6841358	0.0000001	0.0000000
0.65	0.7602034	0.7602031	0.0000004	0.0000000
0.70	0.8422873	0.8422869	0.0000004	0.0000000
0.75	0.9315951	0.9315947	0.0000004	0.0000000
0.80	1.0296364	1.0296364	0.0	0.0000000
0.85	1.1383305	1.1383295	0.0000010	0.0000000
0.90	1.2601557	1.2601547	0.0000010	0.0000000
0.95	1.3983793	1.3983793	0.0	0.0000000
1.00	1.5574045	1.5574036	0.0000010	0.0000000
1.05	1.7433081	1.7433071	0.0000010	0.0000000
1.10	1.9647474	1.9647465	0.0000010	0.0000000
1.15	2.2344770	2.2344761	0.0000010	0.0000000
1.20	2.5721207	2.5721178	0.0000029	0.0000000
1.25	3.0095215	3.0095167	0.0000048	0.0000001
1.30	3.6020241	3.6020155	0.0000086	0.0000019
1.35	4.4550886	4.4550056	0.0000830	0.0000272
1.40	5.7976360	5.7962351	0.0014009	0.0003592
1.45	8.2375507	8.2125196	0.0250311	0.0043386
1.50	14.0997047	13.6012173	0.4984875	0.0481603
1.55	48.0567932	29.7764587	18.2803345	0.4940262

Figure 8.6 Printout of the computer program of Figure 8.5.

expansion of the given differential equation about the point x_0. We shall use this expansion to calculate the value of the solution at any point x which satisfies

$$|e_{19}(x)| = |a_{19}(x - x_0)^{19}| = |\mathrm{A}(20)(x - x_0)^{19}| \leq E.$$

Notice that we have again approximated the true error $\epsilon_{19}(x)$ by $e_{19}(x)$. Solving the above inequality for $|x - x_0|$, we find

$$|x - x_0| \leq (E/|\mathrm{A}(20)|)^{1/19}.$$

If we let

$$\Delta x = \max |x - x_0| = \left(\frac{E}{|\mathrm{A}(20)|}\right)^{1/19} \tag{79}$$

and define $I = [x_0 - \Delta x, x_0 + \Delta x]$, then we will have $|e_{19}(x)| \leq E$ for $x \in I$. If we order the points $x_1, x_2, \ldots, x_m$ at which we desire a solution in ascending order—that is, $x_1 < x_2 < \cdots < x_m$—and if $x_0 < x_1$, then when necessary we can always reinitialize at $x_0 + \Delta x$. For this example $x_0 = 0$, $x_1 = .05$, $x_2 = .10, \ldots, x_{31} = 1.55$. In the computer program to solve the given initial value problem using the third power series method we shall again store the power series coefficients $a_0, a_1, \ldots, a_{19}$ in locations named A(1), A(2), ..., A(20), respectively, and we shall store $x_1 = .05$, $x_2 = .10, \ldots, x_{31} = 1.55$ in locations named XI(1), XI(2), ..., XI(31), respectively. A flowchart of the computer program to solve the given initial value problem using the method just described is shown in Figure 8.7. The number of the corresponding program statement or statements which accomplishes each of the operations indicated by the flowchart is also displayed in Figure 8.7. The computer program is shown in Figure 8.8.

The printout of the computer program is shown in Figure 8.9. Notice that the problem was reinitialized at $x_0^{(1)} = .72271$, $x_0^{(2)} = 1.11449, \ldots,$

$x_0^{(7)} = 1.54130$. Thus, the computer program used a power series expansion of 20 terms to successively calculate solutions to the following initial value problems:

(80a) $\quad y' = 1 + y^2; \quad y(0) = 0;$

(80b) $\quad y' = 1 + y^2; \quad y(.722\ 71) = .881\ 877\ 2;$

(80c) $\quad y' = 1 + y^2; \quad y(1.114\ 49) = 2.037\ 251\ 5;$

$\vdots \qquad \vdots$

(80h) $\quad y' = 1 + y^2; \quad y(1.541\ 30) = 33.889\ 236\ 5.$

The 20-term power series expansion for (80a) was used to produce values for the solution to the given initial value problem at $x_1 = .05$, $x_2 = .10, \ldots$, $x_{14} = .70$; the 20-term power series expansion for (80b) was used to produce values for the solution to the given initial value problem at $x_{15} = .75$, $x_{16} = .80, \ldots, x_{22} = 1.10; \ldots$; and the 20-term power series expansion for (80h) was used to produce a value for the solution to the given initial value problem at $x_{31} = 1.55$. Observe that the magnitude of the largest true error is $.0046 > E = 5 \times 10^{-7}$—see the column labeled DIFFERENCE of Figure 8.9. However, the magnitude of the largest error estimate is $.273 \times 10^{-6} < E = 5 \times 10^{-7}$—see the column labeled ERROR EST of Figure 8.9.

In general, the error depends upon the differential equation, the initial conditions, the number of terms used in the power series expansion, the value chosen for the maximum desirable error E, and the computer that is used. The error can most readily be controlled by the choice of the number of terms used in the expansion and the value chosen for the maximum desirable error. The third method requires more computer programming, more computer storage, and a longer running time than either the first or the second method. However, this is a small price to pay for the ability to be able to prescribe *both* the desired number of terms and the desired accuracy.

EXAMPLE Use the third method of employing a power series expansion to produce a numerical solution to the initial value problem

(81) $$y'' + xy = 0; \qquad y(0) = 2, \quad y'(0) = 1$$

at the points 0., .5, 1.0, . . ., 10 and estimate the error.

The differential equation of equation (81) is Airy's equation. If we assume that equation (81) has a power series solution of the form $y = \sum_{n=0}^{\infty} a_n x^n$, then from the given initial conditions we know that $y(0) = a_0 = 2$ and $y'(0) = a_1 = 1$. We previously developed the recursion formula for Airy's equation—see equation (34), Section 8.3.1—and found that $a_2 = 0$ and that the remaining coefficients satisfy

(82) $$a_{n+2} = \frac{-a_{n-1}}{(n+2)(n+1)} \quad \text{for } n \geq 1.$$

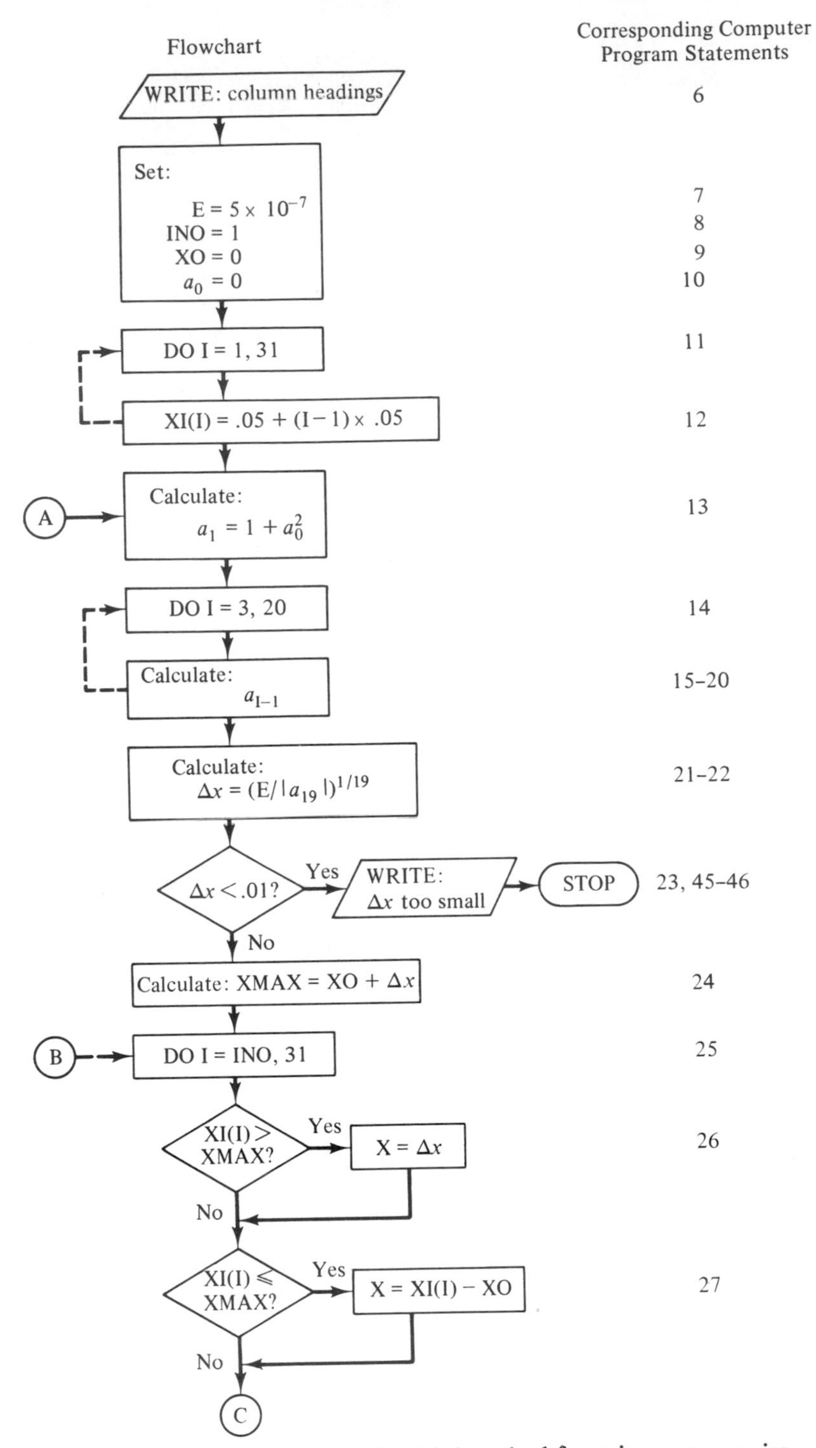

Figure 8.7 Flowchart of the third method for using power series expansions to produce numerical solutions to the IVP: $y' = 1 + y^2$; $y(0) = 0$.

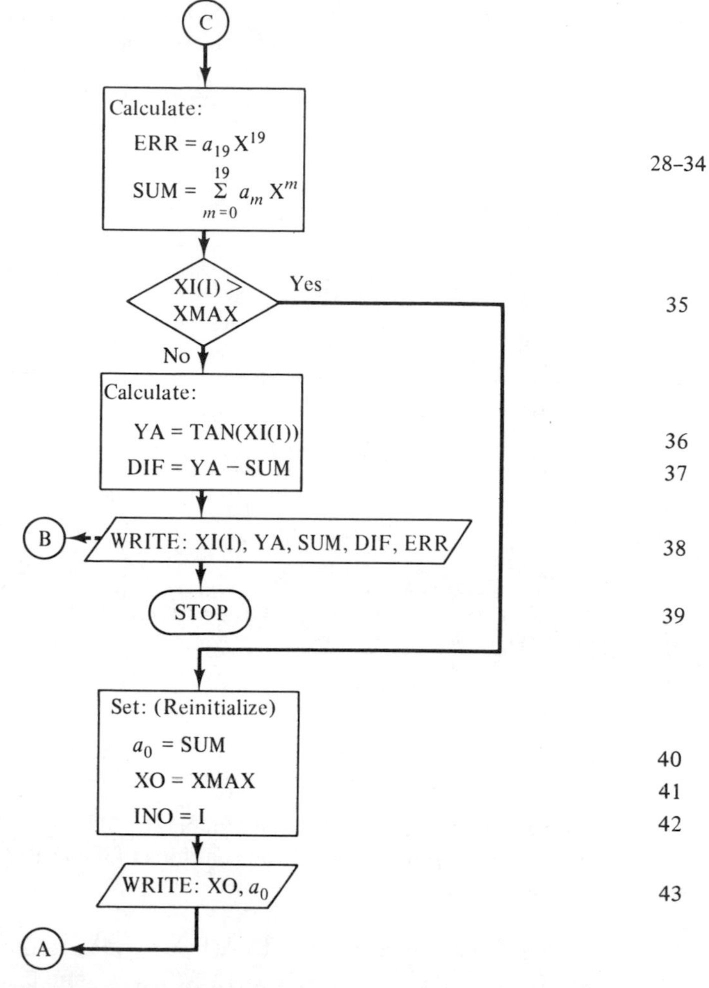

Figure 8.7 (continued)

Let us choose the number of terms of the power series solution to be 20. In the computer program we will store a_0 in A(1), a_1 in A(2), . . ., and a_{19} in A(20). Let us also choose the maximum desired error E to be 5×10^{-7} as before. If we shift the index of the coefficients in equation (82) by -3, we see that the coefficients must satisfy the recursion

$$a_{n-1} = \frac{-a_{n-4}}{(n-1)(n-2)} \quad \text{for } n \geq 4,$$

or

$$\text{A(J)} = \frac{-\text{A(J}-3)}{(\text{J}-1)(\text{J}-2)} \quad \text{for J} \geq 4, \tag{83}$$

```
              C--TWENTY TERM POWER SERIES SOLUTION TO THE IVP Y'=1+Y*Y; Y(0) = 0 ON (0,1.55)
0001                DIMENSION A(20), XI(31)
0002             22 FORMAT(1H1,12X,'X',9X,'TAN(X)',9X,'APPROX',7X,'DIFFERENCE',5X,'ERR
                  1OR EST')
0003             40 FORMAT(11X,F4.2,3(4X,F11.7),4X,E11.4)
0004             50 FORMAT('OREINITIALIZATION AT X = ',F7.5,'    Y(X) = ',F11.7)
0005             60 FORMAT('ORADIUS OF CONVERGENCE TOO SMALL TO CONTINUE, VALUE = ',
                  1 F7.5)
0006                WRITE(3,22)
              C--INITIALIZE VALUES
0007                E = 5.*10.**(-7.)
0008                INO = 1
0009                XO = 0.
0010                A(1) = 0.
0011                DO 5 I = 1,31
0012              5 XI(I) = .05 + (I-1)*.05
              C--RECURSIVELY CALCULATE THE SERIES COEFFICIENTS
0013             10 A(2) = 1. + A(1)*A(1)
0014                DO 20 I = 3,20
0015                N = I - 1
0016                SUM = 0.
0017                DO 15 K = 1,N
0018                L = I - K
0019             15 SUM = SUM + A(K)*A(L)
0020             20 A(I) = SUM/N
              C--CALCULATE THE APPROXIMATE 'NUMERICAL' RADIUS OF CONVERGENCE
0021                POWER = 1./19.
0022                DELX = (E/ABS(A(20)))**POWER
0023                IF(DELX.LT..01)  GO TO 55
0024                XMAX = XO + DELX
0025                DO 30 I = INO,31
              C--DETERMINE IF XI(I) IS WITHIN THIS INTERVAL OF CONVERGENCE
0026                IF(XI(I).GT.XMAX)  X = DELX
0027                IF(XI(I).LE.XMAX)  X = XI(I) - XO
              C--EVALUATE THE POWER SERIES EXPANSION AT XI OR REINITIALIZE AT XMAX
              C--CALCULATE THE VALUE OF THE LAST TERM OF THE POWER SERIES AT XI
0028                SUM = 0.
0029                ERR = A(20)
0030                DO 25 J = 1,19
0031                ERR = ERR*X
0032                L = 21 - J
0033             25 SUM = (SUM + A(L))*X
0034                SUM = SUM + A(1)
0035                IF(XI(I).GT.XMAX) GO TO 45
              C--EVALUATE THE ACTUAL SOLUTION AT XI
0036                YA = TAN(XI(I))
              C--FIND THE DIFFERENCE BETWEEN THE ACTUAL SOLUTION AND THE POWER SERIES SOLUTION
0037                DIF = YA - SUM
0038             30 WRITE(3,40) XI(I), YA, SUM, DIF, ERR
0039                CALL EXIT
              C--REINITIALIZE
0040             45 A(1) = SUM
0041                XO = XMAX
0042                INO = I
0043                WRITE(3,50) XO, A(1)
0044                GO TO 10
0045             55 WRITE(3,60) DELX
0046                CALL EXIT
0047                END
```

Figure 8.8 Computer program for the third method of using a power series expansion to produce numerical solutions to the IVP: $y' = 1 + y^2$; $y(0) = 0$.

where J corresponds to n and a_{n-i} corresponds to A(J $-$ i + 1). In the computer program we set X0 = .0, A(1) = 2., A(2) = 1., A(3) = 0. and calculate A(4), A(5), ..., A(20) recursively using equation (83). The interval over which we will use the expansion

$$y_N(\mathrm{X}) = \sum_{J=1}^{20} \mathrm{A(J)(X - X0)}^{J-1} \tag{84}$$

to approximate the solution of the initial value problem (81) will be I = [X0 $- \Delta x$, X0 $+ \Delta x$], where Δx is determined by equation (79)

$$\Delta x = \left(\frac{E}{|\mathrm{A}(20)|}\right)^{1/19}.$$

In this case when we reinitialize at X0 + Δx we will need the two initial

```
        X           TAN(X)          APPROX        DIFFERENCE       ERROR EST
      0.05        0.0500417       0.0500417       0.0000000       0.4561E-28
      0.10        0.1003346       0.1003346       0.0             0.2391E-22
      0.15        0.1511351       0.1511350       0.0000001       0.5301E-19
      0.20        0.2027100       0.2027099       0.0000002       0.1254E-16
      0.25        0.2553418       0.2553416       0.0000002       0.8699E-15
      0.30        0.3093362       0.3093359       0.0000003       0.2779E-13
      0.35        0.3650284       0.3650283       0.0000002       0.5199E-12
      0.40        0.4227930       0.4227931      -0.0000001       0.6573E-11
      0.45        0.4830549       0.4830546       0.0000003       0.6161E-10
      0.50        0.5463021       0.5463021       0.0             0.4561E-09
      0.55        0.6131049       0.6131049      -0.0000001       0.2789E-08
      0.60        0.6841366       0.6841363       0.0000003       0.1457E-07
      0.65        0.7602041       0.7602036       0.0000005       0.6668E-07
      0.70        0.8422881       0.8422874       0.0000007       0.2726E-06

REINITIALIZATION AT X = 0.72271   Y(X) =    0.8818772
      0.75        0.9315962       0.9315954       0.0000007       0.5183E-28
      0.80        1.0296383       1.0296373       0.0000010       0.2020E-19
      0.85        1.1383324       1.1383314       0.0000010       0.2643E-15
      0.90        1.2601576       1.2601557       0.0000019       0.1433E-12
      0.95        1.3983812       1.3983803       0.0000010       0.1608E-10
      1.00        1.5574064       1.5574055       0.0000010       0.7029E-09
      1.05        1.7433119       1.7433100       0.0000019       0.1640E-07
      1.10        1.9647522       1.9647484       0.0000038       0.2443E-06

REINITIALIZATION AT X = 1.11449   Y(X) =    2.0372515
      1.15        2.2344885       2.2344885       0.0             0.1868E-20
      1.20        2.5721426       2.5721416       0.0000010       0.3335E-13
      1.25        3.0095596       3.0095587       0.0000010       0.2100E-09
      1.30        3.6020775       3.6020765       0.0000010       0.8199E-07

REINITIALIZATION AT X = 1.31852   Y(X) =    3.8794193
      1.35        4.4551878       4.4551888      -0.0000010       0.2662E-16
      1.40        5.7978334       5.7978354      -0.0000019       0.1873E-08

REINITIALIZATION AT X = 1.42785   Y(X) =    6.9481583
      1.45        8.2380161       8.2380352      -0.0000191       0.2863E-14

REINITIALIZATION AT X = 1.48798   Y(X) =   12.0473766
      1.50       14.1012211      14.1013613      -0.0001402       0.1430E-14

REINITIALIZATION AT X = 1.52183   Y(X) =   20.4054565

REINITIALIZATION AT X = 1.54130   Y(X) =   33.8892365
      1.55       48.0744629      48.0789948      -0.0045319       0.2869E-08
```

Figure 8.9 Printout of the computer program of Figure 8.8.

condition $y(\text{X0} + \Delta x)$, and $y'(\text{X0} + \Delta x)$, since Airy's equation is a second order differential equation. Substituting $X = \text{X0} + \Delta x$ into equation (84), we see that we can approximate $y(\text{X0} + \Delta x)$ by

$$y_N(\text{X0} + \Delta x) = \sum_{J=1}^{20} \text{A(J)}\, \Delta x^{J-1}.$$

Differentiating equation (84) with respect to X and substituting $X = \text{X0} + \Delta x$, we see that we can approximate $y'(\text{X0} + \Delta x)$ by

$$y_N'(\text{X0} + \Delta x) = \sum_{J=1}^{19} \text{JA(J + 1)}\, \Delta x^{J-1}.$$

Hence, the new initial conditions will be

$$\text{A(1)} = y_N(\text{X0} + \Delta x) \doteq y(\text{X0} + \Delta x)$$

and

$$\text{A(2)} = y_N'(\text{X0} + \Delta x) \doteq y'(\text{X0} + \Delta x).$$

Upon reinitialization we also change the value of X0 to $\text{X0} + \Delta x$, and leave $\text{A(3)} = 0$. We then recursively calculate new coefficients A(4), A(5), ..., A(20), calculate a new interval of convergence, and evaluate the solution at each point at which we desire a solution that lies within this interval. We proceed in this manner until we have values for the solution at all the points at which we want a value for the solution or until some other prescribed condition is met—in our case we have instructed the computer to stop if Δx is ever less than .01.

The computer program for generating the required solution to the given initial value problem is displayed in Figure 8.10. The basic differences between this computer program and the one of Figure 8.8 are the initial conditions—program statements 10–12, the recursive calculation of the coefficients—program statements 15 and 16, and the calculation of the value of the derivative at X0 + Δx when reinitialization occurs—program statements 35–39. The resulting 20-term power series solution to the given initial value

```
          C-TWENTY TERM POWER SERIES SOLUTION TO THE IVP
          C--Y''+X*Y=0;Y(0)=2,Y'(0)=1 ON (0,10)
0001            DIMENSION A(20),XI(21)
0002         22 FORMAT(1H1, 12X, 'X', 9X, 'Y(X)', 9X, 'ERROR EST')
0003         40 FORMAT(11X, F4.1, 4X, F11.7, 4X, E11.4)
0004         50 FORMAT('OREINITIALIZATION AT X = ',F7.5,'   Y(X) = ',F11.7,11H   Y
               1'(X) = ,F11.7)
0005         60 FORMAT('ORADIUS OF CONVERGENCE TOO SMALL TO CONTINUE, VALUE = ',
               1 F7.5)
0006            WRITE(3,22)
          C--INITIALIZE VALUES
0007            E = 5.*10.**(-7.)
0008            INO = 1
0009            XO = 0.
0010            A(1) = 2.
0011            A(2) = 1.
0012            A(3) = 0.
0013            DO 5 I = 1,21
0014          5 XI(I) = 0. + (I-1)*.5
          C--RECURSIVELY CALCULATE THE SERIES COEFFICIENTS
0015         10 DO 20 J = 4,20
0016         20 A(J) = -A(J-3)/((J-1)*(J-2))
          C--CALCULATE THE APPROXIMATE 'NUMERICAL' RADIUS OF CONVERGENCE
0017            POWER = 1./19.
0018            DELX = (E/ABS(A(20)))**POWER
0019            IF(DELX.LT..01)  GO TO 55
0020            XMAX = XO + DELX
0021            DO 30 I = INO,21
          C--DETERMINE IF XI(I) IS WITHIN THIS INTERVAL OF CONVERGENCE
0022            IF(XI(I).GT.XMAX) X = DELX
0023            IF(XI(I).LE.XMAX) X = XI(I) - XO
          C--EVALUATE THE POWER SERIES EXPANSION AT XI OR REINITIALIZE AT XMAX
          C--CALCULATE THE VALUE OF THE LAST TERM OF THE POWER SERIES AT XI
0024            ERR = A(20)
0025            SUM = 0.
0026            DO 25 J = 1,19
0027            ERR = ERR*X
0028            L = 21 - J
0029         25 SUM = (SUM + A(L))*X
0030            SUM = SUM + A(1)
0031            IF(XI(I).GT.XMAX) GO TO 45
0032         30 WRITE(3,40) XI(I), SUM, ERR
0033            CALL EXIT
          C--REINITIALIZE
0034         45 A(1) = SUM
          C--CALCULATE VALUE OF DERIVATIVE AT XMAX
0035            SUM = 0.
0036            DO 46 J = 1, 18
0037            L = 21 - J
0038         46 SUM = (SUM + (L-1)*A(L))*X
0039            A(2) = SUM + A(2)
0040            XO = XMAX
0041            INO = I
0042            WRITE(3,50) XO, A(1), A(2)
0043            GO TO 10
0044         55 WRITE(3,60) DELX
0045            CALL EXIT
0046            END
```

Figure 8.10 Computer program for the third method of using a power series expansion to produce numerical solutions to Airy's equation.

```
     X          Y(X)          ERROR EST
    0.0       2.0000000      0.0
    0.5       2.4533129      0.3284E-17
    1.0       2.5962534      0.1722E-11
    1.5       2.1065607      0.3817E-08

REINITIALIZATION AT X = 1.93877    Y(X) =  1.0542431    Y'(X) =  -2.9622612
    2.0       0.8728335     -0.4567E-34
    2.5      -0.6147490     -0.8734E-16
    3.0      -1.9868574     -0.1577E-10
    3.5      -2.8195076     -0.2418E-07

REINITIALIZATION AT X = 3.76984    Y(X) = -2.8498240    Y'(X) =   0.5617027
    4.0      -2.7148848      0.7312E-24
    4.5      -2.2703657      0.2457E-14
    5.0      -1.4306479      0.4952E-10
    5.5      -0.1839447      0.3229E-07

REINITIALIZATION AT X = 5.76837    Y(X) =  0.5438862    Y'(X) =   2.6328011
    6.0       1.1519651      0.3868E-23
    6.5       2.3728027      0.1197E-13
    7.0       3.1445961      0.2374E-09
    7.5       2.9656477      0.1538E-06

REINITIALIZATION AT X = 7.61084    Y(X) =  2.7592154    Y'(X) =  -2.1459198
    8.0       1.9011459     -0.6027E-19
    8.5       0.6452847     -0.3965E-12
    9.0      -0.7237864     -0.1905E-08

REINITIALIZATION AT X = 9.47324    Y(X) = -1.7797031    Y'(X) =  -1.6323709
    9.5      -1.8233824     -0.3729E-41
   10.0      -2.5859919     -0.1444E-16
```

Figure 8.11 Printout of the computer program of Figure 8.10.

problem with a maximum desirable error of $E = 5 \times 10^{-7}$ is shown in Figure 8.11.

The power series method can also be used to produce numerical solutions to systems of equations. Let us formally develop the power series expansions for the solution of the Volterra prey-predator initial value problem

$$\begin{aligned} y_1' &= ay_1 - by_1y_2 \\ y_2' &= -cy_2 + dy_1y_2 \end{aligned} \quad ; \quad y_1(0) = c_{10}, \quad y_2(0) = c_{20}. \tag{85}$$

(See Section 9.2 for a general development and discussion of Volterra prey-predator initial value problems.) We assume that both $y_1(x)$ and $y_2(x)$ have power series expansions about $x = 0$. Thus, we assume that

$$y_1(x) = \sum_{n=0}^{\infty} a_n x^n \tag{86a}$$

and

$$y_2(x) = \sum_{n=0}^{\infty} b_n x^n. \tag{86b}$$

Differentiating (86a) and (86b) and shifting the index of the resulting summations by 1, we obtain

$$y_1'(x) = \sum_{n=1}^{\infty} na_n x^{n-1} = \sum_{n=0}^{\infty} (n+1)a_{n+1}x^n \tag{87a}$$

and

$$y_2'(x) = \sum_{n=1}^{\infty} nb_n x^{n-1} = \sum_{n=0}^{\infty} (n+1)b_{n+1}x^n. \tag{87b}$$

Substituting equations (86) and (87) into the differential equation of (85), we see that the coefficients a_n and b_n must be chosen to simultaneously satisfy

$$\text{(88a)} \qquad \sum_{n=0}^{\infty} (n+1)a_{n+1}x^n = a\sum_{n=0}^{\infty} a_n x^n - b\sum_{n=0}^{\infty}\left(\sum_{k=0}^{n} a_{n-k}b_k\right)x^n$$

and

$$\text{(88b)} \qquad \sum_{n=0}^{\infty} (n+1)b_{n+1}x^n = -c\sum_{n=0}^{\infty} b_n x^n + d\sum_{n=0}^{\infty}\left(\sum_{k=0}^{n} a_{n-k}b_k\right)x^n.$$

Equating the coefficients of x^n and solving for a_{n+1} and b_{n+1}, we see that the coefficients must satisfy the recursive formulas

$$\text{(89a)} \qquad a_{n+1} = \frac{aa_n - b\sum_{k=0}^{n} a_{n-k}b_k}{n+1} \qquad \text{for } n = 0, 1, \ldots$$

and

$$\text{(89b)} \qquad b_{n+1} = \frac{cb_n + d\sum_{k=0}^{n} a_{n-k}b_k}{n+1} \qquad \text{for } n = 0, 1, \ldots.$$

In order to satisfy the initial conditions of (85), we set $x = 0$ in equations (86) and find

$$\text{(90a)} \qquad y_1(0) = a_0 = c_{10}$$

and

$$\text{(90b)} \qquad y_2(0) = b_0 = c_{20}.$$

Thus, to obtain the power series solution of the given initial value problem, we initiate the recursive formulas (89) with $a_0 = c_{10}$ and $b_0 = c_{20}$. Notice that the summations appearing in (89a) and (89b) are the same. This will make the computer programming of these recursions somewhat simpler than one might anticipate at first glance. We shall not develop a computer program to solve the given Volterra prey-predator initial value problem. However, the interested reader may wish to do so and in so doing discover some of the additional problems which one encounters when solving a system of differential equation as opposed to solving a single differential equation.

In summary, we would like to point out that the power series method for obtaining numerical solutions to initial value problems should only be considered for use when a high degree of accuracy is required, when the same differential equation with different initial conditions is to be solved many times, and when the differential equation itself is not too complicated. Initial value problems of this type often occur in various branches of the physical sciences. For example, in celestial mechanics problems of this nature occur in the solution of satellite trajectory problems and in the solution of the n-body problem when n is small ($n \leq 5$). If a particular differential equation is to be solved only for a few different initial conditions and the accuracy desired can be obtained using some other numerical technique, then the other technique should normally be employed since an entirely new computer program must be written for the power series solution each time the differential equation is changed, whereas only those few statements which define the differential equation must be changed when using other numerical techniques. If the differential equation is such that a recursive formula or formulas for the calculation of the power series coefficients cannot be derived, then obviously one cannot use the power series method. The third method for using the power series expansion to produce a numerical solution can be thought of as a single-step, variable-stepsize method. Since a larger stepsize can be taken with the power series method than with other single- or multistep methods and the same accuracy still be achieved, the power series method is often competitive with such methods from the standpoint of computing time required to produce a numerical solution.

EXERCISES

1. (a) Find a recursive formula for the coefficients of the power series solution to the differential equation $y' = 2y$.

(b) Write a computer program to calculate values of the 20-term power series solution to the initial value problem $y' = 2y$; $y(0) = 1$ at the points $x = .5, 1.0, 1.5, \ldots, 5.0$ using the first method and compare the results with the actual solution. (*Hint:* See Figure 8.1.)

(c) Derive an explicit formula for the coefficients a_n of the power series solution to the initial value problem of part (b) and estimate the number of terms required to achieve the accuracy $E = 5 \times 10^{-7}$ on the interval $[0, 5]$, if we assume that the error made when using $N + 1$ terms of the power series is $\epsilon_N = a_N x^N$.

2. The error function erf x satisfies the initial value problem

$$y' = \frac{2}{\sqrt{\pi}} e^{-x^2}; \qquad y(0) = 0.$$

(a) Expand e^{-x^2} in a power series about $x = 0$, integrate the series term by term, impose the initial condition, and thereby obtain a power series expansion for the error function.

(b) Assuming that the error made by truncating the series after any particular term is less than or equal to the last term retained in the series, how many terms would be required to ensure four-decimal-place accuracy ($E = 5 \times 10^{-5}$) on the interval $[0, 4]$?

(c) Derive a recursive formula for the coefficients of the error function.

(d) Modify the computer program of Figure 8.5 so that it will calculate the power series coefficients of the error function and determine the number of terms necessary to ensure four-decimal-place accuracy on the interval [0, 4].

(e) Use the modified computer program to calculate the value of the error function at $x = .5, 1.0, \ldots, 4.0$ and compare your results with published tables for the error function.

3. Consider the initial value problem

$$xy' = 1; \qquad y(1) = 0.$$

(a) Verify that $y(x) = \ln x$ is the solution. (Notice that $x = 0$ is a singular point of the given differential equation and that the initial condition is specified at $x = 1$.)

(b) What is the new initial value problem which is obtained by making the change of variable $w = x - 1$?

(c) Where is the singular point of the new initial value problem?

(d) Find a recursive formula for the coefficients of the power series solution,

$$y(x) = \sum_{n=0}^{\infty} a_n w^n,$$

of the new initial value problem.

(e) What is the radius of convergence of the power series solution of part (d)? [*Hint:* See part (c).]

(f) Find an explicit (not recursive) formula for the series coefficient a_n.

(g) Assuming that the error made by truncating the series after any particular term is less than or equal to the last term retained in the series (which it is not in this case), how many terms would be required for four-decimal-place accuracy for $0 \leq w \leq .8$?

(h) Modify the computer program of Figure 8.5 so that it will calculate the power series coefficients of the given initial value problem [these coefficients are the same as the coefficients of part (d)] and determine the number of terms necessary for four-decimal-place accuracy for $1 \leq x \leq 1.8$ $(0 \leq w \leq .8)$.

(i) Use the modified computer program to calculate the value of $\ln x$ at $x = 1.1, 1.2, \ldots, 1.8$ and compare your results closely with the value of $\ln x$ at these points.

4. (a) Find a recursive formula for the coefficients of the power series solution to the differential equation $y'' + y = 0$.

(b) Modify the computer program of Figure 8.10 to solve the initial value problem $y'' + y = 0$; $y(0) = 1$, $y'(0) = 0$ on the interval [0, 6.5] with $E = 5 \times 10^{-7}$. Compute values of the solution for $x = .25, .50, .75, \ldots, 6.50$ and compare the results with the actual solution.

(c) Modify the computer program of part (b) to solve the initial value problem $y'' + y = 0$; $y(0) = 0$, $y'(0) = 1$ on the interval $[0, 6.5]$ with $E = 5 \times 10^{-7}$. (All you need change are the initial conditions.) Compute values of the solution for $x = .25, .50, .75, \ldots, 6.50$ and compare the results with the actual solution.

Notice that since both $\cos x$ and $\sin x$ are periodic with period 2π—that is, $\cos(x + 2\pi) = \cos x$ and $\sin(x + 2\pi) = \sin x$—all that one needs to do in order to be able to compute $\cos x$ or $\sin x$ for any value of x is to find the nonnegative remainder r of x divided by 2π. Then $\cos x = \cos r$ and $\sin x = \sin r$.

5. Modify the computer program of Figure 8.10 to calculate Bessel's function of the first kind of order zero, $J_0(x)$, for $x = .5, 1.0, \ldots, 10.0$. [*Hint:* See equations (47), (49), and (56).] Compare your results with a published table for $J_0(x)$.

9

SYSTEMS OF FIRST ORDER EQUATIONS: METHODS OF SOLUTION

In Chapter 5 we presented the basic theory for the following system of n first order differential equations

$$
\begin{aligned}
y_1' &= f_1(x, y_1, y_2, \ldots, y_n) \\
y_2' &= f_2(x, y_1, y_2, \ldots, y_n) \\
&\vdots \qquad\qquad \vdots \\
y_n' &= f_n(x, y_1, y_2, \ldots, y_n),
\end{aligned} \tag{1}
$$

where y_i $(i = 1, 2, \ldots, n)$ are all real-valued functions of the independent variable x and f_i $(i = 1, 2, \ldots, n)$ are real-valued functions of x and y_i. We also saw that the system of equations (1) could be represented more compactly by the vector differential equation

$$\mathbf{y}' = \mathbf{f}(x, \mathbf{y}) \tag{2}$$

where the vectors

$$\mathbf{y}' = \begin{pmatrix} y_1' \\ y_2' \\ \vdots \\ y_n' \end{pmatrix}, \quad \mathbf{y} = \begin{pmatrix} y_1 \\ y_2 \\ \vdots \\ y_n \end{pmatrix}, \quad \text{and} \quad \mathbf{f}(x, \mathbf{y}) = \begin{pmatrix} f_1(x, \mathbf{y}) \\ f_2(x, \mathbf{y}) \\ \vdots \\ f_n(x, \mathbf{y}) \end{pmatrix}.$$

In this chapter we shall first discuss the form of the solution to system (2) when the system is linear—that is, when each $f_i(x, \mathbf{y})$ is a linear function of the dependent variables $y_1, y_2, \ldots, y_n$. The results we obtain will be similar to the results obtained for nth order linear differential equations. Next we will show how to find the general solution of a homogeneous linear system with constant coefficients and how to find a particular solution to a non-homogeneous linear system. Then we present some applications of linear systems of differential equations. And finally we show how to produce numerical solutions to the general vector initial value problem

$$\mathbf{y}' = \mathbf{f}(x, \mathbf{y}); \qquad \mathbf{y}(x_0) = \mathbf{c}_0. \tag{3}$$

9.1 SYSTEMS OF LINEAR FIRST ORDER EQUATIONS

The general form of a linear system of n first order differential equations is

$$\begin{aligned} y_1' &= a_{11}(x)y_1 + a_{12}(x)y_2 + \cdots + a_{1n}(x)y_n + b_1(x) \\ y_2' &= a_{21}(x)y_1 + a_{22}(x)y_2 + \cdots + a_{2n}(x)y_n + b_2(x) \\ &\vdots \\ y_n' &= a_{n1}(x)y_1 + a_{n2}(x)y_2 + \cdots + a_{nn}(x)y_n + b_n(x). \end{aligned} \tag{4}$$

This system may be represented more concisely in matrix-vector notation as

$$\mathbf{y}' = A(x)\mathbf{y} + \mathbf{b}(x), \tag{5}$$

where the vectors

$$\mathbf{y}' = \begin{pmatrix} y_1' \\ y_2' \\ \vdots \\ y_n' \end{pmatrix}, \qquad \mathbf{y} = \begin{pmatrix} y_1 \\ y_2 \\ \vdots \\ y_n \end{pmatrix}, \qquad \mathbf{b} = \begin{pmatrix} b_1(x) \\ b_2(x) \\ \vdots \\ b_n(x) \end{pmatrix}$$

and the matrix

$$A(x) = \begin{pmatrix} a_{11}(x) & a_{12}(x) & \cdots & a_{1n}(x) \\ a_{21}(x) & a_{22}(x) & \cdots & a_{2n}(x) \\ \vdots & \vdots & & \vdots \\ a_{n1}(x) & a_{n2}(x) & \cdots & a_{nn}(x) \end{pmatrix}.$$

We shall assume that $a_{ij}(x)$ and $b_i(x)$ $(i, j = 1, 2, \ldots, n)$ are all continuous on some interval I. Thus, we know from Theorem 5.5 that the initial value problem

$$\mathbf{y}' = A(x)\mathbf{y} + \mathbf{b}; \qquad \mathbf{y}(x_0) = \mathbf{c}_0, \tag{6}$$

where $x_0 \in I$ and $\mathbf{c}_0$ is a constant vector, has a unique solution on the interval I. If $b_i(x) \not\equiv 0$ for all $x \in I$ and $i = 1, 2, \ldots, n$, then system (5) is said to be *nonhomogeneous* and the associated *homogeneous* system is

$$\mathbf{y}' = A(x)\mathbf{y}. \tag{7}$$

The development of the theory for linear systems of n first order equations closely parallels that for nth order linear equations. We have the following superposition theorem.

Theorem 9.1 If $\mathbf{y}_1$ and $\mathbf{y}_2$ are solutions of the homogeneous linear system (7), then $\mathbf{y}_3 = c_1\mathbf{y}_1 + c_2\mathbf{y}_2$ where c_1 and c_2 are arbitrary constants, is a solution of (7).

Proof

$$\begin{aligned}\mathbf{y}_3' &= c_1\mathbf{y}_1' + c_2\mathbf{y}_2' = c_1A(x)\mathbf{y}_1 + c_2A(x)\mathbf{y}_2\\ &= A(x)(c_1\mathbf{y}_1 + c_2\mathbf{y}_2) = A(x)\mathbf{y}_3.\end{aligned}$$

One can easily extend this theorem and show that if $\mathbf{y}_1, \mathbf{y}_2, \ldots, \mathbf{y}_m$ are solutions of (7), then any linear combination $\mathbf{y} = \sum_{i=1}^m c_i\mathbf{y}_i$, where c_i are any arbitrary constants is also a solution of (7).

Now let us consider the set of n vector functions $\{\mathbf{y}_1, \mathbf{y}_2, \ldots, \mathbf{y}_n\}$.

Here

$$\mathbf{y}_1 = \begin{pmatrix} y_{11}(x)\\ y_{21}(x)\\ \vdots\\ y_{n1}(x)\end{pmatrix}, \qquad \mathbf{y}_2 = \begin{pmatrix} y_{12}(x)\\ y_{22}(x)\\ \vdots\\ y_{n2}(x)\end{pmatrix}, \qquad \ldots, \qquad \mathbf{y}_n = \begin{pmatrix} y_{1n}(x)\\ y_{2n}(x)\\ \vdots\\ y_{nn}(x)\end{pmatrix}.$$

Notice that $y_{ij}(x)$ is the ith component of the jth vector. The set of vectors $\{\mathbf{y}_1, \mathbf{y}_2, \ldots, \mathbf{y}_n\}$ is said to be *linearly dependent on an interval I*, if there exist constants $c_1, c_2, \ldots, c_n$ not all zero such that $c_1\mathbf{y}_1 + c_2\mathbf{y}_2 + \cdots + c_n\mathbf{y}_n = \mathbf{0}$ for all $x \in I$. The set of vectors is said to be *linearly independent on an interval I* if $c_1\mathbf{y}_1 + c_2\mathbf{y}_2 + \cdots + c_n\mathbf{y}_n = \mathbf{0}$ for all $x \in I$ implies that $c_1 = c_2 = \cdots = c_n = 0$. The *Wronskian* of the n vectors $\mathbf{y}_1, \mathbf{y}_2, \ldots, \mathbf{y}_n$ is defined by

$$W(\mathbf{y}_1, \mathbf{y}_2, \ldots, \mathbf{y}_n, x) = \begin{vmatrix} y_{11} & y_{12} & \cdots & y_{1n}\\ y_{21} & y_{22} & \cdots & y_{2n}\\ \vdots & \vdots & & \vdots\\ y_{n1} & y_{n2} & \cdots & y_{nn}\end{vmatrix}.$$

Thus, the Wronskian is the determinant of a matrix whose columns are the vectors $\mathbf{y}_1, \mathbf{y}_2, \ldots, \mathbf{y}_n$.

The following theorems delineate the relationship between the linear independence of the vectors $\mathbf{y}_1, \mathbf{y}_2, \ldots, \mathbf{y}_n$, the value of their Wronskian, and solutions of the homogeneous linear system (7).

Theorem 9.2 If the Wronskian is not zero for any $x \in I$, then the vector functions $\mathbf{y}_1, \mathbf{y}_2, \ldots, \mathbf{y}_n$ are linearly independent on I.

Proof Consider the equation

$$c_1\mathbf{y}_1 + c_2\mathbf{y}_2 + \cdots + c_n\mathbf{y}_n = \mathbf{0}.$$

Viewing $c_1, c_2, \ldots, c_n$ as unknowns, we may rewrite this equation in the following matrix-vector form:

$$\begin{pmatrix} y_{11} & y_{12} & \cdots & y_{1n} \\ y_{21} & y_{22} & \cdots & y_{2n} \\ \vdots & \vdots & & \vdots \\ y_{n1} & y_{n2} & \cdots & y_{nn} \end{pmatrix} \begin{pmatrix} c_1 \\ c_2 \\ \vdots \\ c_n \end{pmatrix} = \begin{pmatrix} 0 \\ 0 \\ \vdots \\ 0 \end{pmatrix}. \tag{8}$$

One knows from linear algebra that if the determinant of the matrix of (8) is nonzero, then the vector of unknowns must be zero. By hypothesis the Wronskian of $\mathbf{y}_1, \mathbf{y}_2, \ldots, \mathbf{y}_n$, which is the determinant of the $n \times n$ matrix of (8), is nonzero on I; therefore, $c_1 = c_2 = \cdots = c_n = 0$ for all $x \in I$. Hence $\mathbf{y}_1, \mathbf{y}_2, \ldots, \mathbf{y}_n$ are linearly independent on I.

Theorem 9.3 If $\mathbf{y}$ is a solution of the homogeneous linear system (7) $\mathbf{y}' = A(x)\mathbf{y}$ on an interval I and if $\mathbf{y}(x_0) = \mathbf{0}$, that is, $y_1(x_0) = 0, y_2(x_0) = 0, \ldots, y_n(x_0) = 0$ for some $x_0 \in I$, then $\mathbf{y}(x) \equiv \mathbf{0}$ for $x \in I$.

Proof Consider the initial value problem $\mathbf{y}' = A(x)\mathbf{y}$; $\mathbf{y}(x_0) = \mathbf{0}$. Clearly, $\mathbf{z}(x) \equiv \mathbf{0}$ for $x \in I$ satisfies this initial value problem and since the solution of this problem is unique, we must have $\mathbf{y} = \mathbf{z} = \mathbf{0}$.

Theorem 9.4 If $\mathbf{y}_1, \mathbf{y}_2, \ldots, \mathbf{y}_n$ are linearly independent solutions of the homogeneous linear system (7) $\mathbf{y}' = A(x)\mathbf{y}$ for $x \in I$, then the Wronskian of $\mathbf{y}_1, \mathbf{y}_2, \ldots, \mathbf{y}_n$ is nonzero for all $x \in I$.

Proof We shall prove this theorem by contradiction. Thus, we assume that for some $x_0 \in I$, $W(\mathbf{y}_1, \mathbf{y}_2, \ldots, \mathbf{y}_n, x_0) = 0$. Consider the following linear algebraic system of equations in the unknowns $c_1, c_2, \ldots, c_n$:

$$\begin{pmatrix} y_{11}(x_0) & y_{12}(x_0) & \cdots & y_{1n}(x_0) \\ y_{21}(x_0) & y_{22}(x_0) & \cdots & y_{2n}(x_0) \\ \vdots & \vdots & & \vdots \\ y_{n1}(x_0) & y_{n2}(x_0) & \cdots & y_{nn}(x_0) \end{pmatrix} \begin{pmatrix} c_1 \\ c_2 \\ \vdots \\ c_n \end{pmatrix} = \begin{pmatrix} 0 \\ 0 \\ \vdots \\ 0 \end{pmatrix}. \tag{9}$$

This system is equivalent to the vector equation

$$c_1\mathbf{y}_1(x_0) + c_2\mathbf{y}_2(x_0) + \cdots + c_n\mathbf{y}_n(x_0) = \mathbf{0}. \tag{10}$$

Since the determinant of the $n \times n$ matrix of (9)—the Wronskian of $\mathbf{y}_1, \mathbf{y}_2, \ldots, \mathbf{y}_n$ at x_0—is assumed to be zero, there exist constants c_1, $c_2, \ldots, c_n$ not all zero which satisfy (9) and (10).

Now let us consider the vector function defined on I by

$$\mathbf{y}(x) = c_1\mathbf{y}_1(x) + c_2\mathbf{y}_2(x) + \cdots + c_n\mathbf{y}_n(x).$$

Since $\mathbf{y}_1, \mathbf{y}_2, \ldots, \mathbf{y}_n$ are solutions of (7), the linear combination $\mathbf{y}$ is also a solution of (7). And since $\mathbf{y}(x_0) = \mathbf{0}$, $\mathbf{y}(x) \equiv \mathbf{0}$ for $x \in I$ by Theorem 9.3. Hence, there exist constants $c_1, c_2, \ldots, c_n$ not all zero such that

$$c_1\mathbf{y}_1(x) + c_2\mathbf{y}_2(x) + \cdots + c_n\mathbf{y}_n(x) \equiv \mathbf{0}$$

for $x \in I$. Thus, $\mathbf{y}_1, \mathbf{y}_2, \ldots, \mathbf{y}_n$ are linearly dependent on I, which contradicts a hypothesis of the theorem.

Notice that Theorem 9.2 applies to any vector functions $\mathbf{y}_1, \mathbf{y}_2, \ldots, \mathbf{y}_n$ whether they are solutions of the same linear system of first order differential equations, or not. However, Theorem 9.4 does not apply to vector functions which are not solutions of the same homogeneous linear system of equations of the form (7). It can be shown, for instance, that the vector functions

$$\mathbf{y}_1 = \begin{pmatrix} x \\ 0 \end{pmatrix} \quad \text{and} \quad \mathbf{y}_2 = \begin{pmatrix} x^2 \\ 0 \end{pmatrix}$$

are not solutions of any linear system of the form (7). (See Petrovski [16], pp. 110–111.) However, $\mathbf{y}_1$ and $\mathbf{y}_2$ are clearly linearly independent on the interval $(-\infty, \infty)$, but

$$W(\mathbf{y}_1, \mathbf{y}_2, x) = \begin{vmatrix} x & x^2 \\ 0 & 0 \end{vmatrix} = 0 \quad \text{for all } x \in (-\infty, \infty).$$

The following theorem, which establishes a test for the linear independence of a set of n solutions of a homogeneous linear system of differential equations of the form (7), follows directly from Theorems 9.2 and 9.4.

Theorem 9.5 Let $\mathbf{y}_1, \mathbf{y}_2, \ldots, \mathbf{y}_n$ be solutions of the homogeneous system of linear differential equations $\mathbf{y}' = A(x)\mathbf{y}$ on some interval I. The set of solutions $\{\mathbf{y}_1, \mathbf{y}_2, \ldots, \mathbf{y}_n\}$ is linearly independent on I if and only if $W(\mathbf{y}_1, \mathbf{y}_2, \ldots, \mathbf{y}_n, x) \neq 0$ for all $x \in I$.

EXAMPLE Verify that

$$\mathbf{y}_1 = \begin{pmatrix} e^x \\ e^x \end{pmatrix} \quad \text{and} \quad \mathbf{y}_2 = \begin{pmatrix} 2e^{-x} \\ e^{-x} \end{pmatrix}$$

are linearly independent solutions of the homogeneous linear system

$$\mathbf{y}' = \begin{pmatrix} -3 & 4 \\ -2 & 3 \end{pmatrix}\mathbf{y}. \tag{11}$$

Differentiating $\mathbf{y}_1$ and $\mathbf{y}_2$ we find, respectively,

$$\mathbf{y}_1' = \begin{pmatrix} e^x \\ e^x \end{pmatrix} \quad \text{and} \quad \mathbf{y}_2' = \begin{pmatrix} -2e^{-x} \\ -\ e^{-x} \end{pmatrix}.$$

Substituting $\mathbf{y} = \mathbf{y}_1$ into (11), we find

$$\mathbf{y}_1' = \begin{pmatrix} e^x \\ e^x \end{pmatrix} = \begin{pmatrix} -3 & 4 \\ -2 & 3 \end{pmatrix}\begin{pmatrix} e^x \\ e^x \end{pmatrix} = \begin{pmatrix} -3 & 4 \\ -2 & 3 \end{pmatrix}\mathbf{y}_1.$$

Hence, $\mathbf{y}_1$ is a solution of (11) for all real x. Substituting $\mathbf{y} = \mathbf{y}_2$ into (11), we get

$$\mathbf{y}_2' = \begin{pmatrix} -2e^{-x} \\ -e^{-x} \end{pmatrix} = \begin{pmatrix} -3 & 4 \\ -2 & 3 \end{pmatrix}\begin{pmatrix} 2e^{-x} \\ e^{-x} \end{pmatrix} = \begin{pmatrix} -3 & 4 \\ -2 & 3 \end{pmatrix}\mathbf{y}_2.$$

Therefore, $\mathbf{y}_2$ is also a solution of (11) for all real x. In order to verify the linear independence of $\mathbf{y}_1$ and $\mathbf{y}_2$, we calculate the Wronskian of these functions.

$$W(\mathbf{y}_1, \mathbf{y}_2, x) = \begin{vmatrix} e^x & 2e^{-x} \\ e^x & e^{-x} \end{vmatrix} = -1 \neq 0,$$

so $\mathbf{y}_1$ and $\mathbf{y}_2$ are linearly independent.

Theorem 9.6 There exist n linearly independent solutions of the homogeneous system of n linear first order differential equations

$$\mathbf{y}' = A(x)\mathbf{y}. \tag{7}$$

Proof Suppose that the elements, $a_{ij}(x)$, of the matrix $A(x)$ are all continuous on some interval I and let $x_0 \in I$. Define the constant vectors $\mathbf{e}_1, \mathbf{e}_2, \ldots, \mathbf{e}_n$ as follows:

$$\mathbf{e}_1 = \begin{pmatrix} 1 \\ 0 \\ \vdots \\ 0 \end{pmatrix}, \quad \mathbf{e}_2 = \begin{pmatrix} 0 \\ 1 \\ \vdots \\ 0 \end{pmatrix}, \quad \ldots, \quad \mathbf{e}_n = \begin{pmatrix} 0 \\ 0 \\ \vdots \\ 1 \end{pmatrix}.$$

That is, $\mathbf{e}_i$ is a vector with ith component 1 and all other components 0. Now consider the n vector initial value problems

$$\mathbf{y}' = A(x)\mathbf{y}; \qquad \mathbf{y}(x_0) = \mathbf{e}_i \qquad (i = 1, 2, \ldots, n). \tag{12}$$

By Theorem 5.5 there exist n unique solutions $\mathbf{y}_1, \mathbf{y}_2, \ldots, \mathbf{y}_n$ of the n initial value problems (12). Suppose that there exist n real constants $c_1, c_2, \ldots, c_n$ such that $c_1\mathbf{y}_1 + c_2\mathbf{y}_2 + \cdots + c_n\mathbf{y}_n = \mathbf{0}$ for all $x \in I$. This last vector equation is equivalent to the following system of algebraic equations:

$$\begin{pmatrix} y_{11}(x) & y_{12}(x) & \cdots & y_{1n}(x) \\ y_{21}(x) & y_{22}(x) & \cdots & y_{2n}(x) \\ \vdots & \vdots & & \vdots \\ y_{n1}(x) & y_{n2}(x) & \cdots & y_{nn}(x) \end{pmatrix} \begin{pmatrix} c_1 \\ c_2 \\ \vdots \\ c_n \end{pmatrix} = \begin{pmatrix} 0 \\ 0 \\ \vdots \\ 0 \end{pmatrix}. \tag{13}$$

At $x_0 \in I$, the determinant of the $n \times n$ matrix in equation (11) is

$$W(\mathbf{y}_1, \mathbf{y}_2, \ldots, \mathbf{y}_n, x_0) = \begin{vmatrix} 1 & 0 & \cdots & 0 \\ 0 & 1 & \cdots & 0 \\ \vdots & \vdots & & \vdots \\ 0 & 0 & \cdots & 1 \end{vmatrix} = 1 \neq 0.$$

So $c_1 = c_2 = \cdots = c_n = 0$, and therefore the functions $\mathbf{y}_1, \mathbf{y}_2, \ldots, \mathbf{y}_n$ are linearly independent on I.

Theorem 9.7 If $\mathbf{y}_1, \mathbf{y}_2, \ldots, \mathbf{y}_n$ are linearly independent solutions of the homogeneous system of n linear first order differential equations (7), $\mathbf{y}' = A(x)\mathbf{y}$, on some interval I and if $\mathbf{y}$ is also a solution of (7) on I, then there exist constants $c_1, c_2, \ldots, c_n$ such that $\mathbf{y} = \sum_{i=1}^{n} c_i\mathbf{y}_i$.

Proof Let $x_0 \in I$ and let

$$\mathbf{y}(x_0) = \mathbf{k} = \begin{pmatrix} k_1 \\ k_2 \\ \vdots \\ k_n \end{pmatrix}.$$

Consider the linear algebraic system

$$\begin{pmatrix} y_{11}(x_0) & y_{12}(x_0) & \cdots & y_{1n}(x_0) \\ y_{21}(x_0) & y_{22}(x_0) & \cdots & y_{2n}(x_0) \\ \vdots & \vdots & & \vdots \\ y_{n1}(x_0) & y_{n2}(x_0) & \cdots & y_{nn}(x_0) \end{pmatrix} \begin{pmatrix} c_1 \\ c_2 \\ \vdots \\ c_n \end{pmatrix} = \begin{pmatrix} k_1 \\ k_2 \\ \vdots \\ k_n \end{pmatrix}. \tag{14}$$

Since $\mathbf{y}_1, \mathbf{y}_2, \ldots, \mathbf{y}_n$ are linearly independent solutions of (7),

$$W(\mathbf{y}_1, \mathbf{y}_2, \ldots, \mathbf{y}_n, x_0) \neq 0,$$

and therefore there exists a unique solution vector

$$\mathbf{c} = \begin{pmatrix} c_1 \\ c_2 \\ \vdots \\ c_n \end{pmatrix} \quad \text{of (14).}$$

Define $\mathbf{z} = \sum_{i=1}^{n} c_i \mathbf{y}_i$. Since $\mathbf{y}_1, \mathbf{y}_2, \ldots, \mathbf{y}_n$ are solutions of (7), $\mathbf{z}$ is a solution of (7). Observe from equation (14) that

$$\mathbf{z}(x_0) = \sum_{i=1}^{n} c_i \mathbf{y}_i(x_0) = \mathbf{k} = \mathbf{y}(x_0).$$

Since $\mathbf{z}$ is the unique solution of the initial value problem $\mathbf{y}' = A(x)\mathbf{y}$; $\mathbf{y}(x_0) = \mathbf{k}$ on the interval I, and since $\mathbf{y}$ satisfies the same initial value problem, we find that $\mathbf{y}(x) \equiv \mathbf{z}(x) = \sum_{i=1}^{n} c_i \mathbf{y}_i$.

Theorem 9.6 shows that there are at least n linearly independent solutions to the homogeneous linear system (7), $\mathbf{y}' = A(x)\mathbf{y}$, and Theorem 9.7 shows that there are at most n linearly independent solutions to that system. If $\mathbf{y}_1, \mathbf{y}_2, \ldots, \mathbf{y}_n$ are linearly independent solutions of (7), then the linear combination $\mathbf{y} = \sum_{i=1}^{n} c_i \mathbf{y}_i$, where the c_i's are arbitrary constants, is called the *general solution* of (7).

Now, let us consider the nonhomogeneous system of linear first order differential equations

$$\mathbf{y}' = A(x)\mathbf{y} + \mathbf{b}(x), \tag{15}$$

where $\mathbf{y}$, $\mathbf{y}'$, and $\mathbf{b}(x)$ are n-dimensional vectors of functions and $A(x)$ is an $n \times n$ matrix of functions, where the functions comprising $A(x)$ and $\mathbf{b}(x)$ are assumed to be continuous on some interval I, and where $\mathbf{b}(x) \not\equiv \mathbf{0}$ for $x \in I$. The associated homogeneous linear system is

$$\mathbf{y}' = A(x)\mathbf{y}. \tag{16}$$

Theorem 9.8 If $\mathbf{y}_p$ is any solution of (15) on the interval I and if $\mathbf{y}_1, \mathbf{y}_2, \ldots, \mathbf{y}_n$ are n linearly independent solutions of (16) on I, then every solution of (15) on I has the form

$$\mathbf{y} = c_1\mathbf{y}_1 + c_2\mathbf{y}_2 + \cdots + c_n\mathbf{y}_n + \mathbf{y}_p, \tag{17}$$

where $c_1, c_2, \ldots, c_n$ are suitably chosen constants.

Proof Let $\mathbf{z}(x)$ be any solution of (15). Consider $\mathbf{w}(x) = \mathbf{z}(x) - \mathbf{y}_p(x)$.

$$\begin{aligned}\mathbf{w}'(x) &= \mathbf{z}'(x) - \mathbf{y}_p'(x)\\ &= A(x)\mathbf{z}(x) + \mathbf{b}(x) - [A(x)\mathbf{y}_p(x) + \mathbf{b}(x)]\\ &= A(x)[\mathbf{z}(x) - \mathbf{y}_p(x)]\\ &= A(x)\mathbf{w}(x).\end{aligned}$$

Hence, $\mathbf{w}$ satisfies (16) and by Theorem 9.7 there exist constants $c_1, c_2, \ldots, c_n$ such that $\mathbf{w}(x) = c_1\mathbf{y}_1(x) + c_2\mathbf{y}_2(x) + \cdots + c_n\mathbf{y}_n$ for $x \in I$. But since $\mathbf{w}(x) = \mathbf{z}(x) - \mathbf{y}_p(x)$, we find that $\mathbf{z}(x) = c_1\mathbf{y}_1(x) + c_2\mathbf{y}_2 + \cdots + c_n\mathbf{y}_n(x) + \mathbf{y}_p(x)$ for $x \in I$.

The *general solution* of the nonhomogeneous system (15) on an interval I is

$$\mathbf{y} = c_1\mathbf{y}_1 + c_2\mathbf{y}_2 + \cdots + c_n\mathbf{y}_n + y_p, \tag{18}$$

where $\mathbf{y}_p$ is any *particular* solution of (15) on I; $\mathbf{y}_1, \mathbf{y}_2, \ldots, \mathbf{y}_n$ are any linearly independent solutions of the associated homogeneous system (16); and $c_1, c_2, \ldots, c_n$ are arbitrary constants.

EXAMPLE Verify that

$$\mathbf{y}_p = \begin{pmatrix} 2x \\ x - 1 \end{pmatrix}$$

is a particular solution of the nonhomogeneous linear system

$$\mathbf{y}' = \begin{pmatrix} -3 & 4 \\ -2 & 3 \end{pmatrix}\mathbf{y} + \begin{pmatrix} 2x + 6 \\ x + 4 \end{pmatrix} \tag{19}$$

and write the general solution.

Differentiating, we find

$$\mathbf{y}_n' = \begin{pmatrix} 2 \\ 1 \end{pmatrix}.$$

Substituting $\mathbf{y} = \mathbf{y}_p$ in equation (19), we get

$$\begin{pmatrix} 2 \\ 1 \end{pmatrix} = \begin{pmatrix} -3 & 4 \\ -2 & 3 \end{pmatrix}\begin{pmatrix} 2x \\ x - 1 \end{pmatrix} + \begin{pmatrix} 2x + 6 \\ 2 + 4 \end{pmatrix}.$$

Hence, $\mathbf{y}_p$ is a particular solution of (19). As we verified earlier,

$$\mathbf{y}_1 = \begin{pmatrix} e^x \\ e^x \end{pmatrix} \quad \text{and} \quad \mathbf{y}_2 = \begin{pmatrix} 2e^{-x} \\ e^{-x} \end{pmatrix}$$

are linearly independent solutions of the associated homogeneous linear system (11). So the general solution of (19) is

$$\mathbf{y} = c_1\begin{pmatrix} e^x \\ e^x \end{pmatrix} + c_2\begin{pmatrix} 2e^{-x} \\ e^{-x} \end{pmatrix} + \begin{pmatrix} 2x \\ x - 1 \end{pmatrix}.$$

EXERCISES

1. Use the Wronskian to show that the following sets of vector functions are linearly independent.

(a) $\begin{pmatrix} x \\ 3 \end{pmatrix}, \begin{pmatrix} 2 \\ x - 1 \end{pmatrix}$

(b) $\begin{pmatrix} 2 \\ 1 \end{pmatrix}, \begin{pmatrix} 1 \\ 0 \end{pmatrix}$

(c) $\begin{pmatrix} 2e^x \\ e^x \end{pmatrix}, \begin{pmatrix} e^{-x} \\ 0 \end{pmatrix}$

(d) $\begin{pmatrix} 1 \\ 0 \\ 0 \end{pmatrix}, \begin{pmatrix} 0 \\ \sin x \\ \cos x \end{pmatrix}, \begin{pmatrix} 0 \\ -\cos x \\ \sin x \end{pmatrix}$

(e) $\begin{pmatrix} e^{2x} \\ -3e^{2x} \\ 2e^{2x} \end{pmatrix}, \begin{pmatrix} 2 \\ 3 \\ 1 \end{pmatrix}, \begin{pmatrix} 3x \\ -x \\ -2x \end{pmatrix}$

2. Show that the following sets of vector functions are linearly dependent by finding constants $c_1, c_2, \ldots, c_n$ not all zero such that

$$c_1\mathbf{y}_1 + c_2\mathbf{y}_2 + \cdots + c_n\mathbf{y}_n = \mathbf{0}.$$

(a) $\begin{pmatrix} 2 \\ -1 \end{pmatrix}, \begin{pmatrix} -6 \\ 3 \end{pmatrix}$

(b) $\begin{pmatrix} x - 1 \\ 2x \end{pmatrix}, \begin{pmatrix} -2x + 2 \\ -4x \end{pmatrix}$

(c) $\begin{pmatrix} 2e^{-x} \\ e^{-x} \\ -3e^{-x} \end{pmatrix}, \begin{pmatrix} e^{-x} \\ -4e^{-x} \\ 6e^{-x} \end{pmatrix}, \begin{pmatrix} -e^{-x} \\ -2e^{-x} \\ 4e^{-x} \end{pmatrix}$

(d) $\begin{pmatrix} \cos x \\ \sin x - \cos x \\ 2\sin x \end{pmatrix}, \begin{pmatrix} 2\sin x - \cos x \\ -\sin x + 3\cos x \\ 2\cos x \end{pmatrix}, \begin{pmatrix} \sin x + \cos x \\ \sin x \\ \cos x + 3\sin x \end{pmatrix}$

3. Verify that

$$\mathbf{y}_1 = \begin{pmatrix} e^x \\ 2e^x \end{pmatrix} \quad \text{and} \quad \mathbf{y}_2 = \begin{pmatrix} e^{-x} \\ -2e^{-x} \end{pmatrix}$$

are linearly independent solutions of

$$\mathbf{y}' = \begin{pmatrix} 0 & \frac{1}{2} \\ 2 & 0 \end{pmatrix} \mathbf{y}.$$

Why are

$$\begin{pmatrix} \cosh x \\ 2 \sinh x \end{pmatrix} \quad \text{and} \quad \begin{pmatrix} \sinh x \\ 2 \cosh x \end{pmatrix}$$

also solutions? [*Hint:* Recall $\sinh x = (e^x - e^{-x})/2$ and $\cosh x = (e^x + e^{-x})/2$.]

4. Verify that

$$\mathbf{y}_p = \begin{pmatrix} 1 - e^x \\ 2x \end{pmatrix}$$

is a particular solution of

$$\mathbf{y}' = \begin{pmatrix} 0 & \frac{1}{2} \\ 2 & 0 \end{pmatrix} \mathbf{y} + \begin{pmatrix} -x - e^x \\ 2e^x \end{pmatrix}$$

and write the general solution.

9.1.1 Homogeneous Linear Systems with Constant Coefficients

In this section we shall consider homogeneous linear first order systems with constant coefficients. That is, we shall consider systems of the form

$$\mathbf{y}' = A\mathbf{y}, \tag{20}$$

where $\mathbf{y}$ and $\mathbf{y}'$ are n dimensional vectors and A is an $n \times n$ matrix of real numbers. In order to solve nth order linear homogeneous differential equations with constant coefficients, we assumed that there were solutions of the form $y = e^{rx}$ and succeeded in finding an algebraic equation, which we called the auxiliary equation, which r had to satisfy. By analogy we shall assume that (20) has a solution of the form

$$\mathbf{y} = \mathbf{c}e^{rx}, \tag{21}$$

where $\mathbf{c}$ is an unknown constant n-dimensional vector and r is an unknown scalar constant. Differentiating (21) and substituting into (20), we find that $\mathbf{c}$ and r must satisfy

$$r\mathbf{c}e^{rx} = A\mathbf{c}e^{rx}$$

or

$$(A\mathbf{c} - r\mathbf{c})e^{rx} = \mathbf{0}.$$

Since the solution is to be valid for all real x, $\mathbf{c}$ and r must satisfy the following system of algebraic equations:

$$(A - rI)\mathbf{c} = \mathbf{0} \tag{22}$$

where I is the $n \times n$ identity matrix

$$\begin{pmatrix} 1 & 0 & \cdots & 0 & 0 \\ 0 & 1 & \cdots & 0 & 0 \\ \vdots & \vdots & & \vdots & \vdots \\ 0 & 0 & \cdots & 1 & 0 \\ 0 & 0 & \cdots & 0 & 1 \end{pmatrix}.$$

The system of equations (22) has a solution $\mathbf{c} \neq \mathbf{0}$ if and only if the determinant of the matrix $(A - rI)$ is zero. Thus, (22) has a nontrivial (nonzero) solution $\mathbf{c}$ if and only if

$$|A - rI| = 0. \tag{23}$$

Equation (23) is a polynomial of degree n in the unknown r and is called the *characteristic equation* of the matrix A. The roots r of the characteristic equation are called the *characteristic roots* or *eigenvalues* of the matrix A. If r_1 is a characteristic root of A, then any nonzero vector $\mathbf{c}_1$ which satisfies

$$(A - r_1 I)\mathbf{c}_1 = \mathbf{0} \tag{24}$$

—equation (22) with $r = r_1$—is called a *characteristic vector* or *eigenvector* of A associated with the characteristic root r_1. Notice that if $\mathbf{c}_1$ satisfies (24), then $k\mathbf{c}_1$, where k is an arbitrary constant, also satisfies (24). So characteristic vectors are not unique but are said to be unique except for a multiplicative constant. Of course, the characteristic polynomial has n roots; however, since a root may be a repeated root, more than one characteristic vector may be associated with a characteristic root. Thus, as was the case for nth order linear homogeneous differential equations with constant coefficients, the form of the general solution to (20) depends upon whether the characteristic roots are real and distinct, complex and distinct, or repeated. We shall now consider each case separately.

Distinct real roots It can be shown that if the characteristic roots of (23), $r_1, r_2, \ldots, r_n$ are all real, and no two roots are equal, then the vectors

$$\mathbf{y}_1 = \mathbf{c}_1 e^{r_1 x}, \qquad \mathbf{y}_2 = \mathbf{c}_2 e^{r_2 x}, \qquad \ldots, \qquad \mathbf{y}_n = \mathbf{c}_n e^{r_n x},$$

where $\mathbf{c}_1, \mathbf{c}_2, \ldots, \mathbf{c}_n$ are characteristic vectors of A associated with $r_1, r_2, \ldots, r_n$, respectively, are linearly independent solutions of (20) and the general solution of (20) is

$$\mathbf{y} = k_1 \mathbf{c}_1 e^{r_1 x} + k_2 \mathbf{c}_2 e^{r_2 x} + \cdots + k_n \mathbf{c}_n e^{r_n x}, \tag{25}$$

where $k_1, k_2, \ldots, k_n$ are arbitrary scalar constants.

EXAMPLE Find the general solution of the following system:

$$\mathbf{y}' = \begin{pmatrix} y_1' \\ y_2' \end{pmatrix} = \begin{pmatrix} 2 & -3 \\ 1 & -2 \end{pmatrix} \begin{pmatrix} y_1 \\ y_2 \end{pmatrix} = A\mathbf{y}. \tag{26}$$

The characteristic equation of A is

$$|A - rI| = \left| \begin{pmatrix} 2 & -3 \\ 1 & -2 \end{pmatrix} - r \begin{pmatrix} 1 & 0 \\ 0 & 1 \end{pmatrix} \right| = \begin{vmatrix} 2 - r & -3 \\ 1 & -2 - r \end{vmatrix} = r^2 - 1 = 0.$$

Hence, the characteristic roots are $r_1 = 1$ and $r_2 = -1$. Any characteristic vector $\mathbf{c}_1$ of A associated with $r_1 = 1$ must satisfy (24). Consequently, in this instance $\mathbf{c}_1$ must satisfy

$$(A - r_1 I)\mathbf{c}_1 = \left(\begin{pmatrix} 2 & -3 \\ 1 & -2 \end{pmatrix} - (1) \begin{pmatrix} 1 & 0 \\ 0 & 1 \end{pmatrix} \right) \begin{pmatrix} c_{11} \\ c_{21} \end{pmatrix} = \begin{pmatrix} 1 & -3 \\ 1 & -3 \end{pmatrix} \begin{pmatrix} c_{11} \\ c_{21} \end{pmatrix} = \begin{pmatrix} 0 \\ 0 \end{pmatrix}.$$

Thus, $c_{11} = 3c_{21}$ and c_{21} is arbitrary. Choosing $c_{21} = 1$, we find that one characteristic vector corresponding to $r_1 = 1$ is $\mathbf{c}_1 = \begin{pmatrix} 3 \\ 1 \end{pmatrix}$. Hence, a solution of (26) is

$$\mathbf{y}_1 = \mathbf{c}_1 e^{r_1 x} = \begin{pmatrix} 3 \\ 1 \end{pmatrix} e^x.$$

A characteristic vector $\mathbf{c}_2$ of A associated with $r_2 = -1$ must satisfy

$$(A - r_2 I)\mathbf{c}_2 = \left(\begin{pmatrix} 2 & -3 \\ 1 & -2 \end{pmatrix} - (-1) \begin{pmatrix} 1 & 0 \\ 0 & 1 \end{pmatrix} \right) \begin{pmatrix} c_{12} \\ c_{22} \end{pmatrix} = \begin{pmatrix} 3 & -3 \\ 1 & -1 \end{pmatrix} \begin{pmatrix} c_{12} \\ c_{22} \end{pmatrix} = \begin{pmatrix} 0 \\ 0 \end{pmatrix}.$$

Hence, $c_{12} = c_{22}$ and c_{22} is arbitrary. Choosing $c_{22} = 1$, we find that $\mathbf{c}_2 = \begin{pmatrix} 1 \\ 1 \end{pmatrix}$ is a characteristic vector associated with $r_2 = -1$ and that

$$\mathbf{y}_2 = \mathbf{c}_2 e^{r_2 x} = \begin{pmatrix} 1 \\ 1 \end{pmatrix} e^{-x}$$

is a second solution of (26). Hence, the general solution of the given system is

$$\mathbf{y} = k_1\begin{pmatrix}3\\1\end{pmatrix}e^{x} + k_2\begin{pmatrix}1\\1\end{pmatrix}e^{-x},$$

where k_1 and k_2 are arbitrary constants.

Distinct complex roots If the characteristic equation has a root $r_1 = a + bi$, where a and b are real numbers and $b \neq 0$, then $r_2 = a - bi$ is also a root of the characteristic equation, since A is a real matrix. Corresponding to r_1 and r_2 there are two complex characteristic vectors $\mathbf{c}_1^*$ and $\mathbf{c}_2^*$ and two linearly independent complex solutions

$$\mathbf{y}_1^* = \mathbf{c}_1^* e^{r_1 x} = \mathbf{c}_1^* e^{(a+bi)x} = \mathbf{c}_1^* e^{ax}(\cos bx + i \sin bx)$$

and

$$\mathbf{y}_2^* = \mathbf{c}_2^* e^{r_2 x} = \mathbf{c}_2^* e^{(a-bi)x} = \mathbf{c}_2^* e^{ax}(\cos bx - i \sin bx).$$

Since A is a real matrix, we will always be able to replace the complex linearly combination $j_1\mathbf{c}_1^*\mathbf{y}_1^* + j_2\mathbf{c}_2^*\mathbf{y}_2^*$, which could appear in the general solution of (20) by a linear combination of real-valued vector functions.

EXAMPLE Find the general solution of the following system expressed as a linear combination of real-valued vector functions:

$$\mathbf{y}' = \begin{pmatrix}y_1'\\y_2'\end{pmatrix} = \begin{pmatrix}1 & -2\\1 & 3\end{pmatrix}\begin{pmatrix}y_1\\y_2\end{pmatrix} = A\mathbf{y}. \tag{27}$$

The characteristic equation of A is

$$|A - rI| = \left|\begin{pmatrix}1 & -2\\1 & 3\end{pmatrix} - r\begin{pmatrix}1 & 0\\0 & 1\end{pmatrix}\right| = \begin{vmatrix}1-r & -2\\1 & 3-r\end{vmatrix} = r^2 - 4r + 5 = 0.$$

The characteristic roots are easily found to be $r_1 = 2 + i$ and $r_2 = 2 - i$.

A characteristic vector $\mathbf{c}_1^*$ of A associated with $r_1 = 2 + i$ must satisfy

$$(A - r_1 I)\mathbf{c}_1^* = \left(\begin{pmatrix}1 & -2\\1 & 3\end{pmatrix} - (2+i)\begin{pmatrix}1 & 0\\0 & 1\end{pmatrix}\right)\begin{pmatrix}c_{11}^*\\c_{21}^*\end{pmatrix}$$

$$= \begin{pmatrix}-1-i & -2\\1 & 1-i\end{pmatrix}\begin{pmatrix}c_{11}^*\\c_{21}^*\end{pmatrix} = \begin{pmatrix}0\\0\end{pmatrix}.$$

Hence, $c_{11}^* = -(1 - i)c_{21}^*$ and c_{21}^* is arbitrary. Choosing $c_{21}^* = 1$, we obtain the following solution of (27) associated with the characteristic root $r_1 = 2 + i$:

$$\mathbf{y}_1^* = \mathbf{c}_1^* e^{r_1 x} = \begin{pmatrix}i-1\\1\end{pmatrix}e^{(2+i)x} = \begin{pmatrix}i-1\\1\end{pmatrix}e^{2x}(\cos x + i \sin x)$$

$$= \begin{pmatrix}-\cos x - \sin x + i(\cos x - \sin x)\\ \cos x + i \sin x\end{pmatrix}e^{2x}.$$

Likewise, any solution of (27) of the form $\mathbf{y}_2^* = \mathbf{c}_2^* e^{(2-i)x}$ associated with the characteristic root $r_2 = 2 - i$ must satisfy

$$(A - r_2 I)\mathbf{c}_2^* = \left(\begin{pmatrix} 1 & -2 \\ 1 & 3 \end{pmatrix} - (2 - i)\begin{pmatrix} 1 & 0 \\ 0 & 1 \end{pmatrix}\right)\begin{pmatrix} c_{12}^* \\ c_{22}^* \end{pmatrix}$$

$$= \begin{pmatrix} -1 + i & -2 \\ 1 & i + 1 \end{pmatrix}\begin{pmatrix} c_{12}^* \\ c_{22}^* \end{pmatrix} = \begin{pmatrix} 0 \\ 0 \end{pmatrix}.$$

Thus, $c_{12}^* = -(1 + i)c_{22}^*$ and c_{22}^* is arbitrary. Choosing $c_{22}^* = 1$, we find a second solution of (27) to be

$$\mathbf{y}_2^* = \begin{pmatrix} -1 - i \\ 1 \end{pmatrix} e^{(2-i)x} = \begin{pmatrix} -1 - i \\ 1 \end{pmatrix} e^{2x}(\cos x - i \sin x)$$

$$= \begin{pmatrix} -\cos x - \sin x + i(\sin x - \cos x) \\ \cos x - i \sin x \end{pmatrix} e^{2x}.$$

Since $\mathbf{y}_1^*$ and $\mathbf{y}_2^*$ are complex solutions of $\mathbf{y}' = A\mathbf{y}$ and since A is a real matrix, both the real and imaginary parts of $\mathbf{y}_1^*$ and $\mathbf{y}_2^*$ must be real solutions of $\mathbf{y}' = A\mathbf{y}$. Notice that the real parts of $\mathbf{y}_1^*$ and $\mathbf{y}_2^*$ are equal and the imaginary part of $\mathbf{y}_1^*$ is equal to minus one times the imaginary part of $\mathbf{y}_2^*$. Thus, the characteristic roots and the associated vector solutions appear in complex conjugate pairs. Hence, corresponding to the two complex linearly independent solutions $\mathbf{y}_1^*$ and $\mathbf{y}_2^*$ there are two real linearly independent solutions, namely

$$\mathbf{y}_1 = \operatorname{Re} \mathbf{y}_1^* = \begin{pmatrix} -\cos x - \sin x \\ \cos x \end{pmatrix} e^{2x} = \operatorname{Re} \mathbf{y}_2^*$$

and

$$\mathbf{y}_2 = \operatorname{Im} \mathbf{y}_1^* = \begin{pmatrix} \cos x - \sin x \\ \sin x \end{pmatrix} e^{2x} = -\operatorname{Im} y_2^*.$$

Consequently, the general solution of the given system expressed in real terms is

$$\mathbf{y} = k_1\mathbf{y}_1 + k_2\mathbf{y}_2,$$

where k_1 and k_2 are arbitrary real scalar constants.

We see from this example that all one need do when the matrix A has nonrepeated complex conjugate characteristic roots is to find a solution corresponding to one of the complex conjugate roots. Then the real and imaginary parts of this complex vector solution will provide two linearly independent real vector solutions of $\mathbf{y}' = A\mathbf{y}$.

EXAMPLE Find the general solution of

$$\mathbf{y}' = \begin{pmatrix} 1 & 0 & -1 \\ 4 & 1 & 1 \\ 3 & -1 & 2 \end{pmatrix} \mathbf{y} \tag{28}$$

expressed as a linear combination of real-valued vector functions.

The characteristic equation

$$\begin{vmatrix} 1-r & 0 & -1 \\ 4 & 1-r & 1 \\ 3 & -1 & 2-r \end{vmatrix} = -r^3 + 4r^2 - 9r + 10 = 0$$

has roots 2, $1 + 2i$, and $1 - 2i$. A solution of (28), $\mathbf{y}_1 = \mathbf{c}_1 e^{2x}$, associated with the characteristic root 2 must satisfy

$$(A - 2I)\mathbf{c} = \begin{pmatrix} 1-2 & 0 & -1 \\ 4 & 1-2 & 1 \\ 3 & -1 & 2-2 \end{pmatrix} \begin{pmatrix} c_{11} \\ c_{21} \\ c_{31} \end{pmatrix}$$

$$= \begin{pmatrix} -1 & 0 & -1 \\ 4 & -1 & 1 \\ 3 & -1 & 0 \end{pmatrix} \begin{pmatrix} c_{11} \\ c_{21} \\ c_{31} \end{pmatrix} = \begin{pmatrix} 0 \\ 0 \\ 0 \end{pmatrix}.$$

Hence, $c_{31} = -c_{11}$, $c_{21} = 3c_{11}$, and c_{11} is arbitrary. Choosing $c_{11} = 1$, we find that

$$\mathbf{y}_1 = \begin{pmatrix} 1 \\ 3 \\ -1 \end{pmatrix} e^{2x}$$

is a solution of (28). For $\mathbf{y}_2^* = \mathbf{c}_2^* e^{(1+2i)x}$ to be a solution of (28), $\mathbf{c}_2^*$ must satisfy

$$[A - (1 + 2i)I]\mathbf{c}_2^* = \begin{pmatrix} 1-(1+2i) & 0 & -1 \\ 4 & 1-(1+2i) & 1 \\ 3 & -1 & 2-(1+2i) \end{pmatrix} \begin{pmatrix} c_{12}^* \\ c_{22}^* \\ c_{32}^* \end{pmatrix}$$

$$= \begin{pmatrix} -2i & 0 & -1 \\ 4 & -2i & 1 \\ 3 & -1 & 1-2i \end{pmatrix} \begin{pmatrix} c_{12}^* \\ c_{22}^* \\ c_{32}^* \end{pmatrix} = \begin{pmatrix} 0 \\ 0 \\ 0 \end{pmatrix}.$$

Hence, $c_{32}^* = -2ic_{12}^*$, $c_{22}^* = (-1 - 2i)c_{12}^*$, and c_{12}^* is arbitrary. Choosing $c_{12}^* = 1$, we find the following solution associated with the characteristic root $1 + 2i$:

$$\mathbf{y}_2^* = \begin{pmatrix} 1 \\ -1 - 2i \\ -2i \end{pmatrix} e^{(1+2i)x} = \begin{pmatrix} 1 \\ -1 - 2i \\ -2i \end{pmatrix} e^x(\cos 2x + i \sin 2x)$$

$$= \begin{pmatrix} \cos 2x + i \sin 2x \\ -\cos 2x + 2 \sin 2x + i(-2 \cos 2x - \sin 2x) \\ 2 \sin 2x + i(-2 \cos 2x) \end{pmatrix} e^x.$$

So the general solution of the given system in real form is

$$\mathbf{y} = k_1 \begin{pmatrix} 1 \\ 3 \\ -1 \end{pmatrix} e^{2x} + k_2 \begin{pmatrix} \cos 2x \\ -\cos 2x + 2 \sin 2x \\ 2 \sin 2x \end{pmatrix} e^x$$

$$+ k_3 \begin{pmatrix} \sin 2x \\ -2 \cos 2x - \sin 2x \\ -2 \cos 2x \end{pmatrix} e^x.$$

Repeated roots When the $n \times n$ matrix A of real constants has a root r of multiplicity m—that is, when r is a root of the characteristic equation of A m times—where $1 < m \leq n$, then the form of the general solution of the homogeneous system of linear equations (20) $\mathbf{y}' = A\mathbf{y}$ depends upon the number, j, of linearly independent characteristic vectors $\mathbf{c}_1, \mathbf{c}_2, \ldots, \mathbf{c}_j$ associated with the characteristic root r. It can be shown that $1 \leq j \leq m$. Hence, there are two distinct cases to consider: (i) $j = m$ and (ii) $j < m$.

When $j = m$ there are m linearly independent characteristic vectors $\mathbf{c}_1, \mathbf{c}_2, \ldots, \mathbf{c}_m$ associated with r and, consequently, the vectors $\mathbf{c}_1e^{rx}, \mathbf{c}_2e^{rx}, \ldots, \mathbf{c}_me^{rx}$ are linearly independent solutions of equation (20). These solutions will naturally be linearly independent of all other solutions which are associated with other roots of the characteristic equation. Therefore, the general solution will contain the linear combination $k_1\mathbf{c}_1e^{rx} + k_2\mathbf{c}_2e^{rx} + \cdots + k_m\mathbf{c}_me^{rx}$ due to the root r of multiplicity m.

When $j < m$, then there will be j linearly independent characteristic vectors of A, $\mathbf{c}_1, \mathbf{c}_2, \ldots, \mathbf{c}_j$ and j linearly independent solutions $\mathbf{c}_1e^{rx}, \mathbf{c}_2e^{rx}, \ldots, \mathbf{c}_je^{rx}$ of equation (20) corresponding to these characteristic vectors. The method employed to obtain the other $m - j$ linearly independent solutions to (20) associated with the root r depends upon j and the general situation will not be considered in this text. The comments made in this section apply to both repeated real roots and repeated complex roots.

EXAMPLE Find the general solution of the homogeneous linear system

$$\mathbf{y}' = \begin{pmatrix} 1 & 0 & 0 \\ 1 & 2 & 2 \\ -2 & -2 & -3 \end{pmatrix} \mathbf{y} = A\mathbf{y}. \tag{29}$$

We find the characteristic equation of A to be

$$|A - rI| = \begin{vmatrix} 1 - r & 0 & 0 \\ 1 & 2 - r & 2 \\ -2 & -2 & -3 - r \end{vmatrix} = -r^3 + 3r - 2 = 0.$$

The roots of this equation are -2, 1, 1. Thus, there is one solution of (29) of the form $\mathbf{y}_1 = \mathbf{c}_1 e^{-2x}$ and one solution or two linearly independent solutions of the form $\mathbf{y}_2 = \mathbf{c}_2 e^{x}$. The vector $\mathbf{c}_1$ must satisfy $(A - r_1 I)\mathbf{c}_1 = \mathbf{0}$ with $r_1 = -2$ in order for $\mathbf{y}_1$ to be a solution of (29). Hence, $\mathbf{c}_1$ must satisfy

$$\begin{pmatrix} 1 - (-2) & 0 & 0 \\ 1 & 2 - (-2) & 2 \\ -2 & -2 & -3 - (-2) \end{pmatrix} \begin{pmatrix} c_{11} \\ c_{21} \\ c_{31} \end{pmatrix} = \begin{pmatrix} 0 \\ 0 \\ 0 \end{pmatrix}.$$

Consequently, $c_{11} = 0$, $c_{31} = -2c_{21}$, and c_{21} is arbitrary. Choosing $c_{21} = 1$, we find one solution of (29) to be

$$\mathbf{y}_1 = \begin{pmatrix} 0 \\ 1 \\ -2 \end{pmatrix} e^{-2x}.$$

For $\mathbf{y}_2 = \mathbf{c}_2 e^{x}$ to be a solution $\mathbf{c}_2$ must satisfy

$$(A - I)\mathbf{c}_2 = \begin{pmatrix} 1 - 1 & 0 & 0 \\ 1 & 2 - 1 & 2 \\ -2 & -2 & -3 - 1 \end{pmatrix} \begin{pmatrix} c_{12} \\ c_{22} \\ c_{32} \end{pmatrix} = \begin{pmatrix} 0 \\ 0 \\ 0 \end{pmatrix}.$$

This system of equations reduces to the following single equation relating c_{12}, c_{22}, and c_{32}: $c_{12} + c_{22} + 2c_{32} = 0$. Hence, two of the three values c_{12}, c_{22}, c_{32} are arbitrary and the third is determined from the previous equation. Since two components of the vector $\mathbf{c}_2$ may be chosen arbitrarily, we may make two different choices of two components in such a manner that the resultant vectors are linearly independent. For example, choosing $c_{22} = 1$ and $c_{32} = 0$, we see that $c_{12} = -1$. And choosing $c_{22} = 0$ and $c_{32} = 1$, we find that $c_{12} = -2$. We leave it to the reader to show that the two characteristic vectors

$$\mathbf{c}_2^1 = \begin{pmatrix} -1 \\ 1 \\ 0 \end{pmatrix} \quad \text{and} \quad \mathbf{c}_2^2 = \begin{pmatrix} -2 \\ 0 \\ 1 \end{pmatrix}$$

are linearly independent. Since these two vectors are linearly independent, the corresponding solutions will be linearly independent and, consequently, the general solution of (29) is

$$\mathbf{y} = k_1 \begin{pmatrix} 0 \\ 1 \\ -2 \end{pmatrix} e^{-2x} + k_2 \begin{pmatrix} -1 \\ 1 \\ 0 \end{pmatrix} e^{x} + k_3 \begin{pmatrix} -2 \\ 0 \\ 1 \end{pmatrix} e^{x}.$$

EXAMPLE Find the general solution of

$$\mathbf{y}' = \begin{pmatrix} 1 & 2 & 2 \\ -2 & 2 & 0 \\ 0 & -4 & -3 \end{pmatrix} \mathbf{y} = A\mathbf{y}. \tag{30}$$

The characteristic equation of A is

$$\begin{vmatrix} 1-r & 2 & 2 \\ -2 & 2-r & 0 \\ 0 & -4 & -3-r \end{vmatrix} = -r^3 + 3r - 2 = 0.$$

This is the same characteristic equation that we obtained in the previous example. So the characteristic roots are again $-2, 1, 1$, and again there is one solution of (30) of the form $\mathbf{y}_1 = \mathbf{c}_1 e^{-2x}$ and one solution or two linearly independent solutions of the form $\mathbf{y}_2 = \mathbf{c}_2 e^{x}$.

The vector $\mathbf{c}_1$ must satisfy

$$\begin{pmatrix} 1-(-2) & 2 & 2 \\ -2 & 2-(-2) & 0 \\ 0 & -4 & -3-(-2) \end{pmatrix} \begin{pmatrix} c_{11} \\ c_{21} \\ c_{32} \end{pmatrix} = \begin{pmatrix} 0 \\ 0 \\ 0 \end{pmatrix}.$$

Hence, $c_{11} = 2c_{21}$, $c_{31} = -4c_{21}$, and c_{21} is arbitrary. Choosing $c_{21} = 1$, we find one solution of (30) to be

$$\mathbf{y}_1 = \begin{pmatrix} 2 \\ 1 \\ -4 \end{pmatrix} e^{-2x}.$$

The vector $\mathbf{c}_2$ must satisfy

$$\begin{pmatrix} 1-(1) & 2 & 2 \\ -2 & 2-(1) & 0 \\ 0 & -4 & -3-(1) \end{pmatrix} \begin{pmatrix} c_{12} \\ c_{22} \\ c_{32} \end{pmatrix} = \begin{pmatrix} 0 \\ 0 \\ 0 \end{pmatrix}.$$

Thus, $c_{32} = -c_{22}$, $c_{22} = 2c_{12}$, and c_{12} is arbitrary. Choosing $c_{12} = 1$, we find a second solution of (30) to be

$$\mathbf{y}_2 = \begin{pmatrix} 1 \\ 2 \\ -2 \end{pmatrix} e^x.$$

Since only one component of $\mathbf{c}_2$ is arbitrary, there will be only one solution of (30) of the form $\mathbf{y}_2 = \mathbf{c}_2 e^x$ corresponding to the characteristic root $r_2 = 1$. Since $r_2 = 1$ is a characteristic root of multiplicity two, we must seek a second linearly independent solution of (30) which has a different form.

Our experience with repeated roots in the case of nth order linear homogeneous differential equations with constant coefficients suggest that we seek a solution of the form $\mathbf{y} = \mathbf{c}xe^{r_2 x} = \mathbf{c}xe^x$. Differentiating, we find $\mathbf{y}' = \mathbf{c}(x + 1)e^x$. Substituting into (30), we see that $\mathbf{c}$ must satisfy $\mathbf{c}(x + 1)e^x = A\mathbf{c}xe^x$ for all x. Rearranging, we find that $\mathbf{c}$ must satisfy $(I - A)\mathbf{c}x + \mathbf{c} = \mathbf{0}$ for all x. For a polynomial in x—a vector polynomial in this case—to be zero for all values of x, each coefficient must be zero. Hence, we must have $(I - A)\mathbf{c} = \mathbf{0}$ and $\mathbf{c} = \mathbf{0}$. Consequently, there can be no nonzero solution of (30) of the form $\mathbf{y} = \mathbf{c}xe^x$. Thus, we have found one instance in which our analogy between nth order linear homogenous differential equations with constant coefficients and linear homogeneous systems of first order differential equations with constant coefficient breaks down.

When r was a root of order two of an auxiliary equation, we found two linearly independent solutions of the corresponding nth order linear homogeneous differential equation of the form $y_1 = c_1 e^{rx}$ and $y_2 = c_2 xe^{rx}$, where c_1 and c_2 are arbitrary constants. The solution y_2 did not contain a term of the form $c_3 e^{rx}$, where c_3 is an arbitrary constant, since c_3 is always a multiple of c_1. Notice that if $y_2^* = c_2 xe^{rx} + c_3 e^{rx}$, then the term $c_3 e^{rx}$ could be combined with y_1 when writing the general solution as follows:

$$\begin{aligned} y = y_1 + y_2^* + \cdots &= c_1 e^{rx} + c_2 xe^{rx} + c_3 e^{rx} + \cdots = (c_1 + c_3)e^{rx} \\ &+ c_2 xe^{rx} + \cdots = d_1 e^{rx} + c_2 xe^{rx} + \cdots = y_1 + y_2 + \cdots. \end{aligned}$$

So in the scalar case it is not necessary to assume that a second linear independent solution y_2 contains the term $c_3 e^{rx}$. However, since vectors **a** and **b** are not always scalar multiples of one another, it is appropriate to seek a solution of (30) of the form

$$\mathbf{y} = \mathbf{c}xe^{r_2x} + \mathbf{d}e^{r_2x}. \tag{31}$$

Differentiating (31) and substituting into (30), we find

$$r_2\mathbf{c}xe^{r_2x} + \mathbf{c}e^{r_2x} + r_2\mathbf{d}e^{r_2x} = A(\mathbf{c}xe^{r_2x} + \mathbf{d}e^{r_2x}) \quad \text{for all } x$$

or

$$r_2\mathbf{c}x + (\mathbf{c} + r_2\mathbf{d}) = A\mathbf{c}x + A\mathbf{d} \quad \text{for all } x. \tag{32}$$

Hence, $\mathbf{c}$ must satisfy

$$r_2\mathbf{c} = A\mathbf{c} \quad \text{or} \quad A\mathbf{c} - r_2\mathbf{c} = (A - r_2I)\mathbf{c} = \mathbf{0}.$$

That is, $\mathbf{c}$ must be a characteristic vector of A associated with the characteristic root r_2, which is 1 in this example. Thus, we may take

$$\mathbf{c} = \mathbf{c}_2 = \begin{pmatrix} 1 \\ 2 \\ -2 \end{pmatrix}.$$

From (32) we also see that $\mathbf{d}$ must satisfy

$$\mathbf{c} + r_2\mathbf{d} = A\mathbf{d} \quad \text{or} \quad A\mathbf{d} - r_2\mathbf{d} = (A - r_2I)\mathbf{d} = \mathbf{c}.$$

In this case, then, $\mathbf{d}$ must satisfy

$$\begin{pmatrix} 1-(1) & 2 & 2 \\ -2 & 2-(1) & 0 \\ 0 & -4 & -3-(1) \end{pmatrix}\begin{pmatrix} d_1 \\ d_2 \\ d_3 \end{pmatrix} = \begin{pmatrix} 1 \\ 2 \\ -2 \end{pmatrix}.$$

So $2d_2 + 2d_3 = 1$, $-2d_1 + d_2 = 2$, and d_1, d_2, or d_3 may be chosen arbitrarily. Choosing $d_2 = 0$, we find that $d_3 = \frac{1}{2}$ and $d_1 = -1$. So a solution of (30) is

$$\mathbf{y}_3 = \begin{pmatrix} 1 \\ 2 \\ -2 \end{pmatrix} xe^x + \begin{pmatrix} -1 \\ 0 \\ \frac{1}{2} \end{pmatrix} e^x.$$

To verify that $\mathbf{y}_1$, $\mathbf{y}_2$, and $\mathbf{y}_3$ are linearly independent, we calculate the Wronskian

$$W(\mathbf{y}_1, \mathbf{y}_2, \mathbf{y}_3, x) = \begin{vmatrix} 2e^{-2x} & e^x & xe^x - e^x \\ e^{-2x} & 2e^x & 2xe^x \\ -4e^{-2x} & -2e^x & -2xe^x + \frac{1}{2}e^x \end{vmatrix}$$

$$= e^{-2x}e^xxe^x\begin{vmatrix} 2 & 1 & 1 \\ 1 & 2 & 2 \\ -4 & -2 & -2 \end{vmatrix} + e^{-2x}e^xe^x\begin{vmatrix} 2 & 1 & -1 \\ 1 & 2 & 0 \\ -4 & -2 & \frac{1}{2} \end{vmatrix}$$

$$= 0 - \tfrac{9}{2} \neq 0.$$

So $\mathbf{y}_1, \mathbf{y}_2, \mathbf{y}_3$ are linearly independent for all real x and the general solution to (30) on any interval is $\mathbf{y} = k_1\mathbf{y}_1 + k_2\mathbf{y}_2 + k_3\mathbf{y}_3$, where k_1, k_2, and k_3 are arbitrary constants.

The preceeding example illustrates the rule that if r is a characteristic root of the real constant matrix A of multiplicity, $m \geq 2$, and if there is only one (unique except for a multiplicative constant) characteristic vector $\mathbf{c}$ associated with r, then the homogeneous system $\mathbf{y}' = A\mathbf{y}$ has one solution of the form $\mathbf{y}_1 = \mathbf{c}e^{rx}$ and another solution of the form $\mathbf{y}_2 = \mathbf{c}xe^{rx} + \mathbf{d}e^{rx}$, where $\mathbf{d}$ satisfies the equation

$$(A - rI)\mathbf{d} = \mathbf{c}. \tag{33}$$

EXERCISES

1. Show that the solutions which we obtained for equation (26), namely,

$$\mathbf{y}_1 = \begin{pmatrix} 3 \\ 1 \end{pmatrix} e^x, \qquad \mathbf{y}_2 = \begin{pmatrix} 1 \\ 1 \end{pmatrix} e^{-x}$$

are linearly independent.

2. Verify that

$$\mathbf{y}_1 = \begin{pmatrix} -\cos x - \sin x \\ \cos x \end{pmatrix} e^{2x}, \qquad \mathbf{y}_2 = \begin{pmatrix} \cos x - \sin x \\ \sin x \end{pmatrix} e^{2x}$$

are solutions of equation (27) and that they are linearly independent.

3. Verify that

$$\mathbf{y}_1 = \begin{pmatrix} \cos 2x \\ -\cos 2x + 2\sin 2x \\ 2\sin 2x \end{pmatrix} e^x, \qquad \mathbf{y}_2 = \begin{pmatrix} \sin 2x \\ -2\cos 2x - \sin 2x \\ -2\cos 2x \end{pmatrix} e^x$$

are solutions of equation (28) and that $\mathbf{y}_1$, $\mathbf{y}_2$, and $\mathbf{y}_3$ are linearly independent, where $\mathbf{y}_3 = \begin{pmatrix} 1 \\ 3 \\ -1 \end{pmatrix} e^{2x}$.

4. Show that the solutions

$$\mathbf{y}_1 = \begin{pmatrix} 0 \\ 1 \\ -2 \end{pmatrix} e^{-2x}, \qquad \mathbf{y}_2 = \begin{pmatrix} -1 \\ 1 \\ 0 \end{pmatrix} e^x, \qquad \mathbf{y}_3 = \begin{pmatrix} -2 \\ 0 \\ 1 \end{pmatrix} e^x$$

of equation (29) are linearly independent.

In Exercises 5–10 find the general solution of the homogeneous linear system $\mathbf{y}' = A\mathbf{y}$ for the given matrix A.

5. $A = \begin{pmatrix} 1 & 1 \\ 2 & 0 \end{pmatrix}$

6. $A = \begin{pmatrix} 3 & 1 \\ -1 & 1 \end{pmatrix}$

7. $A = \begin{pmatrix} -1 & 5 \\ -1 & 1 \end{pmatrix}$

8. $$A = \begin{pmatrix} 1 & -1 & 1 \\ 2 & 1 & 4 \\ 1 & 1 & 1 \end{pmatrix}$$

9. $$A = \begin{pmatrix} 1 & -1 & 1 \\ -1 & 1 & 1 \\ 1 & 1 & 1 \end{pmatrix}$$

10. $$A = \begin{pmatrix} 1 & 2 & -3 \\ -3 & 4 & -2 \\ 2 & 0 & 1 \end{pmatrix}$$

11. Show that the system

$$\mathbf{y}' = \begin{pmatrix} 4 & 1 & -1 \\ -1 & 3 & 1 \\ 1 & 1 & 2 \end{pmatrix} \mathbf{y} = A\mathbf{y}$$

has solutions of the form $\mathbf{y}_1 = \mathbf{c}e^{3x}$, $\mathbf{y}_2 = \mathbf{c}xe^{3x} + \mathbf{d}e^{3x}$, and $\mathbf{y}_3 = \mathbf{c}x^2e^{3x} + \mathbf{a}xe^{3x} + \mathbf{b}e^{3x}$, where $\mathbf{c}$ is any characteristic vector of A, $\mathbf{d}$ satisfies $(A - 3I)\mathbf{d} = \mathbf{c}$, $\mathbf{a}$ satisfies $(A - 3I)\mathbf{a} = 2\mathbf{c}$, and $\mathbf{b}$ satisfies $(A - 3I)\mathbf{b} = \mathbf{d}$. Verify that $\mathbf{y}_1$, $\mathbf{y}_2$, and $\mathbf{y}_3$ are linearly independent.

9.1.2 Nonhomogeneous Linear Systems

In this section we shall discuss methods for solving the nonhomogeneous linear system of differential equations

$$\mathbf{y}' = A(x)\mathbf{y} + \mathbf{b}(x), \tag{34}$$

where $\mathbf{y}$ is an n-dimensional vector, $A(x)$ is an $n \times n$ matrix of functions $a_{ij}(x)$ $(i, j = 1, 2, \ldots, n)$, which are all continuous on some interval I, and $\mathbf{b}(x) \not\equiv \mathbf{0}$

is an n-dimensional vector of functions $b_i(x)$ ($i = 1, 2, \ldots, n$), which are all continuous on the interval I. The associated homogeneous linear system of equations is

$$\text{(35)} \qquad \mathbf{y}' = A(x)\mathbf{y}.$$

We know from existence and uniqueness theorems already presented that (34) and (35) have unique solutions on the interval I for any specified initial condition $\mathbf{y}(x_0)$ for any $x_0 \in I$.

Assuming that we have been able to find n linearly independent solutions $\mathbf{y}_1, \mathbf{y}_2, \ldots, \mathbf{y}_n$ of (35), we may always employ the method of variation of parameters to find a particular solution $\mathbf{y}_p$ of (34). Thus, we seek a solution of (34) of the form

$$\text{(36)} \qquad \mathbf{y}_p = v_1(x)\mathbf{y}_1 + v_2(x)\mathbf{y}_2 + \cdots + v_n(x)\mathbf{y}_n,$$

where $v_1(x), v_2(x), \ldots, v_n(x)$ are unknown scalar functions which are to be determined so that $\mathbf{y}_p$ is a solution of (34). Differentiating (36), substituting into (34), and rearranging we find

$$\text{(37)} \qquad \begin{aligned} \mathbf{y}_p' &= (v_1'\mathbf{y}_1 + v_2'\mathbf{y}_2 + \cdots + v_n'\mathbf{y}_n) + (v_1\mathbf{y}_1' + v_2\mathbf{y}_2' + \cdots + v_n\mathbf{y}_n') \\ &= A(v_1\mathbf{y}_1 + v_2\mathbf{y}_2 + \cdots + v_n\mathbf{y}_n) + \mathbf{b} = A\mathbf{y}_p + \mathbf{b}. \end{aligned}$$

Since $\mathbf{y}_i$ satisfies (35) for $i = 1, 2, \ldots, n$, $v_i\mathbf{y}_i' = v_iA\mathbf{y}_i = Av_i\mathbf{y}_i$ for $i = 1, 2, \ldots, n$. Therefore, (37) reduces to

$$\text{(38)} \qquad v_1'\mathbf{y}_1 + v_2'\mathbf{y}_2 + \cdots + v_n'\mathbf{y}_n = \mathbf{b}.$$

This equation is a linear algebraic system of equations in the unknown functions $v_1', v_2', \ldots, v_n'$ and may be rewritten as

$$\text{(39)} \qquad \begin{pmatrix} y_{11} & y_{12} & \cdots & y_{1n} \\ y_{21} & y_{22} & \cdots & y_{2n} \\ \vdots & \vdots & & \vdots \\ y_{n1} & y_{n2} & \cdots & y_{nn} \end{pmatrix} \begin{pmatrix} v_1' \\ v_2' \\ \vdots \\ v_n' \end{pmatrix} = \begin{pmatrix} b_1 \\ b_2 \\ \vdots \\ b_n \end{pmatrix}.$$

Since the determinant of the $n \times n$ matrix appearing in equation (39) is the Wronskian of a linearly independent set of solutions of (35), the determinant is nonzero, and therefore (39) may always be solved for the unknown functions $v_1', v_2', \ldots, v_n'$. Integration then produces the desired functions $v_1, v_2, \ldots, v_n$.

Notice that the functions $v_1, v_2, \ldots, v_n$ are completely determined by the nonhomogeneous linear system of equations, whereas, when solving nth order nonhomogeneous linear equations, it was necessary to specify additional conditions that the functions $v_1', v_2', \ldots, v_n'$ had to satisfy.

EXAMPLE Find the general solution of the nonhomogeneous linear system

$$\mathbf{y}' = \begin{pmatrix} y_1' \\ y_2' \end{pmatrix} = \begin{pmatrix} 2 & -3 \\ 1 & -2 \end{pmatrix}\begin{pmatrix} y_1 \\ y_2 \end{pmatrix} + \begin{pmatrix} 4x - 2 \\ 3x \end{pmatrix} = A\mathbf{y} + \mathbf{b}. \tag{40}$$

The associated homogeneous equation is equation (26), so two linearly independent solutions of the associated homogeneous equation are

$$\mathbf{y}_1 = \begin{pmatrix} 3 \\ 1 \end{pmatrix} e^x \quad \text{and} \quad \mathbf{y}_2 = \begin{pmatrix} 1 \\ 1 \end{pmatrix} e^{-x}.$$

Consequently, we seek a particular solution of (40) of the form

$$\mathbf{y}_p = \begin{pmatrix} y_{p1} \\ y_{p2} \end{pmatrix} = v_1(x)\begin{pmatrix} 3 \\ 1 \end{pmatrix} e^x + v_2(x)\begin{pmatrix} 1 \\ 1 \end{pmatrix} e^{-x}. \tag{41}$$

From the proceeding development we know that v_1' and v_2' must satisfy (39), which in this case is

$$\begin{pmatrix} 3e^x & e^{-x} \\ e^x & e^{-x} \end{pmatrix}\begin{pmatrix} v_1' \\ v_2' \end{pmatrix} = \begin{pmatrix} 4x - 2 \\ 3x \end{pmatrix}. \tag{42}$$

Multiplying (42) by the matrix

$$\begin{pmatrix} \frac{1}{2}e^{-x} & -\frac{1}{2}e^{-x} \\ -\frac{1}{2}e^x & \frac{3}{2}e^x \end{pmatrix},$$

which is the inverse of the 2×2 matrix appearing in (42), we obtain

$$\begin{pmatrix} v_1' \\ v_2' \end{pmatrix} = \begin{pmatrix} \frac{1}{2}e^{-x} & -\frac{1}{2}e^{-x} \\ -\frac{1}{2}e^x & \frac{3}{2}e^x \end{pmatrix}\begin{pmatrix} 4x - 2 \\ 3x \end{pmatrix} = \begin{pmatrix} \frac{1}{2}xe^{-x} - e^{-x} \\ \frac{5}{2}xe^x + e^x \end{pmatrix}.$$

Integrating, we find

$$\begin{pmatrix} v_1 \\ v_2 \end{pmatrix} = \begin{pmatrix} -\frac{1}{2}xe^{-x} + \frac{1}{2}e^{-x} \\ \frac{5}{2}xe^x - \frac{3}{2}e^x \end{pmatrix}.$$

Hence,

$$\begin{aligned} \mathbf{y}_p &= (-\tfrac{1}{2}xe^{-x} + \tfrac{1}{2}e^{-x})\begin{pmatrix} 3 \\ 1 \end{pmatrix} e^x + (\tfrac{5}{2}xe^x - \tfrac{3}{2}e^x)\begin{pmatrix} 1 \\ 1 \end{pmatrix} e^{-x} \\ &= \begin{pmatrix} x \\ 2x - 1 \end{pmatrix} \end{aligned}$$

and the general solution of (40) is

$$\mathbf{y} = k_1\mathbf{y}_1 + k_2\mathbf{y}_2 + \mathbf{y}_p.$$

When A is a matrix whose entries are all real constants and when $\mathbf{b}$ has elements that are all polynomials, exponential functions, sinusoidal functions, or the sums or products of such functions, then one may use the method of undetermined coefficients to find a particular solution of (34). The procedure for selecting the form of a particular solution of (34) is similar to the procedures used in selecting the form of a particular solution of an nth order nonhomogeneous linear equation with constant coefficients (Section 6.3.1).

EXAMPLE Use the method of undetermined coefficients to find a particular solution of (40).

Since the roots of the characteristic equation of the associated homogeneous system are 1 and -1 and since the functions $b_1 = 4x - 2$ and $b_2 = 3x$ correspond to a characteristic equation with roots 0, 0, it is appropriate to seek a particular solution of (40) of the form

$$\mathbf{y}_p = \begin{pmatrix} ax + b \\ cx + d \end{pmatrix}, \tag{43}$$

where a, b, c, and d are constants which are to be determined. Differentiating (43) and substituting into (40), we find

$$\begin{aligned}\begin{pmatrix} a \\ c \end{pmatrix} &= \begin{pmatrix} 2 & -3 \\ 1 & -2 \end{pmatrix}\begin{pmatrix} ax + b \\ cx + d \end{pmatrix} + \begin{pmatrix} 4x - 2 \\ 3x \end{pmatrix} \\ &= \begin{pmatrix} (2a - 3c + 4)x + 2b - 3d - 2 \\ (a - 2c + 3)x + b - 2d \end{pmatrix}.\end{aligned}$$

Equating coefficients we see that a, b, c, and d must satisfy the following system of equations:

$$\begin{aligned} 2a - 3c + 4 &= 0 \\ 2b - 3d - 2 &= a \\ a - 2c + 3 &= 0 \\ b - 2d &= c. \end{aligned}$$

The solution of this system is easily found to be $a = 1$, $b = 0$, $c = 2$, and $d = -1$. Consequently,

$$\mathbf{y}_p = \begin{pmatrix} x \\ 2x - 1 \end{pmatrix}.$$

EXERCISES

Find particular solutions of the following nonhomogeneous linear systems of equations.

1. $\mathbf{y}' = \begin{pmatrix} 1 & -2 \\ 1 & 3 \end{pmatrix}\mathbf{y} + \begin{pmatrix} 4\cos x - 2\sin x \\ -3\cos x - 3\sin x \end{pmatrix}$ [See equation (27).]

2.
$$\mathbf{y}' = \begin{pmatrix} 1 & 0 & -1 \\ 4 & 1 & 1 \\ 3 & -1 & 2 \end{pmatrix}\mathbf{y} + \begin{pmatrix} 1 \\ 3 \\ 4 \end{pmatrix} e^{2x} \quad \text{[See equation (28).]}$$

3.
$$\mathbf{y}' = \begin{pmatrix} 1 & 0 & 0 \\ 1 & 2 & 2 \\ -2 & -2 & -3 \end{pmatrix}\mathbf{y} + \begin{pmatrix} 0 \\ -1 \\ 2 \end{pmatrix} e^{x} \quad \text{[See equation (29).]}$$

4.
$$\mathbf{y}' = \begin{pmatrix} 1 & 2 & 2 \\ -2 & 2 & 0 \\ 0 & -4 & -3 \end{pmatrix}\mathbf{y} + \begin{pmatrix} -2 \\ 0 \\ 4 \end{pmatrix} e^{x} \quad \text{[See equation (30).]}$$

5.
$$\mathbf{y}' = \begin{pmatrix} 1 & 1 \\ 2 & 0 \end{pmatrix}\mathbf{y} + \begin{pmatrix} \cos 2x - \sin 2x - \cos x - 2\sin x \\ -4\sin 2x - \cos x \end{pmatrix} \quad \text{(See Exercise 5, Section 9.1.)}$$

6.
$$\mathbf{y}' = \begin{pmatrix} 3 & 1 \\ -1 & 1 \end{pmatrix}\mathbf{y} + \begin{pmatrix} -x + 1 \\ x \end{pmatrix} e^{2x} \quad \text{(See Exercise 6, Section 9.1.)}$$

7.
$$\mathbf{y}' = \begin{pmatrix} -1 & 5 \\ -1 & 1 \end{pmatrix}\mathbf{y} + \begin{pmatrix} 4x^2 - 2x + 20 + 11e^{x} \\ 4x^2 - 2x + 6 + 3e^{x} \end{pmatrix} \quad \text{(See Exercise 7, Section 9.1.)}$$

For each of the following problems, verify that the given vectors are linearly independent solutions of the associated homogeneous system and find a particular solution of the given nonhomogeneous system.

8.
$$\mathbf{y}' = \begin{pmatrix} \dfrac{2}{x} & \dfrac{-1}{x^2} \\ 2 & 0 \end{pmatrix}\mathbf{y} + \begin{pmatrix} -3 + \dfrac{1}{x} - \dfrac{1}{x^2} \\ 1 - 6x \end{pmatrix}, \quad x > 0$$

$$\mathbf{y}_1 = \begin{pmatrix} 1 \\ 2x \end{pmatrix}, \qquad \mathbf{y}_2 = \begin{pmatrix} x \\ x^2 \end{pmatrix}$$

9.
$$\mathbf{y}' = \begin{pmatrix} \dfrac{5}{x} & \dfrac{4}{x} \\ \dfrac{-6}{x} & \dfrac{-5}{x} \end{pmatrix}\mathbf{y} + \begin{pmatrix} -2x \\ 5x \end{pmatrix}, \quad x > 0$$

$$\mathbf{y}_1 = \begin{pmatrix} 1 \\ -1 \end{pmatrix} x, \qquad \mathbf{y}_2 = \begin{pmatrix} -2 \\ 3 \end{pmatrix} x^{-1}$$

10.

$$\mathbf{y}' = \begin{pmatrix} \dfrac{1}{2x} & \dfrac{1}{2x} & \dfrac{-1}{2x} \\ \dfrac{-1}{2x} & \dfrac{3}{2x} & \dfrac{1}{2x} \\ \dfrac{-1}{x} & \dfrac{1}{x} & \dfrac{1}{x} \end{pmatrix} \mathbf{y} + \begin{pmatrix} 2x + \frac{3}{2} \\ -2x + \frac{1}{2} \\ 2x + 2 \end{pmatrix} e^{2x}, \quad x > 0$$

$$\mathbf{y}_1 = \begin{pmatrix} 1 \\ 0 \\ 1 \end{pmatrix}, \qquad \mathbf{y}_2 = \begin{pmatrix} 1 \\ 1 \\ 0 \end{pmatrix} x, \qquad \mathbf{y}_3 = \begin{pmatrix} 0 \\ 1 \\ 1 \end{pmatrix} x^2.$$

9.1.3 Applications of Linear Systems

We shall present a few applications of systems of linear differential equations to some electrical and physical systems. Of course, there are many other biological, chemical, economical, electrical, physical, sociological, and so on, systems that can also be adequately modeled by a system of linear differential equations.

9.1.3.1 Electrical systems

Kirchhoff's laws and the common abbreviations, graphic symbols, and units for the electrical components and quantities that will be used in this section were introduced previously in Section 7.4.

An RLC network Let us consider the *RLC* network depicted schematically by Figure 9.1. A *loop* is any closed path through which current can flow. The network represented by Figure 9.1 has three closed loops: *abcdefa*, *abefa*, and *bcdeb*. A *junction* is any point in the network at which the current can flow in three or more directions. This network has two junctions *b* and *c*. We have arbitrarily assigned the direction of current flow as indicated by the arrows in Figure 9.1. The manner of assigning the current flow does not

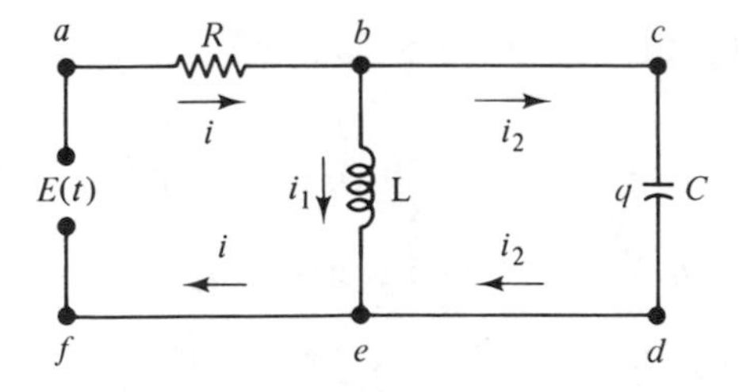

Figure 9.1 *RLC* network.

affect the results one obtains. If the current flows in the opposite direction for some or all the loops of any network, the results will reflect this condition by giving a negative current flow for that particular loop.

Applying Kirchhoff's current law at junction b or e, we find

$$i = i_1 + i_2. \tag{44}$$

Applying Kirchhoff's voltage law to the three loops of this network, we obtain:

for loop *abefa*:

$$Ri + L\frac{di_1}{dt} = E(t), \tag{45}$$

for loop *bcdeb*:

$$\frac{q}{C} - L\frac{di_1}{dt} = 0, \tag{46}$$

and for loop *abcdefa*:

$$Ri + \frac{q}{C} = E(t). \tag{47}$$

We shall also need the following relationship between the charge on the capacitor, q, and the current flowing through the capacitor, i_2:

$$i_2 = \frac{dq}{dt}. \tag{48}$$

Equations (45), (46), and (47) are obviously not independent since (47) may be obtained by adding (45) and (46). Substituting (44) into (45) and solving for di_1/dt, we get

$$\frac{di_1}{dt} = -\frac{L}{R}i_1 - \frac{L}{R}i_2 + \frac{E(t)}{L}. \tag{49}$$

Differentiating (47) and substituting i_2 for dq/dt, we find

$$R\frac{di}{dt} + \frac{i_2}{C} = \frac{dE}{dt}. \tag{50}$$

Differentiating (44), substituting into (50), and solving for di_2/dt, we obtain

$$\frac{di_2}{dt} = -\frac{di_1}{dt} - \frac{i_2}{RC} + \frac{1}{R}\frac{dE}{dt}. \tag{51}$$

Substituting for di_1/dt from (49) and rearranging, we find

$$\frac{di_2}{dt} = \frac{R}{L}i_1 + \left(\frac{R}{L} - \frac{1}{RC}\right)i_2 - \frac{E(t)}{L} + \frac{1}{R}\frac{dE}{dt}. \tag{52}$$

Viewing equations (49) and (52) as a linear system with components i_1 and i_2, we see that we must solve the system

$$\begin{pmatrix} i_1' \\ i_2' \end{pmatrix} = \begin{pmatrix} -\dfrac{R}{L} & -\dfrac{R}{L} \\ \dfrac{R}{L} & \dfrac{R}{L} - \dfrac{1}{RC} \end{pmatrix} \begin{pmatrix} i_1 \\ i_2 \end{pmatrix} + \begin{pmatrix} \dfrac{E(t)}{L} \\ -\dfrac{E(t)}{L} + \dfrac{1}{R}\dfrac{dE}{dt} \end{pmatrix}. \tag{53}$$

EXAMPLE Find the currents, i, i_1, i_2, and the charge q for the network shown in Figure 9.1 if $E(t) = 30$ V; $R = 10\ \Omega$; $L = .1$ H; $C = 1.6 \times 10^{-4}$ F; and at $t = 0$, $i_1(0) = i_2(0) = 0$.

Substituting these values into (53), we obtain

$$\begin{pmatrix} i_1' \\ i_2' \end{pmatrix} = \begin{pmatrix} -100 & -100 \\ 100 & -525 \end{pmatrix} \begin{pmatrix} i_1 \\ i_2 \end{pmatrix} + \begin{pmatrix} 300 \\ -300 \end{pmatrix} = A\mathbf{i} + \mathbf{b}. \tag{54}$$

The characteristic polynomial of the associated homogeneous system of equations is

$$\begin{vmatrix} -100 - r & -100 \\ 100 & -525 - r \end{vmatrix} = r^2 + 625r + 62\ 500 = 0.$$

The roots of this quadratic equation are $r_1 = -125$ and $r_2 = -500$. Thus, the associated homogeneous system of equations $\mathbf{i}' = A\mathbf{i}$ has two linearly independent solutions of the form $\mathbf{i}_1 = \mathbf{c}_1 e^{-125t}$ and $\mathbf{i}_2 = \mathbf{c}_2 e^{-500t}$. Recall that $\mathbf{c}_1$ must satisfy

$$(A - r_1 I)\mathbf{c}_1 = \mathbf{0}$$

or

$$\begin{pmatrix} 25 & -100 \\ 100 & -400 \end{pmatrix} \begin{pmatrix} c_{11} \\ c_{12} \end{pmatrix} = \begin{pmatrix} 0 \\ 0 \end{pmatrix}.$$

Hence, $c_{11} = 4c_{12}$ and c_{12} is arbitrary. The constant vector $\mathbf{c}_2$ must satisfy

$$(A - r_2 I)\mathbf{c}_2 = \mathbf{0}$$

or

$$\begin{pmatrix} 400 & -100 \\ 100 & -25 \end{pmatrix} \begin{pmatrix} c_{21} \\ c_{22} \end{pmatrix} = \begin{pmatrix} 0 \\ 0 \end{pmatrix}.$$

Hence, $c_{22} = 4c_{21}$ and c_{21} is arbitrary. Thus, the transient solution (complementary solution) for the given system is

$$\mathbf{i}_c = \begin{pmatrix} i_1 \\ i_2 \end{pmatrix}_c = k_1 \begin{pmatrix} 4 \\ 1 \end{pmatrix} e^{-125t} + k_2 \begin{pmatrix} 1 \\ 4 \end{pmatrix} e^{-500t}.$$

A steady state solution (particular solution) is easily found to be

$$\mathbf{i}_p = \begin{pmatrix} i_1 \\ i_2 \end{pmatrix}_p = \begin{pmatrix} 3 \\ 0 \end{pmatrix}.$$

Hence, the general solution of (54) is $\mathbf{i} = \mathbf{i}_c + \mathbf{i}_p$. Imposing the initial conditions, we see that k_1 and k_2 must satisfy

$$\mathbf{i}(0) = \mathbf{i}_c(0) + \mathbf{i}_p(0).$$

That is,

$$\begin{pmatrix} i_1(0) \\ i_2(0) \end{pmatrix} = \begin{pmatrix} 0 \\ 0 \end{pmatrix} = \begin{pmatrix} 4k_1 + k_2 \\ k_1 + 4k_2 \end{pmatrix} + \begin{pmatrix} 3 \\ 0 \end{pmatrix}.$$

Solving this system, we find $k_1 = -\frac{4}{5}$ and $k_2 = \frac{1}{5}$. So

$$\begin{aligned} i_1 &= -\tfrac{16}{5}e^{-125t} + \tfrac{1}{5}e^{-500t} + 3, \\ i_2 &= -\tfrac{4}{5}e^{-125t} + \tfrac{4}{5}e^{-500t}, \end{aligned}$$

and

$$i = i_1 + i_2 = -4e^{-125t} + e^{-500t} + 3.$$

Since $i_2 = dq/dt$,

$$q(t) = \int i_2\, dt = \frac{4}{625}\, e^{-125t} - \frac{4}{2500}\, e^{-500t} + k. \tag{55}$$

Evaluating equation (47) at $t = 0$, we get

$$10i(0) + \frac{q(0)}{1.6 \times 10^{-3}} = 30.$$

Since $i(0) = 0, q(0) = 48 \times 10^{-3}$. Imposing this initial condition on equation (55), we see that k must satisfy

$$48 \times 10^{-3} = \frac{4}{625} - \frac{4}{2500} + k.$$

Hence, $k = 27/625$.

EXERCISES

1. Find the transient solution for the currents i_1 and i_2 of Figure 9.1 if the values for the voltage, resistance, and inductance are the same as those used in the example of the text [$E(t) = 30$ V, $R = 10\ \Omega$, $L = .1$ H], but the capacitance is increased to $C = 2.5 \times 10^{-4}$ F.

2. Find the transient solution for the currents i_1 and i_2 of Figure 9.1, if the values for the voltage, resistance, and inductance are the same as those

used in the example in the text but the capacitance is increased to $C = 5 \times 10^{-4}$ F.

3. Find the steady state solution for the currents i_1 and i_2 of Figure 9.1, if $E(t) = 30 \cos 100t$, $R = 10\ \Omega$, $L = .1$ H, and $C = 2.5 \times 10^{-4}$ F.

For each of the following electrical networks:

(a) Derive the general system of differential equations which the currents i_1 and i_2 must satisfy.

(b) Solve this system for the particular constants and initial conditions given.

4. (a)

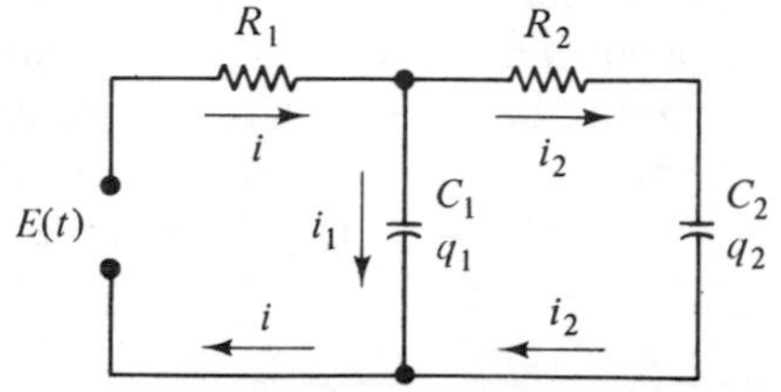

(b) $E(t) = 90$ V, $R_1 = 10\ \Omega$, $R_2 = 20\ \Omega$, $C_1 = 2 \times 10^{-4}$ F, $C_2 = 10^{-4}$ F, $i_1(0) = 6$ A, and $i_2(0) = 0$ A.

(c) What are the initial charges, $q_1(0)$ and $q_2(0)$, on the capacitors?

5. (a)

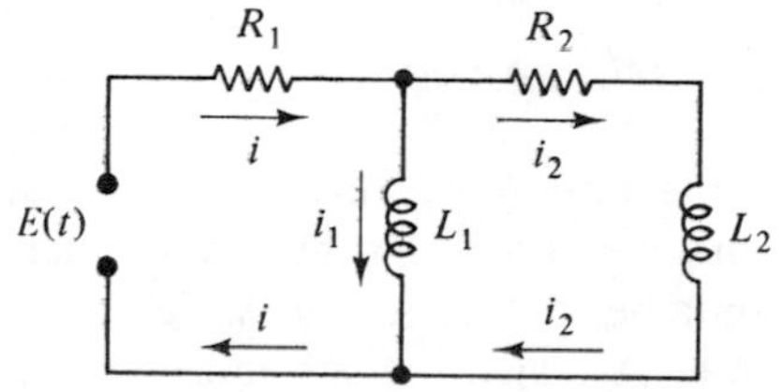

(b) $E(t) = 30$ V, $R_1 = 20\ \Omega$, $R_2 = 10\ \Omega$, $L_1 = .2$ H, $L_2 = .1$ H, and $i_1(0) = i_2(0) = 0$ A.

6. (a)

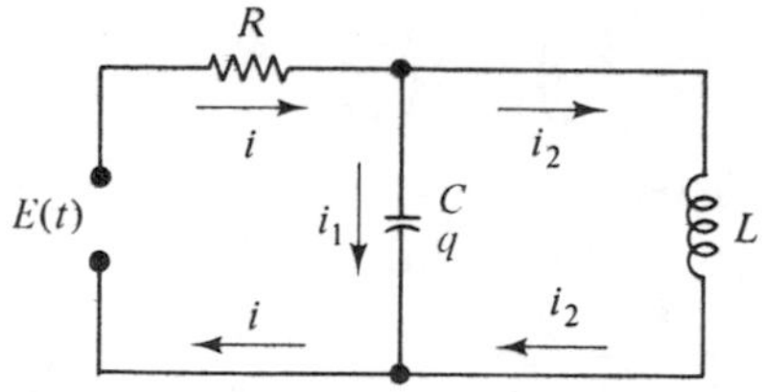

(b) $E(t) = 120$ V, $R = 40\ \Omega$, $L = .2$ H, $C = 2.5 \times 10^{-4}$ F, and $i_1(0) = i_2(0) = 0$ A.

9.1.3.2 Physical systems

In this section we shall develop the systems of differential equations which effectively model two physical systems. We leave it to the reader to calculate specific solutions to these systems.

A mixture system Suppose that at time $t = 0$ a quantity q_1 of a particular substance is present in a solution that fills a container of volume V_1 and a quantity q_2 of the same substance is present in a solution that fills a second container of volume V_2. The two containers are connected by tubes of negligible length as shown in Figure 9.2. Assume that at time $t = 0$, (i) a fluid containing a concentration c_α of the substance is allowed to enter the first container with volume V_1 at a constant rate α; (ii) the fluid in the first container is kept at a uniform, but not constant, concentration and is allowed to flow into the second container at a constant rate β; and (iii) the fluid in the second container is kept at a uniform, but not constant, concentration and is allowed to flow back into the first container at a constant rate γ and out of the system at a constant rate δ. We also assume that both containers are always filled.

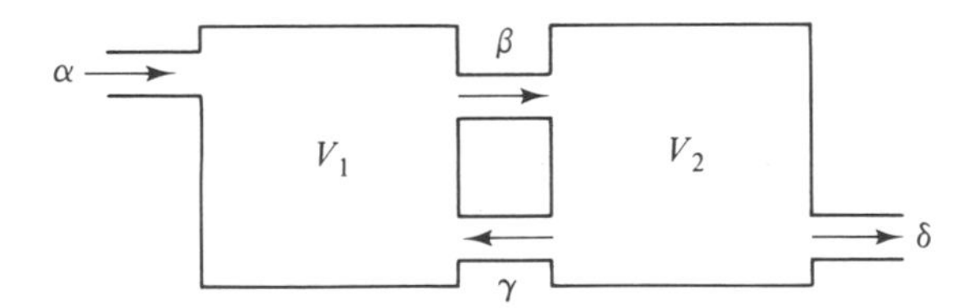

Figure 9.2 Mixture system.

The problem is to determine the amount of substance, $Q_1(t)$ and $Q_2(t)$, in each container as a function of time. Equating the rate of change of the amount of substance in each container to the rate at which the substance enters the container minus the rate at which the substance leaves the container, we obtain the following system:

$$Q_1' = \alpha c_\alpha + \gamma\left(\frac{Q_2}{V_2}\right) - \beta\left(\frac{Q_1}{V_1}\right) \tag{56a}$$

$$Q_2' = \beta\left(\frac{Q_1}{V_1}\right) - \gamma\left(\frac{Q_2}{V_2}\right) - \delta\left(\frac{Q_2}{V_2}\right). \tag{56b}$$

Since we have assumed that the containers are always full, the following relationships between the rates of flow must hold:

$$\alpha + \gamma = \beta \tag{57}$$

and

$$\beta = \gamma + \delta. \tag{58}$$

Substituting (57) into (56a) and (58) into (56b) and rearranging, we obtain

$$\begin{pmatrix} Q_1' \\ Q_2' \end{pmatrix} = \begin{pmatrix} -\dfrac{\alpha + \gamma}{V_1} & \dfrac{\gamma}{V_2} \\[2ex] \dfrac{\delta + \gamma}{V_1} & -\dfrac{\delta + \gamma}{V_2} \end{pmatrix} \begin{pmatrix} Q_1 \\ Q_2 \end{pmatrix} + \begin{pmatrix} \alpha c_\alpha \\ 0 \end{pmatrix}. \tag{59}$$

The system (59) is to satisfy the initial conditions

$$\begin{pmatrix} Q_1(0) \\ Q_2(0) \end{pmatrix} = \begin{pmatrix} q_1 \\ q_2 \end{pmatrix}. \tag{60}$$

A compound spring-mass system A mass m_1 is attached to one end of a spring S_1 whose spring constant is k_1. The other end of this spring is suspended from a fixed support. A second mass m_2 is attached to one end of a spring S_2 whose spring constant is k_2. The other end of this spring is suspended from the mass m_1 and the resulting system is allowed to come to rest in the equilibrium position as shown in Figure 9.3(a). Let y_1 denote the distance of mass m_1 from its equilibrium position and let y_2 denote the distance of mass m_2 from its equilibrium position. In both cases, the positive direction is chosen to be downward. Suppose that at some time t, $y_2(t) > y_1(t) > 0$ as shown in Figure 9.3(b).

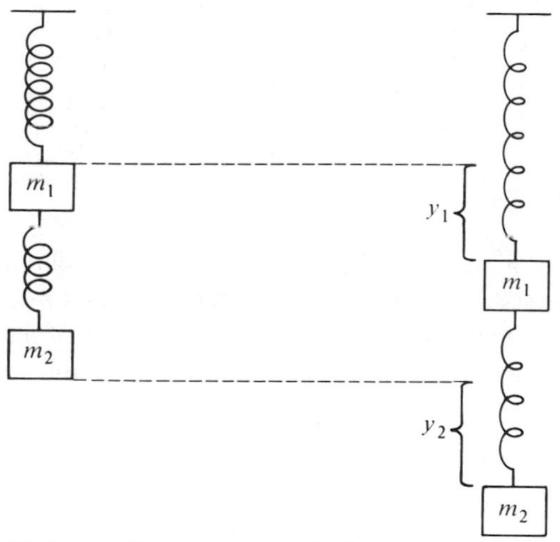

(a) Equilibrium position (b) Both springs stretched position

Figure 9.3 Compound spring-mass system.

Applying Newton's Second Law of Motion ($F = ma$) to mass m_1, we obtain

$$m_1 y_1'' = -k_1 y_1 + k_2(y_2 - y_1). \tag{61}$$

The first term on the right-hand side of this equation is negative, since the spring S_1 is stretched from its equilibrium length ($y_1 > 0$) and the force of the spring acts to restore the mass m_1 to its equilibrium position. Since the spring S_2 is stretched from its equilibrium length ($y_2 - y_1 > 0$), the force of the spring acts to draw the masses m_1 and m_2 together, and therefore the second term on the right-hand side of equation (61) is positive. Applying Newton's Second Law of Motion to mass m_2, we obtain

$$m_2 y_2'' = -k_2(y_2 - y_1). \tag{62}$$

If $y_1 < 0$, then spring S_1 is compressed and it exerts a downward force on mass m_1 of $-k_1 y_1 > 0$. If $y_2 - y_1 < 0$, then spring S_2 is compressed and it exerts an upward force on mass m_1 of $k_2(y_2 - y_1) < 0$ and a downward force on mass m_2 of $-k_2(y_2 - y_1) > 0$. So regardless of the particular physical state of the system at any instant, the system satisfies equations (61) and (62).

Let $v_1 = y_1$, $v_2 = y_1'$, $v_3 = y_2$, and $v_4 = y_2'$. Differentiating and substituting from the equations above and equations (61) and (62), we obtain the following system of first order differential equations:

$$\mathbf{v}' = \begin{pmatrix} v_1' \\ v_2' \\ v_3' \\ v_4' \end{pmatrix} = \begin{pmatrix} 0 & 1 & 0 & 0 \\ -\dfrac{k_1 + k_2}{m_1} & 0 & \dfrac{k_2}{m_1} & 0 \\ 0 & 0 & 0 & 1 \\ \dfrac{k_2}{m_2} & 0 & -\dfrac{k_2}{m_2} & 0 \end{pmatrix} \begin{pmatrix} v_1 \\ v_2 \\ v_3 \\ v_4 \end{pmatrix} = A\mathbf{v}. \tag{63}$$

Notice that v_1 is the position of mass m_1, v_2 is the velocity of mass m_1, v_3 is the position of mass m_2, and v_4 is the velocity of mass m_2. So appropriate initial conditions are the initial positions and velocities of masses m_1 and m_2. That is, appropriate initial conditions for system (63) are

$$\mathbf{v}(0) = \begin{pmatrix} v_1(0) \\ v_2(0) \\ v_3(0) \\ v_4(0) \end{pmatrix} = \begin{pmatrix} y_1(0) \\ y_1'(0) \\ y_2(0) \\ y_2'(0) \end{pmatrix}. \tag{64}$$

EXERCISES

1. Find the solution of system (59) subject to the initial conditions of (60) if $\alpha = 15$ gal/min, $\gamma = 5$ gal/min, $\delta = 15$ gal/min, $V_1 = 200$ gal, $V_2 = 100$ gal, $c_\alpha = 1$ lb/gal, $q_1 = 10$ lb, and $q_2 = 0$ lb.

2. Find the solution of system (63) for $k_1 = 18$ kg/s^2, $k_2 = 3$ kg/s^2, $m_1 = 3$ kg, and $m_2 = .5$ kg subject to the following general initial conditions;
   (a) $v_1(0) = v_3(0) = C$ m, $v_2(0) = v_4(0) = 0$
   (b) $v_1(0) = C$ m, $v_2(0) = v_3(0) = v_4(0) = 0$
   (c) $v_1(0) = v_2(0) = v_3(0) = 0$, $v_4(0) = C$ m/s
   (d) $v_1(0) = C$ m, $v_2(0) = 0$, $v_3(0) = D$ m, $v_4(0) = 0$

9.2 NUMERICAL SOLUTIONS

As one might expect, it is often difficult or impossible to obtain an analytic solution to the vector initial value problem

$$\mathbf{y}' = \mathbf{f}(x, \mathbf{y}); \qquad \mathbf{y}(x_0) = \mathbf{c}_0. \tag{3}$$

Hence, one needs to know some methods for obtaining approximate solutions. The numerical methods discussed in Chapter 4 for solving the scalar initial value problem $y' = f(x, y)$; $y(x_0) = c_0$ can easily be extended to the vector initial value problem (3).

Let $x_n = x_0 + nh$, let $\mathbf{y}_n$ denote an approximation to $\mathbf{y}(x_n)$, and let $\mathbf{f}_n = \mathbf{f}(x_n, \mathbf{y}_n)$. The following recursive formulas are often used to generate approximate solutions to the vector initial value problem (3).

9.2.1 Single-Step Methods

Taylor series expansion of order three:

$$\mathbf{y}_{n+1} = \mathbf{y}_n + h\mathbf{f}_n + \frac{h^2\mathbf{f}'_n}{2} + \frac{h^3\mathbf{f}''_n}{6} \tag{65}$$

Euler's method:

$$\mathbf{y}_{n+1} = \mathbf{y}_n + h\mathbf{f}_n \tag{66}$$

Improved Euler's method:

$$\mathbf{y}_{n+1} = \mathbf{y}_n + \frac{h(\mathbf{f}_n + \mathbf{f}(x_{n+1}, \mathbf{y}_n + h\mathbf{f}_n))}{2} \tag{67}$$

Fourth order Runge-Kutta method

$$\mathbf{y}_{n+1} = \mathbf{y}_n + \frac{h(\mathbf{k}_1 + 2\mathbf{k}_2 + 2\mathbf{k}_3 + \mathbf{k}_4)}{6} \tag{68}$$

where

$$\mathbf{k}_1 = \mathbf{f}_n = \mathbf{f}(x_n, \mathbf{y}_n)$$

$$\mathbf{k}_2 = \mathbf{f}\left(x_n + \frac{h}{2}, \mathbf{y}_n + \frac{h\mathbf{k}_1}{2}\right)$$

$$\mathbf{k}_3 = \mathbf{f}\left(x_n + \frac{h}{2}, \mathbf{y}_n + \frac{h\mathbf{k}_2}{2}\right)$$

$$\mathbf{k}_4 = \mathbf{f}(x_{n+1}, \mathbf{y}_n + h\mathbf{k}_3).$$

Predictor-corrector:

$$\mathbf{y}_{n+1}^0 = \mathbf{y}_n + h\mathbf{f}_n \tag{69p}$$

$$\mathbf{y}_{n+1}^k = \mathbf{y}_n + \frac{h(\mathbf{f}_n + \mathbf{f}(x_{n+1}, \mathbf{y}_{n+1}^{k-1}))}{2}, \qquad k = 1, 2, \ldots. \tag{69c}$$

9.2.2 Multistep Methods

Adams-Bashforth method, $m = 3$:

$$\mathbf{y}_{n+1} = \mathbf{y}_n + \frac{h(55\mathbf{f}_n - 59\mathbf{f}_{n-1} + 37\mathbf{f}_{n-2} - 9\mathbf{f}_{n-3})}{24} \tag{70}$$

Nystrom method, $m = 3$:

$$\mathbf{y}_{n+1} = \mathbf{y}_{n-1} + \frac{h(8\mathbf{f}_n - 5\mathbf{f}_{n-1} + 4\mathbf{f}_{n-2} - \mathbf{f}_{n-3})}{3} \tag{71}$$

Milne's method:

$$\mathbf{y}_{n+1} = \mathbf{y}_{n-3} + \frac{4h(2\mathbf{f}_n - \mathbf{f}_{n-1} + 2\mathbf{f}_{n-2})}{3} \tag{72}$$

Milne predictor-corrector:

$$\mathbf{y}_{n+1}^0 = \mathbf{y}_{n-3} + \frac{4h(2\mathbf{f}_n - \mathbf{f}_{n-1} + 2\mathbf{f}_{n-2})}{3} \tag{73p}$$

$$\mathbf{y}_{n+1}^k = \mathbf{y}_{n-1} + \frac{h(\mathbf{f}(x_{n+1}, \mathbf{y}_{n+1}^{k-1}) + 4\mathbf{f}_n + \mathbf{f}_{n-1})}{3}, \qquad k = 1, 2, \ldots. \tag{73c}$$

Adams-Bashforth, $m = 3$, predictor-Adams-Moulton, $m = 2$, corrector:

$$\mathbf{y}_{n+1}^0 = \mathbf{y}_n + \frac{h(55\mathbf{f}_n - 59\mathbf{f}_{n-1} + 37\mathbf{f}_{n-2} - 9\mathbf{f}_{n-3})}{24} \tag{74p}$$

$$\text{(74c)}\qquad \mathbf{y}_{n+1}^{k} = \mathbf{y}_n + \frac{h(9\mathbf{f}(x_{n+1}, \mathbf{y}_{n+1}^{k-1}) + 19\mathbf{f}_n - 5\mathbf{f}_{n-1} + \mathbf{f}_{n+2})}{24},$$

$$k = 1, 2, \ldots.$$

In order to better understand the vector notation let us use the fourth order Runge-Kutta method to solve the *Volterra prey-predator* system. Volterra (1860–1940) developed his model for the interaction between two different species in the mid-1920s. His model was based upon a statistical analysis made by a friend of fish catches in the Adriatic Sea. His model assumes that two species of fish interact with each other in a prey-predator relationship and that no other factors affect their existence. That is, the model assumes (i) that there is a species of small fish whose population at any time, x, is $y_1(x)$; (ii) that this species eats only plant life; and (iii) that there is always an ample supply of plant life to support the population. It is further assumed (i) that there is a species of large fish whose population at time, x, is $y_2(x)$; and (ii) that this species survives by feeding only on the other species of fish. It was observed that the population of the small fish increased and decreased periodically and that the population of the larger fish did likewise but somewhat out of phase. Volterra's prey-predator model is

$$\text{(75)}\qquad \begin{aligned} y_1' &= ay_1 - by_1y_2 \\ y_2' &= -cy_2 + dy_1y_2, \end{aligned}$$

where a, b, c, and d are positive constants. The model is nonlinear and cannot be solved in terms of elementary functions. The term $-by_1y_2$ represents the rate of population decrease of the small fish due to encounters with the large fish, and the term dy_1y_2 represents the rate of population growth of the large fish due to the same encounter. Notice that if there were no large fish—that is, if $y_2(x) \equiv 0$, then the small fish population would grow exponentially, whereas if there were no small fish, $y_1(x) \equiv 0$, the large fish population would die out exponentially.

EXAMPLE Solve the Volterra prey-predator initial value problem whose differential equation system is (75), where $a = 1$, $b = .5$, $c = 2$, and $d = .25$ subject to the initial conditions $y_1(0) = 10$ and $y_2(0) = 5$ using the fourth order Runge-Kutta method with stepsize $h = .1$.

For this system equation (68) written in detail becomes

$$y_{1\,n+1} = y_{1n} + \frac{h(k_{11} + 2k_{12} + 2k_{13} + k_{14})}{6}$$

$$y_{2\,n+1} = y_{2n} + \frac{h(k_{21} + 2k_{22} + 2k_{23} + k_{24})}{6}$$

where

$$k_{11} = f_1(x_n, y_{1n}, y_{2n}) = y_{1n} - .5y_{1n}y_{2n}$$

$$k_{21} = f_2(x_n, y_{1n}, y_{2n}) = -2y_{2n} + .25y_{1n}y_{2n}$$

$$k_{12} = f_1\left(x_n + \frac{h}{2}, y_{1n} + \frac{hk_{11}}{2}, y_{2n} + \frac{hk_{21}}{2}\right)$$

$$= \left(y_{1n} + \frac{hk_{11}}{2}\right) - .5\left(y_{1n} + \frac{hk_{11}}{2}\right)\left(y_{2n} + \frac{hk_{21}}{2}\right)$$

$$k_{22} = f_2\left(x_n + \frac{h}{2}, y_{1n} + \frac{hk_{11}}{2}, y_{2n} + \frac{hk_{21}}{2}\right)$$

$$= -2\left(y_{2n} + \frac{hk_{21}}{2}\right) + .25\left(y_{1n} + \frac{hk_{11}}{2}\right)\left(y_{2n} + \frac{hk_{21}}{2}\right)$$

$$k_{13} = f_1\left(x_n + \frac{h}{2}, y_{1n} + \frac{hk_{12}}{2}, y_{2n} + \frac{hk_{22}}{2}\right)$$

$$= \left(y_{1n} + \frac{hk_{12}}{2}\right) - .5\left(y_{1n} + \frac{hk_{12}}{2}\right)\left(y_{2n} + \frac{hk_{22}}{2}\right)$$

$$k_{23} = f_2\left(x_n + \frac{h}{2}, y_{1n} + \frac{hk_{12}}{2}, y_{2n} + \frac{hk_{22}}{2}\right)$$

$$= -2\left(y_{2n} + \frac{hk_{22}}{2}\right) + .25\left(y_{1n} + \frac{hk_{12}}{2}\right)\left(y_{2n} + \frac{hk_{22}}{2}\right)$$

$$k_{14} = f_1(x_{n+1}, y_{1n} + hk_{13}, y_{2n} + hk_{23})$$

$$= (y_{1n} + hk_{13}) - .5(y_{1n} + hk_{13})(y_{2n} + hk_{23})$$

$$k_{24} = f_2(x_{n+1}, y_{1n} + hk_{13}, y_{2n} + hk_{23})$$

$$= -2(y_{2n} + hk_{23}) + .25(y_{1n} + hk_{13})(y_{2n} + hk_{23}).$$

A flowchart of the fourth order Runge-Kutta method for solving a vector initial value problem is displayed in Figure 9.4. The initial conditions are x_0 and $\mathbf{y}_0$, the final value of the independent variable for which the solution is to be produced is x_f, and the number of steps of size h required to get from x_0 to or just past x_f is N. The computer program used to produce the fourth order Runge-Kutta solution to the given Volterra prey-predator initial value problem is shown in Figure 9.5. The number of the corresponding computer program statement or statements of Figure 9.5 which accomplishes each of the operations indicated by the flowchart of Figure 9.4 is also shown in

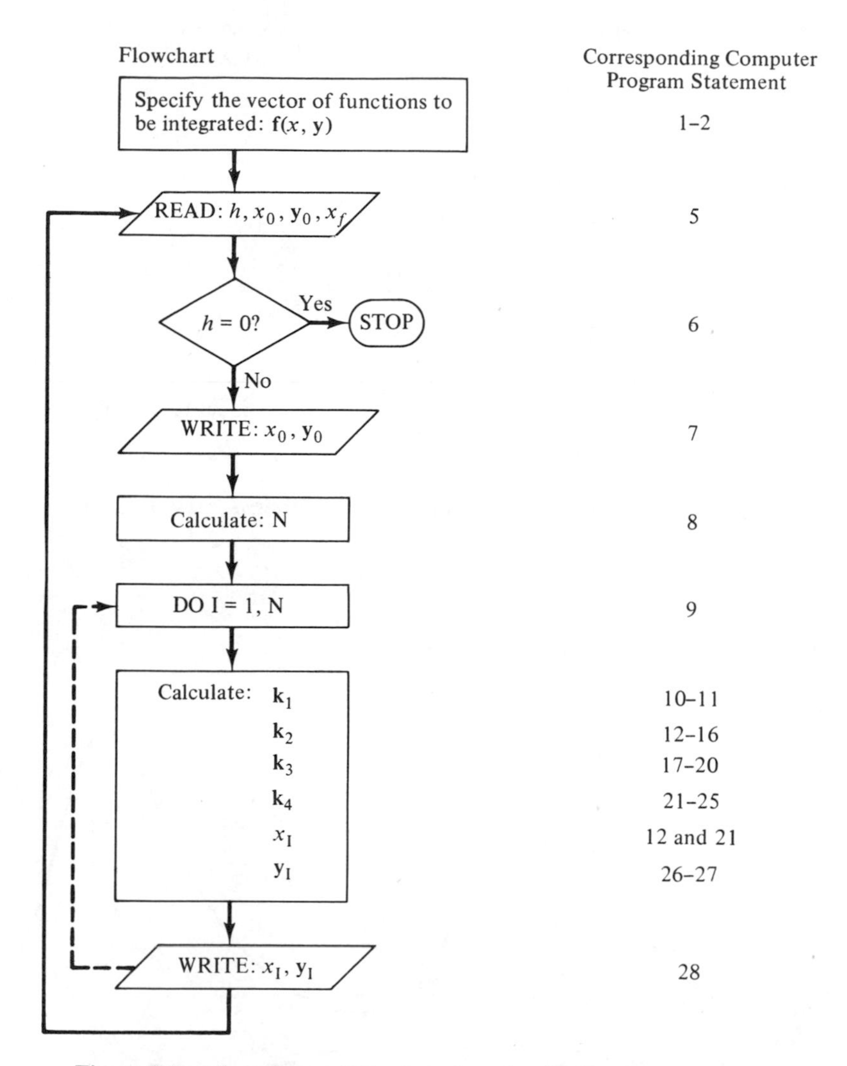

Figure 9.4 Flowchart of the fourth order Runge-Kutta method for a vector initial value problem.

Figure 9.4. Observe that the computer program of Figure 9.5 can easily be expanded to solve larger systems of initial value problems. The reader should also observe that it is probably easier to write the computer program to solve this particular problem than it is to write out the method of solution in non-vector notation. A graph of the solution to the given Volterra prey-predator initial value problem which was produced using the computer program of Figure 9.5 is displayed in Figure 9.6.

```
      C---FOURTH ORDER RUNGE-KUTTA FOR SECOND ORDER SYSTEMS
0001        F1(X,Y1,Y2) = Y1 - .5*Y1*Y2
0002        F2(X,Y1,Y2) =  -2.*Y2 + .25*Y1*Y2
0003    800 FORMAT(3(3X,E18.11))
0004    900 FORMAT(5E16.9)
0005      1 READ(1,900) H,X,Y1,Y2,XF
0006        IF(H.EQ.0.) GO TO 3
0007        WRITE(3,800) X,Y1,Y2
0008        N = (XF - X)/H + .5
0009        DO 2 I = 1,N
0010        XK11 = F1(X,Y1,Y2)
0011        XK21 = F2(X,Y1,Y2)
0012        X = X + .5*H
0013        Z1 = Y1 + .5*H*XK11
0014        Z2 = Y2 + .5*H*XK21
0015        XK12 = F1(X,Z1,Z2)
0016        XK22 = F2(X,Z1,Z2)
0017        Z1 = Y1 + .5*H*XK12
0018        Z2 = Y2 + .5*H*XK22
0019        XK13 = F1(X,Z1,Z2)
0020        XK23 = F2(X,Z1,Z2)
0021        X = X + .5*H
0022        Z1 = Y1 + H*XK13
0023        Z2 = Y2 + H*XK23
0024        XK14 = F1(X,Z1,Z2)
0025        XK24 = F2(X,Z1,Z2)
0026        Y1 = Y1 + H*(XK11 + 2.*XK12 + 2.*XK13 + XK14)/6.
0027        Y2 = Y2 + H*(XK21 + 2.*XK22 + 2.*XK23 + XK24)/6.
0028      2 WRITE(3,800) X,Y1,Y2
0029        GO TO 1
0030      3 CALL EXIT
0031        END
```

Figure 9.5 Computer program for the Volterra prey-predator model.

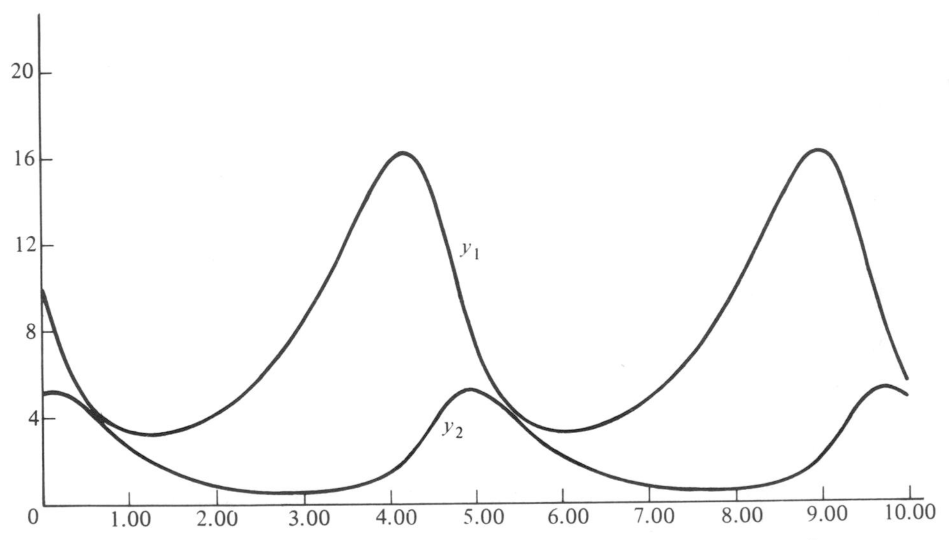

Figure 9.6 Runge-Kutta solution of the Volterra prey-predator model.

EXERCISE

1. Modify the fourth order Runge-Kutta computer program of Figure 9.5 and use it with a stepsize $h = .1$ to solve the initial value problem

$$y_1' = -\frac{1}{y_2}; \qquad y_1(1) = 1$$

$$y_2' = 2y_1y_2; \qquad y_2(1) = 1$$

on the interval [1, 10]. [*Hint:* Statements 1 and 2 of the computer program of Figure 9.5, which specify the functions F1 and F2 which define the system of equations to be integrated, must be changed. And a data card with .1 punched somewhere in columns 1–16, 1. in columns 17–32, 1. in columns 33–48, 1. in columns 49–64, and 10. in columns 65–80 must be included to specify the values of h(H), x_0(X), $\mathbf{y}_0$(Y1, Y2), and x_f(XF), respectively.] Compare the numerical answer you obtain with the exact answer. *Hint:* Verify that the exact solution of the given initial value problem is

$$\begin{pmatrix} y_1(x) \\ y_2(x) \end{pmatrix} = \begin{pmatrix} \dfrac{1}{x} \\ x^2 \end{pmatrix}.$$

10

TWO POINT BOUNDARY VALUE PROBLEMS

Up to this point we have primarily concerned ourselves with the solution of initial value problems. That is, we have mainly been solving differential equations subject to conditions which were all specified for one value of the independent variable. In this chapter we shall show how to use the initial value problem techniques which we have developed previously to solve some particular subcases of the general two point boundary value problems:

$$y'' = f(x, y, y'); \tag{1}$$

$$g(y(a), y'(a)) = 0, \qquad h(y(b), y'(b)) = 0. \tag{2}$$

The general two point boundary value problem consists of solving the differential equation (1) subject to the boundary conditions specified by equations (2). We shall assume that a and b are finite, that $a < b$, and that we desire the solution of (1) subject to the boundary conditions (2) on the interval $[a, b]$.

In this text we shall not consider two point boundary value problems consisting of the differential equation

$$\phi(y'', y', y, x, \lambda) = 0,$$

where λ is a parameter subject to the boundary conditions (2). Problems of this nature are interesting from a physical standpoint and lead to many significant results. However, the solution of such problems depends heavily upon a complete understanding of the concepts of eigenvalue, eigenvector, sets of orthogonal functions, and the convergence of series of functions. Consequently, we believe that the study of this type of two point boundary value problem and the topics just mentioned should be pursued at a later time in a more advanced text on differential equations—in particular, in conjunction with the study of partial differential equations.

EXAMPLE A string is stretched and anchored at two points on a horizontal plane—perhaps a table. Denote the resulting tension in the string by T. Let the string define the x-axis and designate one point at which the string is anchored as the origin of an xy-coordinate system in the plane. Designate the other point at which the string is anchored by $(b, 0)$. A transverse load $f(x)$ whose units are force/length is applied to the string resulting in the deflection $y(x)$ of Figure 10.1. Consider a segment of the deflected string between two nearby points x_1 and x_2. The forces acting on this segment of string are the tensile forces F_1 and F_2 and the applied load $\int_{x_1}^{x_2} f(t)\,dt$. See Figure 10.2. Gravity does not enter into the equilibrium equations, since the xy-plane is horizontal. If we assume that there is no friction between the string and the xy-plane, then the equilibrium equations for the x and y components of force are, respectively,

$$F_1 \cos \theta_1 = F_2 \cos \theta_2 = T$$

and

$$F_1 \sin \theta_1 = F_2 \sin \theta_2 + \int_{x_1}^{x_2} f(t)\,dt.$$

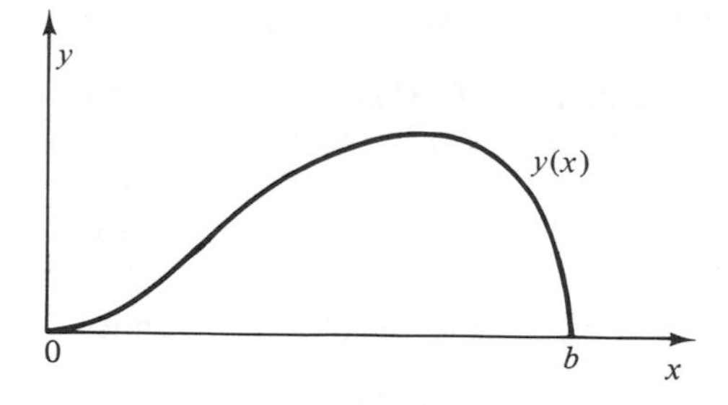

Figure 10.1 Deflection of string.

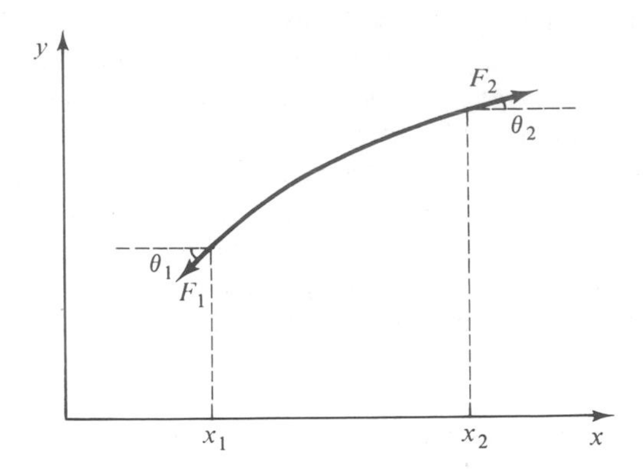

Figure 10.2 Forces acting on string.

Solving the first equation for F_1 and F_2 in terms of T, substituting into the second equation, and rearranging, we find

$$T(\tan\theta_1 - \tan\theta_2) = \int_{x_1}^{x_2} f(t)\,dt.$$

Since $\tan\theta = dy/dx = y'$, we have

$$T(y'(x_1) - y'(x_2)) = \int_{x_1}^{x_2} f(t)\,dt.$$

If $f(x)$ is a continuous function on $[x_1, x_2]$, then by the mean value theorem for integrals there exists a $\xi \in (x_1, x_2)$ such that

$$\int_{x_1}^{x_2} f(t)\,dt = f(\xi)(x_2 - x_1).$$

Therefore,

$$\frac{-T(y'(x_2) - y'(x_1))}{x_2 - x_1} = f(\xi).$$

Taking the limit as $x_2 \to x_1$, we find that y satisfies the differential equation

$$-Ty''(x) = f(x)$$

at each point x of continuity of f. Thus, in this instance one is interested in solving the two points boundary value problem:

$$y''(x) = -\frac{f(x)}{T}; \qquad y(0) = y(b) = 0.$$

Notice that if $f(x) = C$ a constant, then the solution is

$$y = \frac{Cx(b - x)}{2T}.$$

The reader should verify that this equation satisfies the given boundary value problem and that the solution is a parabola with vertex at $(b/2, Cb^2/8T)$ and axis parallel to the y-axis.

If the string is embedded in a medium that exerts a restoring force proportional to the displacement, then one needs to solve the two point boundary value problem

$$y'' - \frac{ky}{T} = -\frac{f(x)}{T}; \qquad y(0) = y(b) = 0,$$

where k is the restoring constant of the medium.

If the string is embedded in a medium that exerts a restoring force proportional to the displacement and a resistive force proportional to the velocity, then one must solve the two point boundary value problem

$$y'' + \frac{cy'}{T} - \frac{ky}{T} = -\frac{f(x)}{T}; \qquad y(0) = y(b) = 0$$

where c is the dissipation constant.

10.1 LINEAR BOUNDARY VALUE PROBLEMS

First let us consider the completely linear boundary value problem consisting of the linear second order differential equation

$$y'' = p(x)y' + q(x)y + r(x) \tag{3}$$

and the linear boundary conditions

$$c_1 y'(a) + c_2 y(a) = k_1, \qquad c_3 y'(b) + c_4 y(b) = k_2, \tag{4}$$

where $p(x)$, $q(x)$, and $r(x)$ are continuous on the interval $[a, b]$; c_1, c_2, c_3, c_4, k_1, and k_2 are real constants; $|c_1| + |c_2| \neq 0$; and $|c_3| + |c_4| \neq 0$. We know from the theory developed previously that on the interval $[a, b]$ there exist two linearly independent solutions $y_1(x)$ and $y_2(x)$ of the homogeneous linear differential equation

$$y'' = p(x)y' + q(x)y. \tag{5}$$

(See Theorem 6.6.) If we can find a particular solution $y_p(x)$ of the non-homogeneous linear equation (3) which is valid on the interval $[a, b]$, then the general solution of (3) on $[a, b]$ is

$$y(x) = d_1 y_1(x) + d_2 y_2(x) + y_p(x), \tag{6}$$

where d_1 and d_2 are arbitrary constants. (See Theorem 6.8.) Consequently, if we can find two linearly independent solutions of the homogeneous equation

(5) and a particular solution of the nonhomogeneous equation (3), then the linear boundary value problem consisting of equations (3) and (4) has a solution on the interval $[a, b]$ if and only if we can find constants d_1 and d_2 of equation (6) which satisfy the boundary conditions of equations (4).

10.1.1 Exact Solutions

When $p(x)$ and $q(x)$ are constant functions, we can always find two linearly independent solutions y_1 and y_2 of the homogeneous linear differential equation (5) $y'' = p(x)y' + q(x)y$ on the interval $[a, b]$. Then, if simpler methods fail, we can at least produce a particular solution to the nonhomogeneous linear differential equation (3), $y'' = p(x)y' + q(x)y + r(x)$, on the interval $[a, b]$ from y_1 and y_2 by the method of variation of parameters. So if $p(x)$ and $q(x)$ are constant functions, the solution of the linear boundary value problem (3)–(4) reduces to the problem of determining whether or not there exist constants d_1 and d_2 of (6) which satisfy the boundary conditions (4).

EXAMPLE Find the solution of the linear boundary value problem

$$y'' = -y + x; \tag{7}$$

$$y'(0) - 2y(0) = 0, \qquad y'\left(\frac{\pi}{2}\right) + 2y\left(\frac{\pi}{2}\right) = 2 + \pi. \tag{8}$$

We easily find two linearly independent solutions of the associated homogeneous equation $y'' = -y$ to be $y_1(x) = \sin x$ and $y_2(x) = \cos x$. And we observe that $y_p(x) = x$ is a particular solution of (7). So the general solution of (7) is

$$y(x) = d_1 y_1(x) + d_2 y_2(x) + x = d_1 \sin x + d_2 \cos x + x, \tag{9}$$

where d_1 and d_2 are arbitrary constants. Differentiating, we get

$$y'(x) = d_1 \cos x - d_2 \sin x + 1. \tag{10}$$

Evaluating y and y' at 0 and $\pi/2$, we find

$$y(0) = d_2, \qquad y'(0) = d_1 + 1, \tag{11}$$

$$y\left(\frac{\pi}{2}\right) = d_1 + \frac{\pi}{2}, y'\left(\frac{\pi}{2}\right) = -d_2 + 1. \tag{12}$$

Substituting from these equations into the boundary conditions (8), we see that d_1 and d_2 must simultaneously satisfy

$$d_1 + 1 - 2d_2 = 0,$$

$$-d_2 + 1 + 2\left(d_1 + \frac{\pi}{2}\right) = 2 + \pi.$$

Thus, d_1 and d_2 must simultaneously satisfy

$$\begin{aligned} d_1 - 2d_2 &= -1, \\ 2d_1 - d_2 &= 1. \end{aligned}$$

Solving this simple linear system of equations, we find that $d_1 = d_2 = 1$. So the solution of the given boundary value problem is

$$y(x) = \sin x + \cos x + x.$$

EXAMPLE Find the solution of the linear boundary value problem

$$y'' = -y + x, \tag{13}$$

$$y'(0) = y(0) = 0, \; -y'\left(\frac{\pi}{2}\right) + y\left(\frac{\pi}{2}\right) = \frac{\pi}{2}. \tag{14}$$

Since the differential equation (13) is the same as (7), the general solution is again (9) $y(x) = d_1 \sin x + d_2 \cos x + x$. Evaluating y and y'—equation (10)—at 0 and $\pi/2$, we again obtain equations (11) and (12). Substituting from equations (11) and (12) into the boundary conditions (14), we see that the constants d_1 and d_2 must simultaneously satisfy

$$\begin{aligned} d_1 + 1 + d_2 &= 0 \\ d_2 - 1 + d_1 + \frac{\pi}{2} &= \frac{\pi}{2}. \end{aligned}$$

So d_1 and d_2 must simultaneously satisfy

$$\begin{aligned} d_1 + d_2 &= -1, \\ d_1 + d_2 &= 1. \end{aligned}$$

Obviously, there are no constants d_1 and d_2 which satisfy the preceding linear system of equations. And, consequently, there is no solution to the boundary value problem (13)–(14).

When the functions $p(x)$ and $q(x)$ of equations (3) and (5) are not constant functions, our ability to write an explicit solution for the boundary value problem (3)–(4) and our ability to show that the boundary value problem has no solution depends upon our ability to find two linearly independent solutions of (5). When we are unable to obtain two linearly independent solutions of (5), we must depend upon numerical techniques to produce approximate solutions to the given boundary value problem or to indicate that there is no solution.

EXAMPLE Given that $y_1(x) = x$ is a solution of the homogeneous linear differential equation

$$y'' = \frac{2x}{x^2 + 1} y' - \frac{2}{x^2 + 1} y, \tag{15}$$

find the solution of the linear boundary value problem

$$y'' = \frac{2x}{x^2 + 1} y' - \frac{2}{x^2 + 1} y + x^2 + 1, \tag{16}$$

$$y(0) = 2, \qquad y(1) = \tfrac{5}{3}. \tag{17}$$

In order to find a second linearly independent solution $y_2(x)$ of the homogeneous linear differential equation (15), we employ the method of reduction of order. That is, we assume that there is a solution of (15) of the form

$$y_2 = y_2(x) = v(x)y_1(x) = vx.$$

Differentiating, we find

$$y_2' = v'x + v \quad \text{and} \quad y_2'' = v''x + 2v'.$$

Substituting y_2, y_2', and y_2'' into (15) we see that v must satisfy the following second order linear differential equation:

$$v''x + 2v' = \frac{2x}{x^2 + 1}(v'x + v) - \frac{2}{x^2 + 1} vx$$

or

$$v''x = \left(\frac{2x^2}{x^2 + 1} - 2\right)v' = -\frac{2}{x^2 + 1} v'.$$

Letting $w = v'$, we see that w must satisfy the first order linear differential equation

$$w'x = -\frac{2w}{x^2 + 1}.$$

Separating variables, we obtain

$$\frac{dw}{w} = -\frac{2\,dx}{x(x^2 + 1)} = -\frac{2\,dx}{x} + \frac{2x\,dx}{x^2 + 1}.$$

Integrating, we get

$$\ln|w| = -2\ln|x| + \ln|x^2 + 1| + \ln c.$$

Upon exponentiating and combining terms, we find

$$v' = w = \frac{c(x^2 + 1)}{x^2} = c\left(1 + \frac{1}{x^2}\right).$$

Choosing $c = 1$ and integrating, we find a particular function v to be

$$v = x - \frac{1}{x}.$$

Hence, a second linearly independent solution of (15) is

$$y_2 = vx = x^2 - 1.$$

Next, we use the method of variation of parameters to find a particular solution to the nonhomogeneous differential equation (16). Thus, we seek a particular solution of the form

$$y_p(x) = v_1(x)y_1(x) + v_2(x)y_2(x) = v_1 x + v_2(x^2 - 1).$$

The derivatives of v_1 and v_2 satisfy

$$v_1' = \frac{\begin{vmatrix} 0 & y_2 \\ x^2 + 1 & y_2' \end{vmatrix}}{\begin{vmatrix} y_1 & y_2 \\ y_1' & y_2' \end{vmatrix}} = \frac{\begin{vmatrix} 0 & x^2 - 1 \\ x^2 + 1 & 2x \end{vmatrix}}{\begin{vmatrix} x & x^2 - 1 \\ 1 & 2x \end{vmatrix}}$$

$$= \frac{-(x^2 + 1)(x^2 - 1)}{2x^2 - (x^2 - 1)} = \frac{-(x^2 + 1)(x^2 - 1)}{x^2 + 1} = -x^2 + 1$$

and

$$v_2' = \frac{\begin{vmatrix} y_1 & 0 \\ y_1' & x^2 + 1 \end{vmatrix}}{x^2 + 1} = \frac{\begin{vmatrix} x & 0 \\ 1 & x^2 + 1 \end{vmatrix}}{x^2 + 1} = \frac{x(x^2 + 1)}{x^2 + 1} = x.$$

(See Section 6.5.) Integrating, we find

$$v_1 = -\frac{x^3}{3} + x \quad \text{and} \quad v_2 = \frac{x^2}{2}.$$

So a particular solution of (16) is

$$y_p = v_1 y_1 + v_2 y_2 = \left(-\frac{x^3}{3} + x\right)x + \frac{x^2(x^2 - 1)}{2}$$

$$= \frac{x^4}{6} + \frac{x^2}{2}.$$

And the general solution of (16) is

$$y = d_1 y_1 + d_2 y_2 + y_p = d_1 x + d_2(x^2 - 1) + \frac{x^4}{6} + \frac{x^2}{2},$$

where d_1 and d_2 are arbitrary constants. In order to satisfy the boundary conditions (17), d_1 and d_2 must simultaneously satisfy

$$y(0) = -d_2 = 2,$$
$$y(1) = d_1 + \tfrac{1}{6} + \tfrac{1}{2} = \tfrac{5}{3}.$$

So $d_1 = 1$, $d_2 = -2$, and the solution of the boundary value problem is

$$y = x - 2(x^2 - 1) + \frac{x^4}{6} + \frac{x^2}{2} = \frac{x^4}{6} - \frac{3x^2}{2} + x + 2.$$

10.1.2 Numerical Solutions

We were very fortunate in the last example. We were able to obtain two linearly independent solutions to a second order linear differential with nonconstant coefficients, and, subsequently, we were able to solve the given boundary value problem explicitly. This will not be the usual situation. Normally, we will not be able to obtain even one solution of a second order linear differential equation with nonconstant coefficients expressed explicitly in terms of elementary functions—we may, of course, be able to obtain power series solutions, but these solutions will generally be unsuitable for our purposes. Therefore, in this section we wish to consider some simple numerical methods for obtaining approximate solutions to boundary value problems of the form (3)–(4).

10.1.2.1 Boundary conditions: $y(a) = V_1$, $y(b) = V_2$

First, let us consider the following boundary value problem:

$$y'' = p(x)y' + q(x)y + r(x) \tag{18}$$

$$y(a) = V_1, \qquad y(b) = V_2. \tag{19}$$

This is the special subcase of the boundary value problem (3)–(4) in which $c_1 = c_3 = 0$ and c_2 and c_4 are not zero. Notice that $V_1 = k_1/c_2$ and $V_2 = k_2/c_4$. Since $p(x)$, $q(x)$, and $r(x)$ are assumed to be continuous on the interval $[a, b]$, there exists a unique solution $y_1(x)$ of the initial value problem

$$y'' = p(x)y' + q(x)y + r(x); \tag{18}$$

$$y(a) = V_1, \qquad y'(a) = 0 \tag{20}$$

on the interval $[a, b]$ and there exists a unique solution $y_2(x)$ of the initial value problem

$$y'' = p(x)y' + q(x)y; \tag{21}$$

$$y(a) = 0, \qquad y'(a) = 1 \tag{22}$$

on the interval $[a, b]$. The existence and uniqueness of the solutions y_1 and y_2 of the initial value problems (18)–(20) and (21)–(22) on the interval $[a, b]$ is proven in Theorem 5.8. Now consider the function

$$y_C(x) = y_1(x) + Cy_2(x), \tag{23}$$

where C is any arbitrary constant. Differentiating, we find

$$y_C' = y_1' + Cy_2' \quad \text{and} \quad y_C'' = y_1'' + Cy_2''.$$

Substituting y_C into equation (18), we obtain the identity

$$\begin{aligned} y_C'' &= p(x)y_C' + q(x)y_C + r(x) \\ &= p(x)(y_1' + Cy_2') + q(x)(y_1 + Cy_2) + r(x) \\ &= (p(x)y_1' + q(x)y_1 + r(x)) + C(p(x)y_2' + q(x)y_2). \\ &= y_1'' + Cy_2''. \end{aligned}$$

Hence, y_C is a solution of the differential equation (18) for any choice of the constant C, and

$$y_C(a) = y_1(a) + Cy_2(a) = V_1 + C \cdot 0 = V_1$$

for any choice of the constant C. Thus, y_C satisfies the differential equation (18) and the first boundary condition of equation (19). If we can choose C so that y_C satisfies the second boundary condition of (19), $y(b) = V_2$, then the boundary value problem (18)–(19) has one or more solutions. If we cannot choose C so that y_C satisfies $y(b) = V_2$, then the boundary value problem (18)–(19) has no solution. In order to satisfy the second boundary condition of (19), C must be chosen to satisfy the equation

$$y_C(b) = y_1(b) + Cy_2(b) = V_2. \tag{24}$$

There are three cases:

1. If $y_2(b) \neq 0$, then $C = [V_2 - y_1(b)]/y_2(b)$ and the corresponding function y_C is the unique solution of the boundary value problem (18)–(19).
2. If $y_2(b) = 0$ and $y_1(b) = V_2$, then equation (24) is satisfied for any value C, and there are infinitely many solutions of the boundary value problem (18)–(19)—namely, y_C for any choice of C.
3. If $y_2(b) = 0$ and $y_1(b) \neq V_2$, then no value of C satisfies equation (24), and consequently there is no solution to the boundary value problem (18)–(19).

So the process for generating a numerical solution of the boundary value problem (18)–(19) consists of the following steps:

1. Compute the numerical solution y_1 of the initial value problem (18)–(20) on the interval $[a, b]$.

2. Compute the numerical solution y_2 of the initial value problem (21)–(22) on the interval $[a, b]$.
3. Compare $y_2(b)$ and 0. If $y_2(b) = 0$, compare $y_1(b)$ and V_2. If $y_2(b) = 0$ and $y_1(b) = V_2$, then indicate that $y_C = y_1 + Cy_2$ is a solution for any choice of C. If $y_2(b) = 0$ and $y_1(b) \neq V_2$, then indicate that there is no solution. If $y_2(b) \neq 0$, then indicate that the unique solution is

$$y_C(x) = y_1(x) + \frac{[V_2 - y_1(b)]y_2(x)}{y_2(b)}.$$

We shall use the fourth order Runge-Kutta method for generating numerical solutions of first order systems of differential equations to compute y_1 and y_2 on the interval $[a, b]$. This method was described in Section 9.2 and used to compute the solution of a Volterra prey-predator system of differential equations. In order to use this method, we must write the differential equations (18) and (21) as first order systems of differential equations.

We let $y_{11}(x) = y_1(x)$ and $y_{12}(x) = y_1'(x)$. Differentiating, we find

$$y_{11}'(x) = y_1'(x) = y_{12}(x)$$

and

$$\begin{aligned} y_{12}'(x) &= y_1''(x) = p(x)y_1'(x) + q(x)y_1(x) + r(x) \\ &= p(x)y_{12}(x) + q(x)y_{11}(x) + r(x), \end{aligned}$$

since $y_1(x)$ is to satisfy (18). Also since $y_1(x)$ is to satisfy (20), we have $y_{11}(a) = y_1(a) = V_1$ and $y_{12}(a) = y_1'(a) = 0$. So a first order system initial value problem which is equivalent to the initial value problem (18)–(20) is

$$\begin{aligned} y_{11}' &= y_{12} = f(x, y_{11}, y_{12}) \\ y_{12}' &= p(x)y_{12} + q(x)y_{11} + r(x) = g(x, y_{11}, y_{12}); \end{aligned} \tag{25}$$

$$\begin{aligned} y_{11}(a) &= V_1 \\ y_{12}(a) &= 0. \end{aligned} \tag{26}$$

We will use the fourth order Runge-Kutta method to solve the first order system (25) on the interval $[a, b]$ subject to the initial conditions (26). The solution component $y_{11}(x)$ will be an approximate value of y_1—the actual solution of (18)–(20)—at x and the solution component $y_{12}(x)$ will be an approximate value of y_1' at x.

Letting $y_{21}(x) = y_2(x)$ and $y_{22}(x) = y_2'(x)$ and differentiating, we see that

$$y_{21}'(x) = y_2'(x) = y_{22}(x)$$

and

$$y_{22}'(x) = y_2''(x) = p(x)y_2'(x) + q(x)y_2(x) = p(x)y_{22}(x) + q(x)y_{21}(x),$$

since $y_2(x)$ is to satisfy (21). And since $y_2(x)$ is also to satisfy (22), we have $y_{21}(a) = y_2(a) = 0$ and $y_{22}(a) = y_2'(a) = 1$. Hence, a first order system initial value problem which is equivalent to the initial value problem (21)–(22) is

$$(27)\qquad \begin{aligned} y_{21}' &= y_{22} &&= f(x, y_{21}, y_{22}) \\ y_{22}' &= p(x)y_{22} + q(x)y_{21} &&= h(x, y_{21}, y_{22}); \end{aligned}$$

$$(28)\qquad \begin{aligned} y_{21}(a) &= 0 \\ y_{22}(a) &= 1. \end{aligned}$$

We will use the fourth order Runge-Kutta method to solve the first order system (27) on the interval $[a, b]$ subject to the initial conditions (28). The solution component $y_{21}(x)$ will be an approximate value of y_2—the actual solution of (21)–(22) at x and the solution component $y_{22}(x)$ will be an approximate value of y_2' at x.

After the approximate solutions y_1 and y_2 are computed, the various comparisons of step 3 for numerically solving the boundary value problem (18)–(19) are performed and the appropriate result is indicated based upon these comparisons. As we shall soon discover in the examples we will consider, "numerical equality" seldom occurs after a large number of calculations of this nature have taken place. This is due to truncation and roundoff errors. So presently we will need to relax the comparisons of step 3 from equality to "near equality."

A simplified flowchart for solving the boundary value problem (18)–(19) is displayed in Figure 10.3. Also included in Figure 10.3 is the number of the corresponding computer program statement or statements which accomplishes each of the operations indicated by the flowchart. The computer program itself is displayed in Figure 10.4. Notice that in order to numerically solve a particular boundary value problem of the type (18)–(19) by the technique just described, all one needs to do is specify $p(x)$, $q(x)$, and $r(x)$—statements 2, 3, and 4 of the computer program of Figure 10.4—and specify a, b, V_1, V_2, and N, where $N + 1$ is the number of points equally spaced throughout the interval $[a, b]$ at which values for the solution are to be calculated. The values of a, b, V_1, V_2, and N are placed on a single data card —a in columns 1–10, b in columns 11–20, V_1 in columns 21–30, V_2 in columns 31–40, and N in columns 61–63. Column 63 is the units column for N, 62 is the tens column, and 61 is the hundreds column. The data a, b, V_1, and V_2 may be placed anywhere in the columns indicated so long as a decimal point is included.

EXAMPLE Compute a numerical solution of the boundary value problem

$$(16)\qquad y'' = \frac{2x}{x^2+1}y' - \frac{2}{x^2+1}y + x^2 + 1;$$

$$(17)\qquad y(0) = 2, \qquad y(1) = \tfrac{5}{3}$$

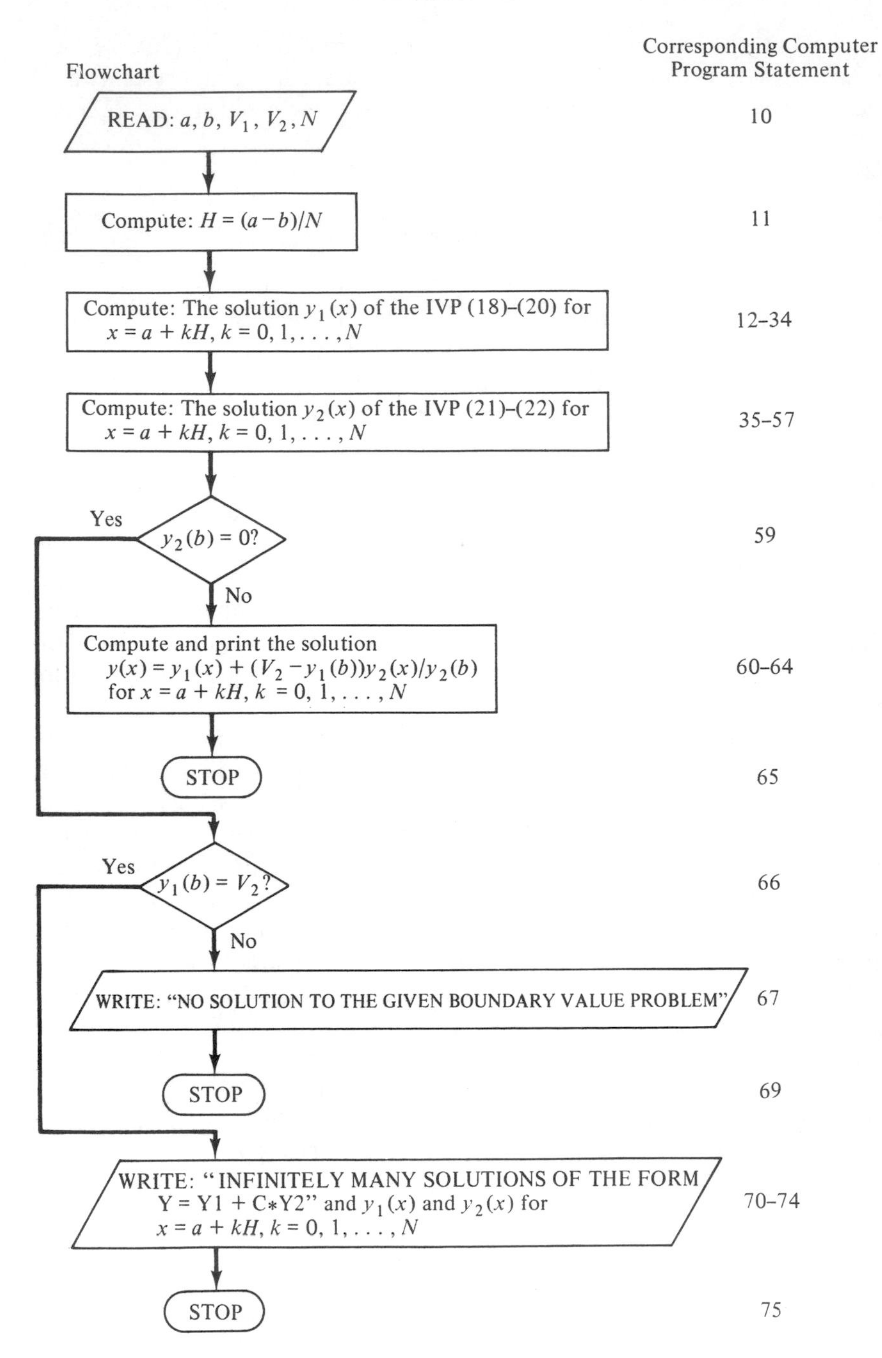

Figure 10.3 Flowchart for the solution of the boundary value problem (18)–(19).

```
      C--FOURTH ORDER RUNGE-KUTTA SOLUTION OF THE BOUNDARY VALUE PROBLEM
      C---   Y'' = P(X)*Y + Q(X)*Y + R(X);  Y(A) = BV1,  Y(B) = BV2
0001         DIMENSION Y11(101),Y21(101)
0002         P(X) = 2.*X/(X*X+1.)
0003         Q(X) = -2./(X*X+1.)
0004         R(X) = X*X+1.
0005         F(X,Y,YP) = YP
0006         G(X,Y,YP) = P(X)*YP + Q(X)*Y + R(X)
0007         HO(X,Y,YP) = P(X)*YP + Q(X)*Y
0008     800 FORMAT(3(2X,E18.11))
0009     900 FORMAT(4E10.5,20X,I3)
0010         READ(1,900) A,B,BV1,BV2,N
0011         H = (B-A)/N
      C--GENERATE AND STORE THE SOLUTION Y1 OF THE IVP
      C---   Y'' = P(X)*Y' + Q(X)*Y + R(X);  Y(A) = BV1,  Y'(A) = 0
0012         X = A
0013         Y11(1) = BV1
0014         Y12 = 0.
0015         DO 2 I = 1,N
0016         XK11 = F(X,Y11(I),Y12)
0017         XK21 = G(X,Y11(I),Y12)
0018         X = X + .5*H
0019         Z1 = Y11(I) + .5*H*XK11
0020         Z2 = Y12    + .5*H*XK21
0021         XK12 = F(X,Z1,Z2)
0022         XK22 = G(X,Z1,Z2)
0023         Z1 = Y11(I) + .5*H*XK12
0024         Z2 = Y12    + .5*H*XK22
0025         XK13 = F(X,Z1,Z2)
0026         XK23 = G(X,Z1,Z2)
0027         X = X + .5*H
0028         Z1 = Y11(I) + .5*H*XK13
0029         Z2 = Y12    + .5*H*XK23
0030         XK14 = F(X,Z1,Z2)
0031         XK24 = G(X,Z1,Z2)
0032         J = I + 1
0033         Y11(J) = Y11(I) + H*(XK11+2.*XK12+2.*XK13+XK14)/6.
0034       2 Y12    = Y12    + H*(XK21+2.*XK22+2.*XK23+XK24)/6.
      C--GENERATE AND STORE THE SOLUTION Y2 OF THE IVP
      C---   Y'' = P(X)*Y' + Q(X)*Y;  Y(A) = 0,  Y'(A) = 1
0035         X = A
0036         Y21(1) = 0.
0037         Y22 = 1.
0038         DO 3 I = 1,N
0039         XK11 = F(X,Y21(I),Y22)
0040         XK21 = HO(X,Y21(I),Y22)
0041         X = X + .5*H
0042         Z1 = Y21(I) + .5*H*XK11
0043         Z2 = Y22    + .5*H*XK21
0044         XK12 = F(X,Z1,Z2)
0045         XK22 = HO(X,Z1,Z2)
0046         Z1 = Y21(I) + .5*H*XK12
0047         Z2 = Y22    + .5*H*XK22
0048         XK13 = F(X,Z1,Z2)
0049         XK23 = HO(X,Z1,Z2)
0050         X = X + .5*H
0051         Z1 = Y21(I) + .5*H*XK13
0052         Z2 = Y22    + .5*H*XK23
0053         XK14 = F(X,Z1,Z2)
0054         XK24 = HO(X,Z1,Z2)
0055         J = I + 1
0056         Y21(J) = Y21(I) + H*(XK11+2.*XK12+2.*XK13+XK14)/6.
0057       3 Y22    = Y22    + H*(XK21+2.*XK22+2.*XK23+XK24)/6.
      C--CHECK TO SEE IF THERE IS A SOLUTION OF THE BVP
0058         M = N + 1
0059         IF(Y21(M).EQ.0.) GO TO 5
0060         C = (BV2 - Y11(M))/Y21(M)
      C--GENERATE AND PRINT THE SOLUTION
0061         DO 4 I = 1,M
0062         X = A + (I-1)*H
0063         Y = Y11(I) + C*Y21(I)
0064       4 WRITE(3,800) X,Y
0065         CALL EXIT
0066       5 IF(Y11(M).EQ.BV2) GO TO 7
      C--INDICATE NO SOLUTION
0067         WRITE(3,6)
0068       6 FORMAT(5X,'NO SOLUTION TO THE GIVEN BOUNDARY VALUE PROBLEM')
0069         CALL EXIT
      C--INDICATE INFINITELY MANY SOLUTIONS OF THE FORM Y = Y1 + C*Y2
      C---AND PRINT THE SOLUTION Y1 AND THE FUNCTION Y2
0070       7 WRITE(3,8)
0071       8 FORMAT(5X,'INFINITELY MANY SOLUTIONS OF THE FORM Y = Y1 + C*Y2'//
            19X,'X',10X,'Y1',10X,'Y2'/)
0072         DO 9 I = 1,M
0073         X = A + (I-1)*H
0074       9 WRITE(3,800) X,Y11(I),Y21(I)
0075         CALL EXIT
0076         END
```

Figure 10.4 Computer program for the solution of the boundary value problem (18)–(19).

for $N = 10$ and $N = 100$ and compare the results with the exact solution at $x = 0., .1, .2, \ldots, 1.0$.

For this problem $p(x) = 2x/(x^2 + 1)$, $q(x) = -2/(x^2 + 1)$, $r(x) = x^2 + 1$, $a = 0$, $b = 1$, $V_1 = 2$, and $V_2 = 5/3$. The exact solution—$y(x) = x^4/6 - 3x^2/2 + x + 2$, the approximate solution for $N = 10$, the difference between this approximate solution and the exact solution, the approximate solution for $N = 100$, and the difference between this approximate solution and the exact solution for $x = 0., .1, .2, \ldots, 1.0$ are shown in Table 10.1.

Table 10.1

x	*Exact solution*	$N = 10$ *Solution*	$N = 10$ *Difference*	$N = 100$ *Solution*	$N = 100$ *Difference*
.0	2.000 00	2.000 00	.000 00	2.000 00	.000 00
.1	2.085 02	2.085 39	.000 37	2.085 07	.000 05
.2	2.140 27	2.141 15	.000 88	2.140 39	.000 12
.3	2.166 35	2.167 79	.001 44	2.166 54	.000 19
.4	2.164 27	2.166 24	.001 97	2.164 51	.000 24
.5	2.135 42	2.137 81	.002 39	2.135 70	.000 28
.6	2.081 60	2.084 21	.002 61	2.081 91	.000 31
.7	2.005 02	2.007 58	.002 56	2.005 31	.000 29
.8	1.908 27	1.910 43	.002 16	1.908 51	.000 24
.9	1.794 35	1.795 68	.001 33	1.794 50	.000 15
1.0	1.666 67	1.666 67	.000 00	1.666 67	.000 00

EXAMPLE Compute a numerical solution of the boundary value problem

$$y'' = -y; \tag{29}$$

$$y(0) = 0, \qquad y(\pi) = 0 \tag{30}$$

for $N = 10$ and $N = 100$ and analyze the results.

For this boundary value problem $p(x) = 0$, $q(x) = -1$, $r(x) = 0$, $a = 0$, $b = \pi$, and $V_1 = V_2 = 0$. We approximated $b = \pi$ by 3.141 593 and ran the appropriate computer program with $N = 10$ and $N = 100$. In both cases the program indicated that the solution was unique and was the function $y(x) = 0$. The general solution of the differential equation (29) is $y(x) = A \sin x + B \cos x$ where A and B are arbitrary constants. Imposing the boundary conditions (30), we find that there are an infinite number of solutions of the boundary value problem (29)–(30) of the form $y = A \sin x$. We need to determine why the computer program did not indicate that the solution was not unique and if possible change the program so that it will do so. First the computer program generates an approximate solution to the initial value problem corresponding to equations (18)–(20). For this example the computer program generates a numerical solution y_1 of the initial value problem

$$y'' = -y;$$
$$y(0) = 0, \qquad y'(0) = 0.$$

The solution generated is $y_1(x) = 0$. Next the program generates an approximate solution to the initial value problem corresponding to equations (21)–(22). In this case the computer program generates a numerical solution y_2 of the initial value problem

$$y'' = -y;$$
$$y(0) = 0, \qquad y'(0) = 1.$$

The solution generated is a numerical approximation of $y(x) = \sin x$. Next the computer compares the value of the approximate solution at b, $y_2(3.141\ 593)$, with the value 0—computer program statement number 59, Figure 10.4—and finds that these values are not identical, so the computer indicates that there is a unique solution and computes the solution to be

$$y(x) = y_1(x) + \frac{[V_2 - y_1(b)]y_2(x)}{y_2(b)}$$

$$= 0 + \frac{(0 - 0)y_2(x)}{y_2(b)} = 0.$$

So our problem lies with the computer program statement number 59 and with the fact that even though the exact solution to the initial value problem (21)–(22) may satisfy $y(b) = 0$, the approximate numerical solution will almost never satisfy the condition, although it may "nearly" satisfy the condition. Due to these numerical considerations we change the hypothesis of program statement number 59 to correspond to the condition $|y_2(b)| < 5 \times 10^{-3}$. Anticipating similar problems for the cases in which there is no solution to the given boundary value problem, we replace computer program statement number 66 of Figure 10.4 with statements which will determine if $y_1(b)$ is "nearly" equal to V_2. We will say that $y_1(b)$ and V_2 are numerically equal if the first five significant digits of these values agree. For $V_2 \neq 0$ the condition is $|V_2 - y_1(b)|/|V_2| = |1 - y_1(b)/V_2| < 10^{-5}$ while for $V_2 = 0$, this condition is $|y_1(b)| < 10^{-5}$. Program statements 58–79 of the revised computer program to solve the general boundary value problem (18)–(19) are shown in Figure 10.5. These statements replace program statements 58–76 of the computer program of Figure 10.4. For $N = 10$ the revised computer program still erroneously indicates that the boundary value problem (29)–(30) has the unique solution $y(x) = 0$, while for $N = 100$, the program correctly indicates that there are an infinite number of solutions to the boundary value problem (29)–(30) of the form $y(x) = y_1(x) + Cy_2(x)$ and prints x, $y_1(x)$, and $y_2(x)$ for $x = 0., .0314, .0628, \ldots, 3.141\ 593$. The results are displayed in Table 10.2.

```
      C--CHECK TO SEE IF THERE IS A SOLUTION OF THE BVP
0058        M = N + 1
0059        IF(ABS(Y21(M)).LT.5.*10.**(-3.)) GO TO 5
0060        C = (BV2 - Y11(M))/Y21(M)
      C--GENERATE AND PRINT THE SOLUTION
0061        DO 4 I = 1,M
0062        X = A + (I-1)*H
0063        Y = Y11(I) + C*Y21(I)
0064      4 WRITE(3,800) X,Y
0065        CALL EXIT
0066      5 IF(BV2.EQ.0.) GO TO 25
0067        IF(ABS(1.-Y11(M)/BV2).LT.10.**(-5.)) GO TO 7
0068        GO TO 26
0069     25 IF(ABS(Y11(M)).LT.10.**(-.5)) GO TO 7
      C--INDICATE NO SOLUTION
0070     26 WRITE(3,6)
0071      6 FORMAT(5X,'NO SOLUTION TO THE GIVEN BOUNDARY VALUE PROBLEM')
0072        CALL EXIT
      C--INDICATE INFINITELY MANY SOLUTIONS OF THE FORM Y = Y1 + C*Y2
      C---AND PRINT THE SOLUTION Y1 AND THE FUNCTION Y2
0073      7 WRITE(3,8)
0074      8 FORMAT(5X,'INFINITELY MANY SOLUTIONS OF THE FORM Y = Y1 + C*Y2'
         1 9X,'X',19X,'Y1',18X,'Y2'/)
0075        DO 9 I = 1,M
0076        X = A + (I-1)*H
0077      9 WRITE(3,800) X,Y11(I),Y21(I)
0078        CALL EXIT
0079        END
```

Figure 10.5 Statements to revise the computer program for the solution of the boundary value problem (18)–(19).

Table 10.2

x	$y_1(x)$	$y_2(x)$
.000	.000 00	.000 00
.314	.000 00	.309 26
.628	.000 00	.588 73
.942	.000 00	.810 99
1.257	.000 00	.954 17
1.571	.000 00	1.004 12
1.885	.000 00	.955 78
2.199	.000 00	.813 74
2.513	.000 00	.591 75
2.827	.000 00	.311 42
3.142	.000 00	.000 13

How "near" any computed value must be to a specified value in order for a certain condition to be satisfied depends upon the differential equation to be solved, the interval of integration, and the boundary conditions. A "nearness" of less than 5×10^{-3} or of five significant digits may not be appropriate for other boundary value problems. What should be noted is that if exactness or near exactness is required, the computer program will indicate that there is a unique solution and print the computed unique solution when in fact there are an infinite number of solutions or no solution. Often one will know from physical considerations that there is one or an infinite number of solutions and consequently will be able to obtain satisfactory and useful numerical results by varying the "nearness" conditions.

EXAMPLE Compute a numerical solution to the boundary value problem

$$y'' = -y; \tag{31}$$

$$y(0) = 0, \qquad y(\pi) = 1 \tag{32}$$

for $N = 10$ and $N = 100$ and analyze the results.

As in the previous example, $p(x) = 0, q(x) = -1, r(x) = 0, a = 0, b = \pi$, and $V_1 = 0$. However, in this example $V_2 = 1$. The general solution of the differential equation (31) is $y(x) = A \sin x + B \cos x$, where A and B are arbitrary constants. Imposing the boundary conditions (32), we find that there is no solution to the given boundary value problem. For $N = 10$ the revised computer program indicates that there is a unique solution. While for $N = 100$ the revised computer program properly indicates that the boundary value problem (31)–(32) has no solution.

10.1.2.2 Boundary conditions: $y(a) = V_1$, $y'(b) + C_2y(b) = V_2$

Now let us consider the linear boundary value problem

$$y'' = p(x)y' + q(x)y + r(x); \tag{33}$$

$$y(a) = V_1, \qquad y'(b) + C_2y(b) = V_2. \tag{34}$$

This is a special subcase of the boundary value problem (3)–(4) in which $c_1 = 0$ and $c_3 \neq 0$. Since $c_1 = 0$ and $|c_1| + |c_2| \neq 0$, $c_2 \neq 0$. Observe that $V_1 = k_1/c_2$, $C_2 = c_4/c_3$, and $V_2 = k_2/c_3$. Since $p(x)$, $q(x)$, and $r(x)$ are assumed to be continuous on $[a, b]$, there exists a unique solution $y_1(x)$ of the initial value problem

$$y'' = p(x)y' + q(x)y + r(x); \tag{33}$$

$$y(a) = V_1, \qquad y'(a) = 0 \tag{35}$$

on the interval $[a, b]$ and there exists a unique solution $y_2(x)$ of the initial value problem

$$y'' = p(x)y' + q(x)y; \tag{36}$$

$$y(a) = 0, \qquad y'(a) = 1 \tag{37}$$

on the interval $[a, b]$. The existence and uniqueness of the solutions y_1 and y_2 on the interval $[a, b]$ is proven in Theorem 5.8. Consider the function

$$y_C(x) = y_1(x) + Cy_2(x), \tag{38}$$

where C is an arbitrary constant. Substituting into the differential equation (33), we see that y_C satisfies (33) for any constant C. Evaluating y_C at a, we see that $y_C(a) = V_1$ for any constant C. Thus, y_C satisfies the first boundary condition of (34). If we can choose C so that the second boundary condition of (34), $y'(b) + C_2 y(b) = V_2$, is satisfied, then the boundary value problem (33)–(34) has one or more solutions. However, if we cannot choose C so that the second boundary condition of (34) is satisfied, then the boundary value problem (33)–(34) has no solution. In order to satisfy

$$y_C'(b) + C_2 y_C(b) = V_2,$$

C must be chosen to satisfy

$$y_1'(b) = Cy_2'(b) + C_2(y_1(b) + Cy_2(b)) = V_2$$

or

$$y_1'(b) + C_2 y_1(b) + C(y_2'(b) + C_2 y_2(b)) = V_2. \tag{39}$$

There are three cases to consider:

1. If $D = y_2'(b) + C_2 y_2(b) \neq 0$, then
$$C = \frac{V_2 - y_1'(b) - C_2 y_1(b)}{y_2'(b) + C_2 y_2(b)}$$
and the corresponding function y_C is the unique solution of the boundary value problem (33)–(34).
2. If $D = 0$ and $y_1'(b) + C_2 y_1(b) = V_2$, then any value of C satisfies equation (39) and there are an infinite number of solutions of the boundary value problem (33)–(34) of the form y_C.
3. If $D = 0$ and $y_1'(b) + C_2 y_1(b) \neq V_2$, then no value of C satisfies equation (39) and, therefore, the boundary value problem (33)–(34) has no solution.

Thus the procedure for producing a numerical solution to the boundary value problem (33)–(34) consists of the following three steps:

1. Compute the numerical solution y_1 of the initial value problem (33)–(35) on the interval $[a, b]$.
2. Compute the numerical solution y_2 of the initial value problem (36)–(37) on the interval $[a, b]$.
3. Compare D and 0. If $D = 0$, compare $y_1'(b) + C_2 y_1(b)$ and V_2. If $D = 0$ and $y_1'(b) + C_2 y_1(b) = V_2$, then indicate that there are an infinite number of solutions of the form $y_C = y_1 + Cy_2$. While if $D = 0$ and $y_1'(b) + C_2 y_1(b) \neq V_2$, then indicate that there is no solution. If $D \neq 0$, then indicate that the unique solution is
$$y_C(x) = y_1(x) + \frac{[V_2 - y_1'(b) - C_2 y_1(b)]y_2(x)}{D}.$$

We observe that the initial value problem (33)–(35) is the same as (18)–(20) and that the initial value problem (36)–(37) is the same as (21)–(22), so steps 1 and 2 of the procedure for generating a numerical solution to the boundary value problem (33)–(34) are identical to steps 1 and 2 of the procedure for generating a numerical solution to (18)–(19). Thus, program statements 1–57 of Figure 10.4 will again provide the desired numerical solutions y_1 and y_2. However, program statements 9 and 10, respectively, must be changed to

```
900 FORMAT (4E10.5, 10X, E10.5, I3)
    READ(1, 900) A, B, BV1, BV2, C2, N
```

and one data card will contain a in columns 1–10, b in columns 11–20, V_1 in columns 21–30, V_2 in columns 31–40, C_2 in columns 51–60, and N in columns 61–63. The comparisons made in step 3 of this procedure differ from those made in step 3 of the previous procedure. However, we can again anticipate that if we make the comparisons of step 3, the conditions tested will never be satisfied exactly, and therefore we should again require only that the conditions be "nearly" satisfied instead of satisfied exactly. A flowchart of the comparisons of step 3 for an accuracy of 5×10^{-3} is shown in Figure 10.6. The number or numbers of the computer program statements which accomplish each of the operations indicated by the flowchart is also given in Figure 10.6. The computer program statements themselves are displayed in Figure 10.7. These statements replace program statements 58–76 of Figure 10.4.

EXAMPLE Compute a numerical solution of the boundary value problem

$$y'' = \frac{2x}{x^2+1}y' - \frac{2}{x^2+1}y + x^2 + 1;$$

$$y(0) = 2, \qquad y'(1) + y(1) = \tfrac{1}{3}$$

for $N = 10$ and $N = 100$ and compare the results with the exact solution

$$y(x) = \frac{x^4}{6} - \frac{3x^2}{2} + x + 2 \quad \text{at } x = 0., .1, .2, \ldots, 1.0.$$

For this problem $p(x) = 2x/(x^2 + 1)$, $q(x) = -2/(x^2 + 1)$, $r(x) = x^2 + 1$, $a = 0$, $b = 1$, $V_1 = 2$, $V_2 = \frac{1}{3}$, and $C_2 = 1$. The exact solution, the approximate solution for $N = 10$, the difference between this approximate solution and the exact solution, the approximate solution for $N = 100$, and the difference between this approximate solution and the exact solution for $x = 0., .1, .2, \ldots, 1.0$ are shown in Table 10.3.

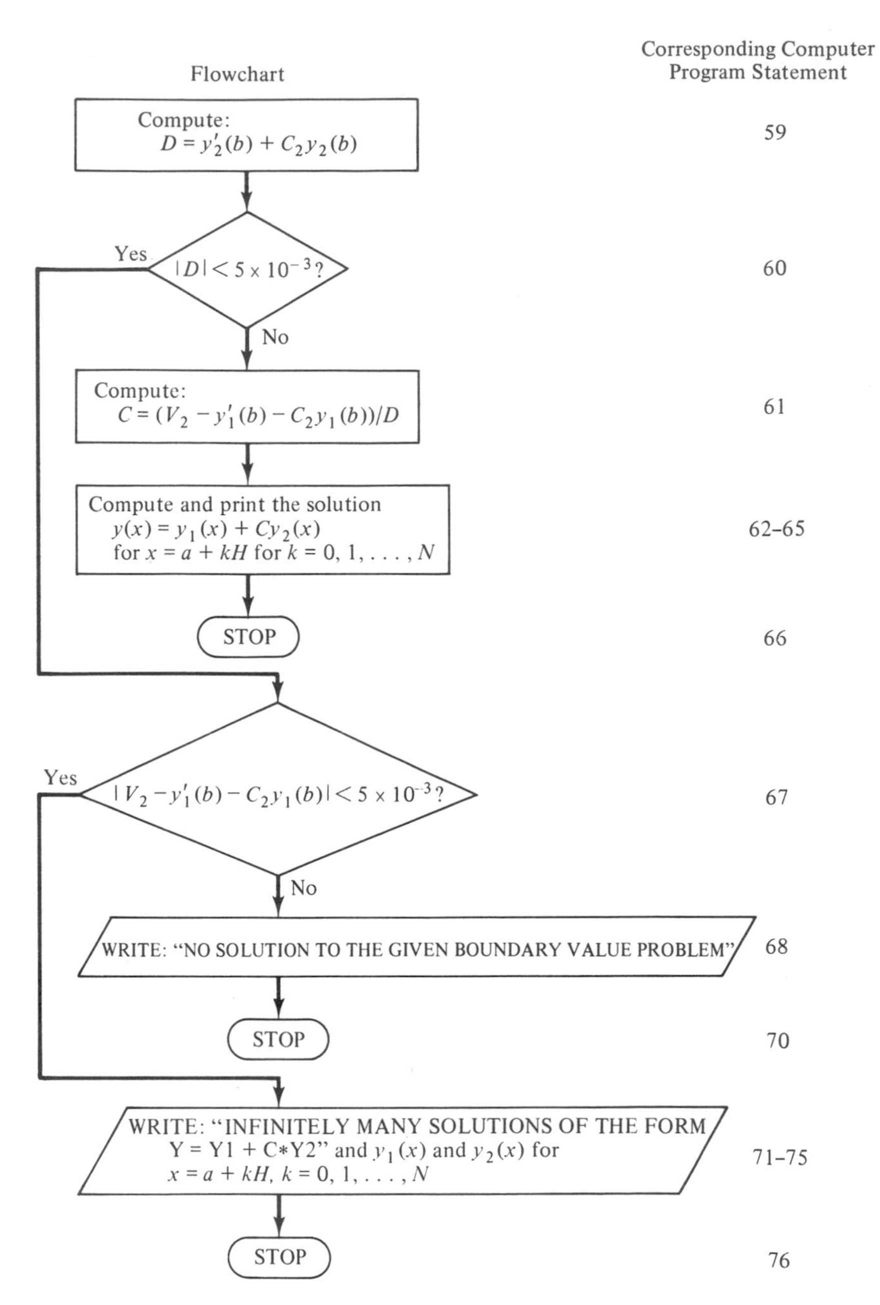

Figure 10.6 Flowchart of the comparisons required for the solution of the boundary value problem (33)–(34).

```
      C--CHECK TO SEE IF THERE IS A SOLUTION OF THE BVP
0058        M = N + 1
0059        D = Y22 + C2*Y21(M)
0060        IF(ABS(D).LT.5.*10.**(-3.)) GO TO 5
0061        C = (BV2 - Y12 - C2*Y11(M))/D
      C--GENERATE AND PRINT THE SOLUTION
0062        DO 4 I = 1,M
0063        X = A + (I-1)*H
0064        Y = Y11(I) + C*Y21(I)
0065      4 WRITE(3,800) X,Y
0066        CALL EXIT
0067      5 IF(ABS(BV2 - Y12 - C2*Y11(M)).LT.5.*10.**(-3.)) GO TO 7
      C--INDICATE NO SOLUTION
0068        WRITE(3,6)
0069      6 FORMAT(5X,'NO SOLUTION TO THE GIVEN BOUNDARY VALUE PROBLEM')
0070        CALL EXIT
      C--INDICATE INFINITELY MANY SOLUTIONS OF THE FORM Y = Y1 + C*Y2
      C---AND PRINT THE SOLUTION Y1 AND THE FUNCTION Y2
0071      7 WRITE(3,8)
0072      8 FORMAT(5X,'INFINITELY MANY SOLUTIONS OF THE FORM Y = Y1 + C*Y2'//
         19X,'X',19X,'Y1',18X,'Y2'/)
0073        DO 9 I = 1,M
0074        X = A + (I-1)*H
0075      9 WRITE(3,800) X,Y11(I),Y21(I)
0076        CALL EXIT
0077        END
```

Figure 10.7 Computer program statements for the comparisons required for the solution of the boundary value problems (33)–(34).

Table 10.3

x	*Exact solution*	$N = 10$		$N = 100$	
		Solution	*Difference*	*Solution*	*Difference*
.0	2.000 00	2.000 00	.000 00	2.000 00	.000 00
.1	2.085 02	2.086 61	.001 59	2.085 20	.000 18
.2	2.140 27	2.143 59	.003 32	2.140 65	.000 38
.3	2.166 35	2.171 46	.005 11	2.166 93	.000 58
.4	2.164 27	2.171 13	.006 86	2.165 03	.000 76
.5	2.135 42	2.143 93	.008 51	2.136 36	.000 94
.6	2.081 60	2.091 57	.009 97	2.082 69	.001 09
.7	2.005 02	2.016 17	.011 15	2.006 22	.001 20
.8	1.908 27	1.920 25	.011 98	1.909 55	.001 28
.9	1.794 35	1.806 74	.012 39	1.795 67	.001 32
1.0	1.666 67	1.678 96	.012 29	1.667 97	.001 30

EXAMPLE Compute a numerical solution to the boundary value problem

$$y'' = \frac{2x}{x^2 + 1} y' - \frac{2}{x^2 + 1} y + x^2 + 1;$$

$$y(0) = 2, \qquad y'(1) - y(1) = -3$$

for $N = 10$ and $N = 100$ and analyze the results.

For $N = 10$ the computer program erroneously indicates that there is a unique solution which is displayed in Table 10.4. For $N = 100$ the computer program properly indicates that there are infinitely many solutions of the form $y(x) = y_1(x) + Cy_2(x)$. The values of $y_1(x)$ and $y_2(x)$ for $x = 0., .1,$

.2, ..., 1.0 are shown in Table 10.5. Notice that $y(x)$ of Table 10.4 is a numerical approximation of a member of the infinite family of solutions $y_1(x) + Cy_2(x)$, where C is approximately 3.82.

Table 10.4

x	$y(x)$
.0	2.000 00
.1	2.366 19
.2	2.703 20
.3	3.011 58
.4	3.292 22
.5	3.546 45
.6	3.775 98
.7	3.982 95
.8	4.169 86
.9	4.339 65
1.0	4.495 63

Table 10.5

x	$y_1(x)$	$y_2(x)$
.0	2.000 00	.000 00
.1	1.985 26	.100 01
.2	1.940 75	.200 03
.3	1.867 05	.300 07
.4	1.765 17	.400 13
.5	1.636 48	.500 20
.6	1.482 79	.600 29
.7	1.306 29	.700 39
.8	1.109 56	.800 51
.9	0.895 61	.900 65
1.0	0.667 82	1.000 80

EXAMPLE Compute a numerical solution to the boundary value problem

$$y'' = \frac{2x}{x^2+1}y' - \frac{2}{x^2+1}y + x^2 + 1;$$

$$y(0) = 2, \qquad y'(1) - y(1) = 1$$

for $N = 10$ and $N = 100$ and analyze the results.

For $N = 10$ the computer program erroneously indicates that there is a unique solution, while for $N = 100$ the computer program properly indicates that there is no solution to the given boundary value problem.

10.1.2.3 Boundary conditions: $y'(a) + C_1y(a) = V_1$, $y'(b) + C_2y(b) = V_2$

When $c_1 \neq 0$ and $c_3 \neq 0$, the linear boundary value problem (3)–(4) may be written as

$$y'' = p(x)y' + q(x)y + r(x); \tag{40}$$

$$y'(a) + C_1y(a) = V_1, \qquad y'(b) + C_2y(b) = V_2, \tag{41}$$

where $C_1 = c_2/c_1$, $V_1 = k_1/c_1$, $C_2 = c_4/c_3$, and $V_2 = k_2/c_4$. Since $p(x)$, $q(x)$, and $r(x)$ are assumed to be continuous on the interval $[a, b]$, we know from Theorem 5.8 that there is a unique solution $y_1(x)$ of the initial value problem

$$y'' = p(x)y' + q(x)y + r(x); \tag{40}$$

$$y(a) = 0, \qquad y'(a) = V_1 \tag{42}$$

on the interval $[a, b]$ and there exists a unique solution $y_2(x)$ of the initial value problem

$$y'' = p(x)y' + q(x)y; \tag{43}$$

$$y(a) = 1, \qquad y'(a) = -C_1 \tag{44}$$

on the interval $[a, b]$. The function

$$y(x) = y_1(x) + Cy_2(x)$$

where C is an arbitrary constant satisfies the differential equation (40) and the first boundary condition of (41), namely, $y'(a) + C_1y(a) = V_1$, for any value of C. When C can be chosen to satisfy the second boundary condition of (41), $y'(b) + C_2y(b) = V_2$, the boundary value problem (40)–(41) has one or more solutions. However, when C cannot be chosen to satisfy the second boundary condition of (41), the boundary value problem (40)–(41) has no solution. In order to satisfy the second boundary condition of (41), C must satisfy equation (39). The three cases to be considered follow equation (39) of Section 10.1.2.2.

The procedure for generating a numerical solution to the boundary value problem (40)–(41) consists of the following three steps:

1. Compute the numerical solution y_1 of the initial value problem (40)–(42) on the interval $[a, b]$.
2. Compute the numerical solution y_2 of the initial value problem (43)–(44) on the interval $[a, b]$.
3. Make the comparisons and perform the operations indicated by the flowchart of Figure 10.6.

The entire computer program for producing a numerical approximation to the solution of the boundary value problem (40)–(41) is shown in Figure 10.8. Notice that all one needs to specify in order to use the computer program are the functions $p(x)$, $q(x)$, and $r(x)$—program statements numbered 2, 3, and 4 respectively—and the constants a, b, V_1, V_2, C_1, C_2, and N. The constants are keypunched in one data card: a in columns 1–10, b in columns 11–20, V_1 in columns 21–30, V_2 in columns 31–40, C_1 in columns 41–50, C_2 in columns 51–60, and N in columns 61–63. (The units column is column 63.)

EXAMPLE Compute a numerical solution to the boundary value problem

$$y'' = \frac{2x}{x^2+1}y' - \frac{2}{x^2+1}y + x^2 + 1; \tag{45}$$

$$3y'(0) - y(0) = 1, \qquad -y'(1) + y(1) = 3 \tag{46}$$

for $N = 10$ and $N = 100$ and compare the results with the exact solution

$$y(x) = \frac{x^4}{6} - \frac{3x^2}{2} + x + 2 \quad \text{at } x = 0., .1, .2, \ldots, 1.0.$$

```
      C--FOURTH ORDER RUNGE-KUTTA SOLUTION OF THE BOUNDARY VALUE PROBLEM
      C---   Y'' = P(X)*Y + Q(X)*Y + R(X); Y'(A)+C1*Y(A)=BV1, Y'(B)+C2*Y(B)=BV2
0001         DIMENSION Y11(101),Y21(101)
0002         P(X) = 2.*X/(X*X+1.)
0003         Q(X) = -2./(X*X+1.)
0004         R(X) = X*X+1.
0005         F(X,Y,YP) = YP
0006         G(X,Y,YP) = P(X)*YP + Q(X)*Y + R(X)
0007         HO(X,Y,YP) = P(X)*YP + Q(X)*Y
0008     800 FORMAT(3(2X,E18.11))
0009     900 FORMAT(6E10.5,I3)
0010         READ(1,900) A,B,BV1,BV2,C1,C2,N
0011         H = (B-A)/N
      C--GENERATE AND STORE THE SOLUTION Y1 OF THE IVP
      C---   Y'' = P(X)*Y' + Q(X)*Y + R(X); Y(A) = 0, Y'(A) = BV1
0012         X = A
0013         Y11(1) = 0.
0014         Y12 = BV1
0015         DO 2 I = 1,N
0016         XK11 = F(X,Y11(I),Y12)
0017         XK21 = G(X,Y11(I),Y12)
0018         X = X + .5*H
0019         Z1 = Y11(I) + .5*H*XK11
0020         Z2 = Y12    + .5*H*XK21
0021         XK12 = F(X,Z1,Z2)
0022         XK22 = G(X,Z1,Z2)
0023         Z1 = Y11(I) + .5*H*XK12
0024         Z2 = Y12    + .5*H*XK22
0025         XK13 = F(X,Z1,Z2)
0026         XK23 = G(X,Z1,Z2)
0027         X = X + .5*H
0028         Z1 = Y11(I) + .5*H*XK13
0029         Z2 = Y12    + .5*H*XK23
0030         XK14 = F(X,Z1,Z2)
0031         XK24 = G(X,Z1,Z2)
0032         J = I + 1
0033         Y11(J) = Y11(I) + H*(XK11+2.*XK12+2.*XK13+XK14)/6.
0034       2 Y12    = Y12    + H*(XK21+2.*XK22+2.*XK23+XK24)/6.
      C--GENERATE AND STORE THE SOLUTION Y2 OF THE IVP
      C---   Y'' = P(X)*Y' + Q(X)*Y; Y(A) = 1, Y'(A) = -C1
0035         X = A
0036         Y21(1) = 1.
0037         Y22 = -C1
0038         DO 3 I = 1,N
0039         XK11 = F(X,Y21(I),Y22)
0040         XK21 = HO(X,Y21(I),Y22)
0041         X = X + .5*H
0042         Z1 = Y21(I) + .5*H*XK11
0043         Z2 = Y22    + .5*H*XK21
0044         XK12 = F(X,Z1,Z2)
0045         XK22 = HO(X,Z1,Z2)
0046         Z1 = Y21(I) + .5*H*XK12
0047         Z2 = Y22    + .5*H*XK22
0048         XK13 = F(X,Z1,Z2)
0049         XK23 = HO(X,Z1,Z2)
0050         X = X + .5*H
0051         Z1 = Y21(I) + .5*H*XK13
0052         Z2 = Y22    + .5*H*XK23
0053         XK14 = F(X,Z1,Z2)
0054         XK24 = HO(X,Z1,Z2)
0055         J = I + 1
0056         Y21(J) = Y21(I) + H*(XK11+2.*XK12+2.*XK13+XK14)/6.
0057       3 Y22    = Y22    + H*(XK21+2.*XK22+2.*XK23+XK24)/6.
      C--CHECK TO SEE IF THERE IS A SOLUTION OF THE BVP
0058         M = N + 1
0059         D = Y22 + C2*Y21(M)
0060         IF(ABS(D).LT.5.*10.**(-3.)) GO TO 5
0061         C = (BV2 - Y12 - C2*Y11(M))/D
      C--GENERATE AND PRINT THE SOLUTION
0062         DO 4 I = 1,M
0063         X = A + (I-1)*H
0064         Y = Y11(I) + C*Y21(I)
0065       4 WRITE(3,800) X,Y
0066         CALL EXIT
0067       5 IF(ABS(BV2 - Y12 - C2*Y11(M)).LT.10.**(-5.)) GO TO 7
      C--INDICATE NO SOLUTION
0068         WRITE(3,6)
0069       6 FORMAT(5X,'NO SOLUTION TO THE GIVEN BOUNDARY VALUE PROBLEM')
0070         CALL EXIT
      C--INDICATE INFINITELY MANY SOLUTIONS OF THE FORM Y = Y1 + C*Y2
      C---AND PRINT THE SOLUTION Y1 AND THE FUNCTION Y2
0071       7 WRITE(3,8)
0072       8 FORMAT(5X,'INFINITELY MANY SOLUTIONS OF THE FORM Y = Y1 + C*Y2'//
            19X,'X',19X,'Y1',18X,'Y2'/)
0073         DO 9 I = 1,M
0074         X = A + (I-1)*H
0075       9 WRITE(3,800) X,Y11(I),Y21(I)
0076         CALL EXIT
0077         END
```

Figure 10.8 Computer program for the solution of the boundary value problem (40)–(41).

First we must rewrite the boundary conditions (46) in the form (41). Dividing the first boundary conditions by 3 and the second by -1, we obtain the equivalent boundary conditions

$$y'(0) - \tfrac{1}{3}y(0) = \tfrac{1}{3}, \qquad y'(1) - y(1) = -3. \tag{47}$$

For the boundary value problem (45)–(47), $p(x) = 2x/(x^2 + 1)$, $q(x) = -2/(x^2 + 1)$, $r(x) = x^2 + 1$, $a = 0$, $b = 1$, $V_1 = 1/3$, $V_2 = -3$, $C_1 = -1/3$, and $C_2 = -1$. The exact solution, the approximate solution for $N = 10$, the difference between this approximate solution and the exact solution, the approximate solution for $N = 100$, and the difference between this approximate solution and the exact solution for $x = 0., .1, .2, \ldots, 1.0$ are shown in Table 10.6.

Table 10.6

x	*Exact solution*	$N = 10$		$N = 100$	
		Solution	*Difference*	*Solution*	*Difference*
.0	2.000 00	1.987 88	−.012 12	1.998 70	−.001 30
.1	2.085 02	2.075 08	−.009 94	2.083 93	−.001 09
.2	2.140 27	2.132 89	−.007 38	2.139 43	−.000 84
.3	2.166 35	2.161 84	−.004 51	2.165 80	−.000 55
.4	2.164 27	2.162 83	−.001 44	2.164 02	−.000 25
.5	2.135 42	2.137 19	.001 77	2.135 48	.000 06
.6	2.081 60	2.086 64	.005 04	2.081 99	.000 39
.7	2.005 02	2.013 30	.008 28	2.005 72	.000 70
.8	1.908 27	1.919 68	.011 41	1.909 27	.001 00
.9	1.794 35	1.808 72	.014 37	1.795 63	.001 28
1.0	1.666 67	1.683 73	.017 06	1.668 20	.001 53

EXAMPLE Compute a numerical solution of the boundary value problem

$$y'' = -y;$$
$$y'(0) = 0, \qquad y'(\pi) = 0$$

for $N = 10$ and $N = 100$ and analyze the results.

For this boundary value problem $p(x) = 0$, $q(x) = -1$, $r(x) = 0$, $a = 0$, $b = \pi$, and $V_1 = V_2 = C_1 = C_2 = 0$. The general solution of the differential equation $y'' = -y$ is $y(x) = A \sin x + B \cos x$, where A and B are arbitrary constants. Imposing the boundary conditions, we find that there are an infinite number of solutions of the given boundary value problem of the form $y(x) = B \cos x$. We approximated π by 3.145 93 and ran the appropriate computer program with $N = 10$ and $N = 100$. For $N = 10$ the program erroneously indicated that $y(x) \equiv 0$ was the unique solution of the given boundary value problem. For $N = 100$ the program correctly indicated that there are an infinite number of solutions of the form $y(x) = y_1(x) + Cy_2(x)$. The function $y_1(x)$ was computed to be the zero function and the function $y_2(x)$ was a numerical approximation of $\cos x$.

EXAMPLE Compute a numerical solution to the boundary value problem

$$y'' = -y;$$
$$y'(0) = 0, \qquad y'(\pi) = 1$$

for $N = 10$ and $N = 100$ and analyze the results.

As in the previous example, $p(x) = 0$, $q(x) = -1$, $r(x) = 0$, $a = 0$, $b = \pi$, and $V_1 = C_1 = C_2 = 0$. However, in this example $V_2 = 1$. The general solution of $y'' = -y$ is $y(x) = A \sin x + B \cos x$, where A and B are arbitrary constants. Imposing the given boundary conditions, we find that there is no solution to the given boundary value problem. For $N = 10$ the computer program erroneously indicates that there is a unique solution, while for $N = 100$ the computer program properly indicates that there is no solution to the given boundary value problem.

10.2 NONLINEAR BOUNDARY VALUE PROBLEMS

The general form of the nonlinear two point boundary value problem is

$$y'' = f(x, y, y'); \tag{48}$$

$$g(y(a), y'(a)) = V_1, \qquad h(y(b), y'(b)) = V_2. \tag{49}$$

We shall consider only one numerical method for solving the boundary value problem (48)–(49)—the "shooting method." In the shooting method, initial conditions $y_0(a)$ and $y_0'(a)$ are chosen so that the first boundary condition $g(y_0(a), y_0'(a)) = 0$ is satisfied. The initial value problem consisting of the differential equation (48) and the initial conditions $y_0(a)$, $y_0'(a)$ is integrated over the interval $[a, b]$. The terminal conditions $y_0(b)$, $y_0'(b)$ are substituted into the terminal boundary condition. That is, $h(y_0(b), y_0'(b)) = C$ is calculated. If the terminal boundary condition is satisfied—that is, if $C = V_2$, then the boundary value problem is solved. If the terminal boundary condition is not satisfied (which is the usual case), the "miss distance," $V_2 - C$, is used to produce new initial conditions and the entire process is repeated. This "hit-or-miss" iterative process is continued until a satisfactory approximate numerical solution is obtained or until a specified number of iterations have taken place. We shall consider two special initial boundary conditions in the following two sections.

10.2.1 Initial Boundary Condition: $y(a) = V_1$

Consider the boundary value problem:

(48) $$y'' = f(x, y, y');$$

(50) $$y(a) = V_1, \qquad h(y(b), y'(b)) = V_2.$$

Since no explicit constraint is placed upon the initial condition $y'(a)$ by the boundary conditions (50), the "shooting method" for producing a numerical solution of the boundary value problem (48)–(50) on the interval $[a, b]$ consists of repeatedly guessing the initial condition $y'(a)$, generating the solution to the resulting initial value problem, and checking to see if the boundary condition at b is satisfied. The specific steps for producing a numerical solution to the BVP (48)–(50) by the shooting method follow:

1. Set $y'(a) = 0$ and calculate a numerical solution y_1 to the initial value problem

(48) $$y'' = f(x, y, y');$$

(51) $$y(a) = V_1, \qquad y'(a) = 0$$

   on the interval $[a, b]$. [We have tacitly assumed that the initial value problem (48)–(51) has a unique solution on the interval $[a, b]$. Sufficient conditions for the existence of a unique solution of (49)–(51) on $[a, b]$ are continuity of f, f_y, and $f_{y'}$ on the domain

$$D = \{(x, y) | a - \epsilon < x < b + \epsilon, c < y < d\},$$

   where $\epsilon > 0$ and c and d are constants chosen so that the range of the solution of (48)–(51) lies within the interval (c, d). See Theorem 5.7.]
2. Check to see if y_1 "nearly" satisfies the terminal boundary condition

$$h(y(b), y'(b)) = V_2.$$

3. If the terminal boundary condition is "nearly" satisfied, then y_1 is a solution.
4. If the terminal boundary condition is not "nearly" satisfied, set $y'(a) = 1$ and calculate a numerical solution y_2 to the initial value problem

$$y'' = f(x, y, y');$$
$$y(a) = V_1, \qquad y'(a) = 1$$

   on the interval $[a, b]$.
5. Check to see if y_2 "nearly" satisfies the terminal boundary condition.
6. If y_2 "nearly" satisfies the terminal boundary condition, then y_2 is a solution.

7. If the terminal boundary condition is not "nearly" satisfied, set $n = 3$, calculate

$$y'_n(a) = y'_{n-2}(a) + \frac{[y'_{n-1}(a) - y'_{n-2}(a)][V_2 - h(y_{n-2}(b), y'_{n-2}(b))]}{[h(y_{n-1}(b), y'_{n-1}(b)) - h(y_{n-2}(b), y'_{n-2}(b))]}, \tag{52}$$

   and calculate the solution y_n of the initial value problem

$$y'' = f(x, y, y'); \tag{48}$$

$$y(a) = V_1, \qquad y'(a) = y'_n(a) \tag{53}$$

   on the interval $[a, b]$. Equation (52) is the formula for the linear interpolation of the initial condition $y'(a)$ from the two previous values, the associated two terminal conditions and the required terminal condition.
8. Check to see if y_n satisfies the terminal boundary condition.
9. If so, y_n is a solution.
10. If not, repeat steps 7–10 for $n = 4, 5, \ldots, N_I$, where N_I is the maximum number of iterations to be performed.

A FORTRAN computer program to perform the procedure given above is displayed in Figure 10.9. The function $f(x, y, y')$ is defined in the third program statement and the terminal boundary function $h(y, y')$ is defined in the fourth program statement. The values of a, b, V_1, V_2, N, and N_I are read from one data card by the fifth program statement. The value of a—the left endpoint of the interval—is punched in columns 1–10, the value of b—the right endpoint of the interval—is punched in columns 11–20, the value of V_1 is punched in columns 21–30, the value of V_2 is punched in 31–40, the value of N—where $N + 1$ is the number of equally spaced points in the interval $[a, b]$ at which solutions of the initial value problems are to be calculated—is punched in columns 41–43 (column 43 is the units column), and the value of N_I—the maximum number of iterations to be performed—is punched in columns 44–45 (column 45 is the units column). As before, the fourth order Runge-Kutta method is used to generate numerical solutions of the required initial value problems. Equation (52)—computer program statement 48—is a linear interpolation formula for a new estimate of the initial condition $y'(a)$ based upon the two most recent estimates and the terminal boundary conditions produced by those estimates. At any one time only the numerical solution y_n of the most recent initial value problem of the form (48)–(53) and the initial and terminal values of y_n—$y'_n(a)$, $y_n(b)$, and $y'_n(b)$—and the initial and terminal values of the next most recent numerical solution y_{n-1} need to be stored in the computer in order to use the linear interpolation formula. The initial conditions $y'_n(a)$ and $y'_{n-1}(a)$ are stored in S(1) and S(2). The terminal conditions $y_n(b)$ and $y_{n-1}(b)$ are stored in E(1) and E(2) and the terminal conditions $y'_n(b)$ and $y'_{n-1}(b)$ are stored in EP(1) and EP(2). The number of initial value problems which have been solved is kept track of in

```
      C--FOURTH-ORDER RUNGE-KUTTA SOLUTION OF THE BOUNDARY VALUE PROBLEM
      C---   Y'' = F(X,Y,Y'); Y(A) = BV1, HB(Y(B),Y'(B)) = BV2
0001        DIMENSION Y(101), S(2), E(2), EP(2)
0002        G(X,Y,YP) = YP
0003        F(X,Y,YP) =  -2.*Y*YP
0004        HB(Y,YP) = Y
0005        READ(1,900) A,B,BV1,BV2,N,NI
0006    900 FORMAT(4E10.5,I3,I2)
0007        H = (B-A)/N
0008        M = N + 1
0009        K = 0
0010        L = 0
0011        S(1) = 0.
0012        S(2) = 1.
      C--GENERATE AND STORE THE SOLUTION OF THE IVP
      C---   Y'' = F(X,Y,Y'); Y(A) = BV1, Y'(A) = S(K)
0013      1 K = K + 1
0014        L = L + 1
0015        X = A
0016        Y(1) = BV1
0017        YP = S(K)
0018        DO 2 I = 1,N
0019        XK11 = G(X,Y(I),YP)
0020        XK21 = F(X,Y(I),YP)
0021        X = X + .5*H
0022        Z1 = Y(I) + .5*H*XK11
0023        Z2 = YP   + .5*H*XK21
0024        XK12 = G(X,Z1,Z2)
0025        XK22 = F(X,Z1,Z2)
0026        Z1 = Y(I) + .5*H*XK12
0027        Z2 = YP   + .5*H*XK22
0028        XK13 = G(X,Z1,Z2)
0029        XK23 = F(X,Z1,Z2)
0030        X = X + .5*H
0031        Z1 = Y(I) + .5*H*XK13
0032        Z2 = YP   + .5*H*XK23
0033        XK14 = G(X,Z1,Z2)
0034        XK24 = F(X,Z1,Z2)
0035        J = I + 1
0036        Y(J) = Y(I) + H*(XK11+2.*XK12+2.*XK13+XK14)/6.
0037      2 YP = YP     + H*(XK21+2.*XK22+2.*XK23+XK24)/6.
0038        E(K) = Y(J)
0039        EP(K) = YP
      C--CHECK TO SEE IF THE BOUNDARY CONDITION HB(Y(B),Y'(B)) = BV2 IS
      C---SATISFIED
0040        IF(BV2.EQ.0.) GO TO 3
0041        IF(ABS(1. - HB(E(K),EP(K))/BV2).LT.10.**(-5.)) GO TO 6
0042        GO TO 23
0043      3 IF(ABS(HB(E(K),EP(K))).LT.10.**(-5.)) GO TO 6
      C--BOUNDARY CONDITION NOT SATISFIED.  DETERMINE ACTION TO TAKE.
0044     23 IF(L.EQ.1) GO TO 1
0045        IF(L.GE.NI) GO TO 4
0046        IF(K.EQ.2) K = K - 2
0047        KK = K + 1
0048        S(KK) = S(1) + (S(2)-S(1))*(BV2-HB(E(1),EP(1)))/(HB(E(2),EP(2))-HB
           1(E(1),EP(1)))
0049        GO TO 1
      C--INDICATE THAT THE SOLUTION CAN NOT BE OBTAINED USING THIS
      C---METHOD IN NI ITERATIONS
0050      4 WRITE(3,5) NI
0051      5 FORMAT(11X,'NO SOLUTION CAN BE OBTAINED USING THIS METHOD IN ',I2,
           1' ITERATIONS'//11X,'THE BEST RESULT WAS')
      C--PRINT SOLUTION
0052      6 WRITE(3,7) L
0053      7 FORMAT(11X,'SOLUTION OBTAINED IN ',I2,' ITERATIONS'//11X,'X',19X,
           1'Y')
0054        DO 8 I = 1,M
0055        X = A + (I-1)*H
0056      8 WRITE(3,9) X,Y(I)
0057      9 FORMAT(2(2X,E18.9))
0058        CALL EXIT
0059        END
```

Figure 10.9 Computer program for the solution of the boundary value problem (48) and (50).

the counter L and where the latest initial and terminal conditions are to be stored is kept track of in the variable K.

Some general comments are in order. First, the method just described can be used to solve boundary value problems of the form

$$y'' = f(x, y, y');$$
$$g(y(a), y'(a)) = V_1, \qquad y(b) = V_2.$$

All that one needs to do is to make the substitution $x = -x$. Then, in effect, the integration proceeds from b to a. Second, linear boundary value problems of the forms (18)–(19) and (33)–(34) may be solved using this method. Third, the first two estimates of $y'(a)$ may be chosen differently from the choices indicated in steps 1 and 4. Different choices for the first two estimates of $y'(a)$ may affect the convergence of the process to a solution and also the rapidity of the convergence. And, finally, some technique other than the fourth order Runge-Kutta method could be used to generate the numerical solutions of the initial value problems.

EXAMPLE Compute a numerical solution to the boundary value problem

$$y'' = -2yy';$$
$$y(0) = 1, \qquad y(1) = .5$$

on the interval [0, 1] with $N = 100$ and compare with the exact solution $y = 1/(x + 1)$ at $x = 0., .1, .2, \ldots, 1.0$.

For this example $f(x, y, y') = -2yy'$, $h(y, y') = y$, $a = 0$, $b = 1$, $V_1 = 1$, and $V_2 = .5$. The computer program of Figure 10.9 was used to generate the solution values shown in the second column of Table 10.7. This solution required seven iterations. Also shown in Table 10.7 is the exact solution and the difference between the computed solution and the exact solution.

Table 10.7

x	*Computed solution*	*Exact solution*	*Difference*
.0	1.000 000	1.000 000	.000 000
.1	.908 975	.909 091	.000 116
.2	.833 162	.833 333	.000 171
.3	.769 040	.769 231	.000 191
.4	.714 098	.714 286	.000 188
.5	.666 497	.666 667	.000 170
.6	.624 857	.625 000	.000 143
.7	.588 124	.588 235	.000 111
.8	.555 480	.555 555	.000 075
.9	.526 277	.526 316	.000 039
1.0	.500 000	.500 000	.000 000

EXAMPLE Compute a numerical solution of the boundary value problem

$$y'' = -2yy';$$
$$y(0) = 1, \qquad (y(1))^2 + y'(1) = 0$$

on the interval [0, 1] with $N = 100$ and compare with the exact solution $y = 1/(x + 1)$ at $x = 0., .1, .2, \ldots, 1.0$.

In this case $f(x, y, y') = -2yy'$, $h(y, y') = y^2 + y'$, $a = 0$, $b = 1$, $V_1 = 1$, and $V_2 = 0$. Four iterations were required to obtain the solution shown in Table 10.8.

Table 10.8

x	*Computed solution*	*Exact solution*	*Difference*
.0	1.000 000	1.000 000	.000 000
.1	.908 920	.909 091	.000 171
.2	.833 061	.833 333	.000 272
.3	.768 899	.769 231	.000 332
.4	.713 921	.714 286	.000 365
.5	.666 287	.666 667	.000 380
.6	.624 617	.625 000	.000 383
.7	.587 855	.588 235	.000 380
.8	.555 184	.555 555	.000 371
.9	.525 955	.526 316	.000 361
1.0	.499 652	.500 000	.000 348

EXAMPLE Use the shooting method to produce a solution of the linear boundary value problem

$$y'' = \frac{2x}{x^2 + 1} y' - \frac{2}{x^2 + 1} y + x^2 + 1;$$

$$y(0) = 2, \qquad y(1) = \tfrac{5}{3}$$

on the interval [0, 1] with $N = 100$ and compare with the exact solution

$$y = \frac{x^4}{6} - \frac{3x^2}{2} + x + 2 \quad \text{at } x = 0., .1, .2, \ldots, 1.0.$$

For this example,

$$f(x, y, y') = \frac{2x}{x^2 + 1} y' - \frac{2}{x^2 + 1} y + x^2 + 1, \quad h(y, y') = y, \ a = 0, \ b = 1,$$

$V_1 = 1$, and $V_2 = \frac{5}{3}$. The solution shown in Table 10.9 was obtained on the third iteration. The solution to linear boundary value problems of this type will be obtained in three or less iterations since the boundary conditions are linear and we are using linear interpolation. The results obtained by using the shooting method are essentially the same as the results obtained by using the computer program designed for linear boundary value problems of this type. See Table 10.1.

Table 10.9

x	*Computed solution*	*Exact solution*	*Difference*
.0	2.000 00	2.000 00	.000 00
.1	2.085 07	2.085 02	.000 05
.2	2.140 39	2.140 27	.000 12
.3	2.166 54	2.166 35	.000 19
.4	2.164 51	2.164 27	.000 24
.5	2.135 71	2.135 42	.000 29
.6	2.081 91	2.081 60	.000 31
.7	2.005 32	2.005 02	.000 30
.8	1.908 52	1.908 27	.000 25
.9	1.794 50	1.794 35	.000 15
1.0	1.666 67	1.666 67	.000 00

EXAMPLE Compute a numerical solution to the boundary value problem

$$y'' = -\frac{(y')^2}{y};$$

$$y(0) = 1, \qquad y'(1) = \tfrac{3}{4}$$

on the interval [0, 1] with $N = 100$.

In this example $f(x, y, y') = -(y')^2/y$. In all our previous examples the functions f, f_y, and $f_{y'}$ have all been defined and continuous on the domain $D = \{(x, y) \mid a - \epsilon < x < b + \epsilon, \epsilon > 0\}$. So the solutions of the initial value problems which were needed to produce a solution to the boundary value problem always existed and were unique on $[a, b]$. The functions $f(x, y, y') = -(y')^2/y, f_y = (y')^2/y^2$, and $f_{y'} = -2y'/y$ are not defined for $y = 0$. Therefore we should not be too surprised if we cannot produce numerical solutions to many boundary value problems of this type. And we should examine the results carefully if y is ever "nearly" zero on $[a, b]$. Also, we should discount the validity of any numerical solution which assumes both positive and negative values on the interval $[a, b]$. That is, the graph of the solution y should never intersect the graph of the line $y = 0$. Returning to the example at hand, we see that $h(y, y') = y'$, $a = 0$, $b = 1$, $V_1 = 1$, and $V_2 = \frac{3}{4}$. The computed solution shown in Table 10.10 was obtained in six iterations. This solution is also compared with the exact solution $y = \sqrt{3x + 1}$ at $x =$ 0., .1, .2, . . . , 1.0 in Table 10.10.

Table 10.10

x	*Computed solution*	*Exact solution*	*Difference*
.0	1.000 00	1.000 00	.000 00
.1	1.140 49	1.140 17	.000 32
.2	1.265 42	1.264 91	.000 51
.3	1.379 05	1.378 40	.000 65
.4	1.483 98	1.483 24	.000 74
.5	1.581 95	1.581 14	.000 81
.6	1.674 18	1.673 32	.000 86
.7	1.761 58	1.760 68	.000 90
.8	1.844 84	1.843 91	.000 93
.9	1.924 50	1.923 54	.000 96
1.0	2.000 98	2.000 00	.000 98

10.2.2 Initial Boundary Condition: $y'(a) + Cy(a) = V_1$

Finally, let us consider the boundary value problem:

$$y'' = f(x, y, y'); \tag{48}$$

$$y'(a) + Cy(a) = V_1, \qquad h(y(b), y'(b)) = V_2. \tag{54}$$

The "shooting method" for producing a numerical solution of the boundary value problem (48)–(54) on the interval $[a, b]$ consists of the following steps:

1. Set $y(a) = 0$, set $y'(a) = V_1$, and calculate a numerical solution y_1 of the initial value problem

$$y'' = f(x, y, y');$$
$$y(a) = 0, \qquad y'(a) = V_1$$

on the interval $[a, b]$.
2. Check to see if y_1 "nearly" satisfies the terminal boundary condition $h(y(b), y'(b)) = V_2$.
3. If so, then y_1 is a solution.
4. If not, set $y(a) = 1$, calculate $y'(a) = V_2 - C$, and calculate a numerical solution y_2 of the initial value problem

$$y'' = f(x, y, y');$$
$$y(a) = 1, \qquad y'(a) = V_1 - C$$

on the interval $[a, b]$.

5. Check to see if y_2 "nearly" satisfies the terminal boundary condition.
6. If so, then y_2 is a solution.
7. If not, set $n = 3$, calculate

$$y_n(a) = y_{n-2}(a) + \frac{[y_{n-1}(a) - y_{n-2}(a)][V_2 - h(y_{n-2}(b), y'_{n-2}(b))]}{[h(y_{n-1}(b), y'_{n-1}(b)) - h(y_{n-2}(b), y'_{n-2}(b))]}, \tag{55}$$

calculate $y'_n(a) = V_1 - Cy_n(a)$, and calculate a numerical solution y_n of the initial value problem

$$y'' = f(x, y, y');$$
$$y(a) = y_n(a), \qquad y'(a) = y'_n(a)$$

on the interval $[a, b]$. Equation (55) is the formula for the linear interpolation of the initial condition $y(a)$ from the two previous values, the associated two terminal conditions, and the required terminal condition.
8. Check to see if y_n satisfies the terminal boundary condition.
9. If so, y_n is a solution.
10. If not, repeat steps 7–10 for $n = 4, 5, \ldots, N_I$, where N_I is the maximum number of iterations to be performed.

A FORTRAN computer program to perform the procedure outlined above is displayed in Figure 10.10. This program is similar to the program of Figure 10.9. The function $f(x, y, y')$ is specified in the third program statement and the terminal boundary function is specified in the fourth program statement. The values of a, b, V_1, V_2, C, N, and N_I are read from one data card by the fifth program statement. The value of a, b, V_1, V_2, and C are punched in columns 1–10, 11–20, 21–30, 31–40, 41–50, respectively; the value of N is punched in columns 61–63 (column 63 is the units column); and the value of N_I is punched in columns 64–65 (column 65 is the units column). Equation (55)—computer program statement 48—is a linear interpolation formula for a new estimate of the initial condition $y(a)$ based upon the last two estimates and the terminal boundary conditions produced by those estimates.

Several comments are appropriate. First, the method described can be used to solve boundary value problems of the form

$$y'' = f(x, y, y');$$
$$g(y(a), y'(a)) = V_1, \qquad y'(b) + Cy(b) = V_2.$$

All that one needs to do is to make the change of variable $x = -x$. Then in effect the integration takes place from b to a. Second, linear boundary value problems of the form (40)–(41) may be solved using this method. Third, the first two estimates of $y(a)$ may be chosen differently from the choices made in steps 1 and 4. Different choices may effect both the convergence and the rapidity of the convergence of the process. And, finally, some integration

```
      C--FOURTH-ORDER RUNGE-KUTTA SOLUTION OF THE BOUNDARY VALUE PROBLEM
      C---   Y'' = F(X,Y,Y'); Y'(A) + C*Y(A) = BV1, HB(Y(B),Y'(B)) = BV2
0001         DIMENSION Y(101), S(2), E(2), EP(2)
0002         G(X,Y,YP) = YP
0003         F(X,Y,YP) = -2.*Y*YP
0004         HB(Y,YP) = Y*Y + YP
0005         READ(1,900) A,B,BV1,BV2,C,N,NI
0006     900 FORMAT(5E10.5,10X,I3,I2)
0007         H = (B-A)/N
0008         M = N + 1
0009         K = 0
0010         L = 0
0011         S(1) = 0.
0012         S(2) = 1.
      C--GENERATE AND STORE THE SOLUTION OF THE IVP
      C---   Y'' = F(X,Y,Y'); Y(A) = S(K), Y'(A) = BV1 - C*Y(A)
0013       1 K = K + 1
0014         L = L + 1
0015         X = A
0016         Y(1) = S(K)
0017         YP = BV1 - C*S(K)
0018         DO 2 I = 1,N
0019         XK11 = G(X,Y(I),YP)
0020         XK21 = F(X,Y(I),YP)
0021         X = X + .5*H
0022         Z1 = Y(I) + .5*H*XK11
0023         Z2 = YP   + .5*H*XK21
0024         XK12 = G(X,Z1,Z2)
0025         XK22 = F(X,Z1,Z2)
0026         Z1 = Y(I) + .5*H*XK12
0027         Z2 = YP   + .5*H*XK22
0028         XK13 = G(X,Z1,Z2)
0029         XK23 = F(X,Z1,Z2)
0030         X = X + .5*H
0031         Z1 = Y(I) + .5*H*XK13
0032         Z2 = YP   + .5*H*XK23
0033         XK14 = G(X,Z1,Z2)
0034         XK24 = F(X,Z1,Z2)
0035         J = I + 1
0036         Y(J) = Y(I) + H*(XK11+2.*XK12+2.*XK13+XK14)/6.
0037       2 YP   = YP   + H*(XK21+2.*XK22+2.*XK23+XK24)/6.
0038         E(K) = Y(J)
0039         EP(K) = YP
      C--CHECK TO SEE IF THE BOUNDARY CONDITION HB(Y(B),Y'(B)) = BV2 IS
      C---SATISFIED
0040         IF(BV2.EQ.0.) GO TO 3
0041         IF(ABS(1. - HB(E(K),EP(K))/BV2).LT.10.**(-3.)) GO TO 6
0042         GO TO 23
0043       3 IF(ABS(HB(E(K),EP(K))).LT.10.**(-3.)) GO TO 6
      C--BOUNDARY CONDITION NOT SATISFIED.  DETERMINE ACTION TO TAKE.
0044      23 IF(L.EQ.1) GO TO 1
0045         IF(L.GE.NI) GO TO 4
0046         IF(K.EQ.2) K = K - 2
0047         KK = K + 1
0048         S(KK) = S(1) + (S(2)-S(1))*(BV2-HB(E(1),EP(1)))/(HB(E(2),EP(2))-HB
            1(E(1),EP(1)))
0049         GO TO 1
      C--INDICATE THAT THE SOLUTION CAN NOT BE OBTAINED USING THIS
      C---METHOD IN NI ITERATIONS
0050       4 WRITE(3,5) NI
0051       5 FORMAT(11X,'NO SOLUTION CAN BE OBTAINED USING THIS METHOD IN ',I2,
            1' ITERATIONS'//11X,'THE BEST RESULT WAS')
      C--PRINT SOLUTION
0052       6 WRITE(3,7) L
0053       7 FORMAT(11X,'SOLUTION OBTAINED IN ',I2,' ITERATIONS'//11X,'X',19X,
            1'Y')
0054         DO 8 I = 1,M
0055         X = A + (I-1)*H
0056       8 WRITE(3,9) X,Y(I)
0057       9 FORMAT(2(2X,E18.9))
0058         CALL EXIT
0059         END
```

Figure 10.10 Computer program for the solution of the boundary value problem (48) and (54).

procedure other than the fourth order Runge-Kutta technique could be used to produce the numerical solutions to the initial value problems.

EXAMPLE Compute a numerical solution to the boundary value problem

$$y'' = -2yy';$$
$$y'(0) + 2y(0) = 1; \qquad (y(1))^2 + y'(1) = 0$$

on the interval [0, 1] with $N = 100$ and compare with the exact solution

$$y = \frac{1}{x+1} \quad \text{at } x = 0., .1, .2, \ldots, 1.0.$$

For this example $f(x, y, y') = -2yy'$, $h(y, y') = y^2 + y'$, $a = 0$, $b = 1$, $V_1 = 1$, $V_2 = 0$, and $C = 2$. The computer program of Figure 10.10 was used to generate the solution values shown in the second column of Table 10.11. This solution only required two iterations since $y'(0) = 1$. Also shown in the table are the exact solution and the difference between the computed solution and the exact solution.

Table 10.11

x	*Computed solution*	*Exact solution*	*Difference*
.0	1.000 000	1.000 000	.000 000
.1	.908 965	.909 091	.000 126
.2	.833 142	.833 333	.000 191
.3	.769 013	.769 231	.000 218
.4	.714 065	.714 286	.000 221
.5	.666 457	.666 667	.000 210
.6	.624 811	.625 000	.000 189
.7	.588 073	.588 235	.000 162
.8	.555 424	.555 555	.000 131
.9	.526 216	.526 316	.000 100
1.0	.499 934	.500 000	.000 066

Readers interested in more advanced techniques for computing numerical solutions to boundary value problems should consult the texts by Roberts and Shipman [18] and Bailey, Shampine, and Waltman [1]. Both of these texts contain numerous references. The first contains an extensive survey of methods of solution while the second deals primarily with existence, uniqueness, and comparison theorems.

EXERCISES

1. Find the exact solution of the boundary value problem

$$y'' = y + x;$$
$$y(0) = 0, \qquad y(1) = 1.$$

Compute a numerical solution on [0, 1] with $N = 100$ and compare with the exact solution at $x = 0., .1, .2, \ldots, 1.0$.

2. Use the fact that x^2 is a solution of the differential equation to solve the boundary value problem

$$y'' = -\frac{y'}{x} + \frac{4y}{x^2};$$

$$y(1) = 2, \qquad y(2) = 4.25.$$

Compute a numerical solution on [1, 2] with $N = 100$ and compare with the exact solution at $x = 1.0, 1.1, \ldots, 2.0$.

3. Compute a numerical solution to the boundary value problem

$$y'' = -xy' - y + 2x;$$

$$y(0) = 1, \qquad y(1) = 0$$

on the interval [0, 1] with $N = 100$.

4. Find the exact solution of the boundary value problem

$$y'' = -y + 2;$$

$$y(0) = 1, \qquad y'(1) + y(1) = 2 + 2 \sin 1 \doteq 3.682\ 941.$$

Compute a numerical solution on [0, 1] with $N = 100$ and compare with the exact solution at $x = 0., .1, .2, \ldots, 1.0$.

5. Compute a numerical solution of the boundary value problem

$$y'' = -xy' + x^2y + \sqrt{1 + x^2};$$

$$y(0) = 1, \qquad 2y'(1) + y(1) = 3$$

on the interval [0, 1] with $N = 100$.

6. Find the exact solution of the boundary value problem

$$y'' = 0;$$

$$y'(0) + y(0) = 2, \qquad y'(1) + y(1) = 3.$$

Compute a numerical solution on [0, 1] with $N = 100$ and compare with the exact solution at $x = 0., .1, .2, \ldots, 1.0$.

7. Compute a numerical solution of the boundary value problem

$$y'' = -xy' - y - 2x;$$

$$y'(0) - 2y(0) = 1, \qquad y'(1) + 2y(1) = 0$$

on the interval [0, 1] with $N = 100$.

8. Verify that $y = 1/x$ is a solution of the boundary value problem

$$y'' = 2y^3;$$
$$y(1) = 1, \qquad y'(2) + y^2(2) = 0.$$

Compute a numerical solution on [1, 2] with $N = 100$ and compare the results with the exact solution at $x = 1.0, 1.1, \ldots 2.0$.

9. Find the exact solution of the boundary value problem

$$y'' = -(y')^3;$$
$$y'(0) + y(0) = \frac{3\sqrt{2}}{2} \doteq 2.121\ 330, \qquad y'(1) = \frac{1}{2}.$$

(*Hint:* Let $v = y'$.) Compute a numerical solution on [0, 1] with $N = 100$ and compare with the exact solution at $x = 0., .1, .2, \ldots, 1.0$.

10. Compute a numerical solution to the boundary value problem

$$y'' = \frac{(y')^2}{y + 1};$$
$$y'(0) + y(0) = 1, \qquad y'(1) - y(1) = 1$$

on the interval [0, 1] with $N = 100$.

REFERENCES

1. Bailey, P. B., Shampine, L. F., and Waltman, P. E. *Nonlinear Two Point Boundary Value Problems*, Academic Press, Inc., New York, 1968.
2. Bell, E. T. *The Development of Mathematics*, 2nd ed., McGraw-Hill, New York, 1945.
3. Birkhoff, G., and Rota, G. C. *Ordinary Differential Equations*, Ginn and Company, Waltham, Mass., 1962.
4. Blum, E. K. *Numerical Analysis and Computation: Theory and Practice*, Addison-Wesley Publishing Company, Inc., Reading, Mass., 1972.
5. Boyce, W. E., and DiPrima, R. C. *Elementary Differential Equations and Boundary Value Problems*, 2nd ed., John Wiley & Sons, Inc., New York, 1969.
6. Conte, S. D., and de Boor, C. *Elementary Numerical Analysis*, 2nd ed., McGraw-Hill Book Company, New York, 1972.
7. Greenspan, D. *Theory and Solution of Ordinary Differential Equations*, Macmillan Publishing Co., Inc., New York, 1960.
8. Hagin, F. G. *A First Course in Differential Equations*, Prentice-Hall, Inc., Englewood Cliffs, N.J., 1975.
9. Henrici, P. *Discrete Variable Methods in Ordinary Differential Equations*, John Wiley & Sons, Inc., New York, 1962.

10. Ince, E. L. *Ordinary Differential Equations*, Dover Publications, Inc., New York, 1956.

11. Kamke, E. *Differentialgleichungen Lösungsmethoden und Lösungen*, Akademische Verlagsgesellschaft, Leipzig, 1943.

12. Kaplan, W. *Ordinary Differential Equations*, Addison-Wesley Publishing Company, Inc., Reading, Mass., 1958.

13. Keller, H. B. *Numerical Methods for Two-Point Boundary-Value Problems*, Blaisdell Publishing Company, Waltham, Mass., 1968.

14. Lapidus, L., and Seinfeld, J. H. *Numerical Solution of Ordinary Differential Equations*, Academic Press, Inc., New York, 1971.

15. Pennisi, L. L. *Elements of Ordinary Differential Equations*, Holt, Rinehart and Winston, Inc., New York, 1972.

16. Petrovski, I. G. *Ordinary Differential Equations*, Prentice-Hall, Inc., Englewood Cliffs, N.J., 1966.

17. Rainville, E. D., and Bedient, P. E. *Elementary Differential Equations*, 5th ed., Macmillan Publishing Co., Inc., New York, 1974.

18. Roberts, S. M., and Shipman, J. S. *Two-Point Boundary Value Problems: Shooting Methods*, American Elsevier Publishing Company, Inc., New York, 1972.

19. Ross, S. *Differential Equations*, 2nd ed., Xerox College Publishing, Lexington, Mass., 1974.

20. Smith, D. E. *History of Mathematics*, Vols. 1 and 2, Dover Publications, Inc., New York, 1958.

ANSWERS TO SELECTED EXERCISES

CHAPTER 1
Page 11

1.

	Linear	*Nonlinear*	*Order*	*Degree*
(a)	×		1	1
(b)		×	1	1
(c)		×	1	2
(d)	×		2	1
(e)		×	1	1
(f)	×		1	1
(g)		×	2	2
(h)	×		4	1

2. (a) No. The function $1/x$ is not defined and therefore not continuous or differentiable at $x = 0 \in [-1, 1]$.
   (b) Yes. The function $1/x$ is continuous, differentiable, and satisfies the differential equation on $(0, \infty)$.

3. (a) No. The function $|x|$ is not differentiable at $x = 0 \in [-1, 1]$.
   (b) Yes. The function $|x|$ is continuous, differentiable, and satisfies the differential equation on $(0, \infty)$.

4. (a) $(n\pi/2, (n+2)\pi/2)$, $n = 0, \pm 1, \pm 2, \ldots$.
(b) $(-\infty, \infty)$; $(-\infty, \infty)$.
(c) $(-\infty, \infty)$; $(-\infty, \infty)$.
(d) $(-\infty, \infty)$; $(-\infty, 0)$, $(0, \infty)$.
(e) $(-\infty, \infty)$; $(0, \infty)$.

14. (a) $y = 0$. (b) $y = 2e^{x^2}$.
(c) $y = e^{x^2+1}$.

15. (a) $y = (e^x + e^{-x})/2$. (b) $y = (e^x - e^{-x})/2$.
(c) $y = (e^x + e^{-x})/2$. (d) $y = (e^x - e^{-x})/2$.

16. (a) $y = \sin x$. (b) $y = c_1 \sin x$, where c_1 is arbitrary.
(c) $y = 0$. (d) No solution.

CHAPTER 2

Page 16

1. (a) $y = \frac{3}{2}x^2 + x - \frac{1}{2}$. (b) $y = -2\cos x - 1$.
(c) $y = 1 + \int_0^x \cos t^2 \, dt$. (d) $y = \sin x - x\cos x$.

2. (a) $y = 1 + \ln|x - 1|$; $I = (1, \infty)$.
(b) $y = 1 + \ln|x - 1|$; $I = (-\infty, 1)$.
(c) $y = 1 + \frac{1}{2}\ln 3 + \frac{1}{2}\ln\left|\frac{x-1}{x+1}\right|$; $I = (1, \infty)$.
(d) $y = 1 + \frac{1}{2}\ln\left|\frac{x-1}{x+1}\right|$; $I = (-\infty, 1)$.
(e) $y = -\ln|\cos x|$; $I = (-\pi/2, \pi/2)$.
(f) $y = -\ln|\cos x|$; $I = (\pi/2, 3\pi/2)$.

3. (a) $y' = \sin x^2$; $y(1) = -2$. (b) $y' = -\sin x^2$; $y(1) = -2$.
(c) $y' = x\tan x$; $y(0) = 0$. (d) $y' = \sin x + x\sec x$; $y(0) = 0$.

Page 28

1. (i) $h \le 5\sqrt{2e} \times 10^{-7} \doteq 1.1658 \times 10^{-6}$; $N \ge 857\ 764$.
(ii) $h \le \sqrt{3} \times 10^{-3} \doteq 1.732 \times 10^{-3}$; $N \ge 578$.
(iii) $h \le \sqrt[4]{750} \times 10^{-2} \doteq 5.233 \times 10^{-2}$; $2N \ge 20$.

2. (a) $y = 1 + \sin x - x\cos x$. (b) $y(1) = 1.301\ 169$.

(c)

	$y(1)$	*Error* Estimated	*Error* True
(i) Rectangular rule			
Left endpoints	1.260 247	.069 089	−.040 922
Right endpoints	1.344 394		.043 225
(ii) Trapezoidal rule	1.302 321	$.000\ 199 \leq E_T \leq .001\ 667$	.001 152
(iii) Simpson's rule	1.301 167	$-.000\ 002\ 2 \leq E_S \leq -.000\ 000\ 7$	−.000 002

3. (a)

	$y(1)$	*Error*
Rectangular rule		
Left endpoints		
$N = 10$	3.953 758 523	$E_R \leq .015\ 058\ 434$
$N = 20$	3.949 983 552	$E_R \leq .007\ 529\ 217$
Right endpoints		
$N = 10$	3.937 905 621	
$N = 20$	3.942 057 101	
(b) Trapezoidal rule		
$N = 5$	3.945 078 780	$-.001\ 111\ 111 \leq E_T \leq -.000\ 797\ 112$
$N = 10$	3.945 832 072	$-.000\ 277\ 777 \leq E_T \leq -.000\ 199\ 278$
$N = 20$	3.946 020 326	$-.000\ 069\ 445 \leq E_T \leq -.000\ 049\ 820$
(c) Simpson's rule		
$2N = 10$	3.946 083 169	$.000\ 000\ 074 \leq E_S \leq .000\ 000\ 111$
$2N = 20$	3.946 083 078	$.000\ 000\ 005 \leq E_S \leq .000\ 000\ 007$

4. .428 134. $.000\ 004 < E_S < .000\ 346$.
1.868 917. $.000\ 006 < E_S < .201\ 700$.

5. $D(.6) = .792\ 932$; $D(1) = .674\ 419$.

6. $J_0(.5) = .938\ 470$; $J_0(3) = -.260\ 060$.

Page 33

1. (i)

	$N = 100$ $h = .01$	$N = 1000$ $h = .001$	$N = 10\ 000$ $h = .0001$
Trapezoidal rule	1.473 15	1.469 70	1.468 06
Simpson's rule	1.465 43	1.469 32	1.468 73

(ii)

	$N = 100$ $h = .01$	$N = 1000$ $h = .001$	$N = 10\ 000$ $h = .0001$
Trapezoidal rule	4.047 38	5.242 98	6.746 65
Simpson's rule	4.084 79	5.340 14	7.093 95

As $N \to \infty (h \to 0)$, $L(1) \to \infty$. That is, the arc length is not finite.

CHAPTER 3

Page 42

1. $y = -e^{3x}; (-\infty, \infty)$.
2. $y = 1; (-\infty, \infty)$.
3. $y = 1 + e^{-x}; (-\infty, \infty)$.
4. $y = 2x; (0, \infty)$.
5. $y = -2x; (-\infty, 0)$.
6. $y = (x^4/4 + x^3/3 + 1)/(x + 1); (-\infty, 1)$.
7. $y = (\sin x)(x - \pi/2); (0, \pi)$
8. $y = -2 + 3e^{x^2/2}; (-\infty, \infty)$.
9. $y = e^{x^2/2}(1 + 2\int_0^x e^{-t^2/2}\, dt); (-\infty, \infty)$.
10. $y = 2e^x - 1; (-\infty, \infty)$.
11. $y = 3e^{(x^2-1)/2} - 1; (-\infty, \infty)$.
12. See Exercise 7.
13. $y = e^{x^2/2}(2e^{-1/2} + \int_1^x e^{-t^2/2}\, dt); (-\infty, \infty)$.
14. $y = x(1 + \int_{-1}^x (1/t) \sin t^2\, dt); (-\infty, \infty)$.
15. $y = x^2[\frac{1}{2} + \int_1^x (e^t/t^2)\, dt]; (0, \infty)$.
16. $y = \sqrt{2(x^2 + 2)}; (-\infty, \infty)$.
17. $y = 2/(1 + e^{2x}); (-\infty, \infty)$.
18. $y = (x + 2/\sqrt{x})^2/9; (0, \infty)$.
19. $y = \sqrt{5}\, x/\sqrt{3 - 2x^5}; (-\infty, 0)$.

Page 49

1. $y^2 = 2x^3/3 + C$.
2. $e^{-y} = -e^x + C$.
3. $\ln |\csc y - \cot y| = x^3/3 + C$.
4. $x^2 + 2xy = C$.
5. $x = y(\ln |y| + C)$.
6. $y = -x/(\ln |x| + C)$.
7. $y = Ce^{x^3/3 + x}$.
8. $y = Cx(x + y)$.

9. $y = e^{-1/x+1}$. **10.** $(x - y)^2 = 1$.

11. $\sin y + \cos x = -1$. **12.** $y = x(\ln|x| - 1)$.

13. $x^2 - 2e^{-y} + 2 = 0$. **14.** $\ln|y| + e^{-x/y} = 1$.

Page 57

1. $xy^2 + 2x^2y + x = C$. **2.** $x^3 + 2y = Cx$.

3. $x \sin y + y^2 = C$. **4.** $x(x + y) = Ce^y$.

5. $x(1 + \ln y) - y^2 = C$. **6.** $xy + y^2 + \ln|x| = C$.

7. $x \sin x + \cos x + xy = C$.

Answers to page 57 continued on page 372.

Page 73

1. 59.9%; 35.8%; 675.7 years. **2.** 96.2%; 5979.5 years.

3. $133\frac{1}{3}$ grams; 2.4 years.

4. ~3804 years before 1977 or ~1827 B.C.

5. 17.3 years; 13.9 years; 11.55 years.

6. 58 years from now. **7.** 2001.85; 2025.94.

8. 4, 8, and 17 years from now.

9. 23.05 million (1850); 76.54 million (1900);
148.40 million (1950); 167.74 million (1970).

10. $r = r_0 - kt$.

11. $(\sqrt{5} - 1)/2$ hours before 8 A.M. or 7:22:55 A.M.

12. 34.5 minutes. **13.** 40°F.

14. 1:20:33 P.M.; 50.15°F. **15.** A.

16. 211 lb.; 400 lb. **17.** 356.9 lb.; 579.1 lb.

18. 7.48%.

19. $Q(t) = (20 - 20e^{-t/20} + te^{-t/20})$ gal; 22.23%.

20. $x^2 + y^2 = k^2$. **21.** $y = \ln|\sin x| + k$.

22. (a) $y^2 = k - 2x^2$. (b) $y^2 = k - 2x$.
(c) $xy^2 = k$. (d) $x^2 + y^2 - \ln y^2 = k$.

23. $$v(t) = \frac{g(e^{2\sqrt{gc}t} - 1) + \sqrt{gc}\, v_0(e^{2\sqrt{gc}t} + 1)}{\sqrt{gc}\,(e^{2\sqrt{gc}t} + 1) + cv_0(e^{2\sqrt{gc}t} - 1)}$$

$v \to \sqrt{g/c}$.

Answers to page 57—*Continued*

	Linear	*Nonlinear*	*Bernoulli*	*Separable*	*Homogeneous*	*Exact*	*Integrating factor*	*Solution*
8.		×	$n = -1$	×	Degree 1	×		$x^2 - y^2 = C$
9.	×			×	Degree 1		$1/x^2$ or $1/y^2$	$y = Cx$
10.		×			Degree 1			$\ln \sqrt{x^2 + y^2} - \text{Arctan}\,(y/x) = C$
11.		×			Degree 1	×		$x^2 - y^2 + 2xy = C$
12.		×				×		$2x^3 - 3y^2 - 6xy = C$
13.	×						$1/x^2$	$y = x(C - x)$
14.		×	$n = 2$	×			$1/y(1 - y)$	$y = 1/(1 + Ce^{-x^2/4})$
15.		×			Degree 1			$\ln \sqrt{x^2 + y^2} - 2\,\text{Arctan}\,(y/x) = C$
16.		×	$n = -2$	×				$y^3 = 1 - Ce^{-x^2/2}$
17.		×				×		$e^y \sin x - x^2 \sin y = C$
18.		×			Degree 1		$1/x^2$ or $1/y^2$	$y[\ln (y/x) - 1] = Cx$
19.	×			×			$((x + 1)/x)^2$	$y = Cx/(x + 1)^3$
20.		×	$n = \frac{1}{2}$					$y = (Ce^{x/2} - x - 2)^2$

24. (a) $v(t) = v_0 e^{-ct}$; $s(t) = v_0(1 - e^{-ct})/c$; v_0/c; never.

(b) $v(t) = v_0/(1 + cv_0 t)$; $s(t) = \dfrac{1}{c}\ln(1 + cv_0 t)$; ∞; never.

(c) $v(t) = (\sqrt{v_0} - ct/2)^2$; $s(t) = v_0 t - \sqrt{v_0}\, ct^2/2 + c^2t^3/12$; $2v_0^{3/2}/3c$; $2\sqrt{v_0}/c$.

25. $y = a \ln\left(\dfrac{a + \sqrt{a^2 - x^2}}{x}\right) - \sqrt{a^2 - x^2}$.

26. $y = (cx^{r+1} - c^{-1}x^{1-r})/2$, where c is a constant of integration and $r = S_R/S_B$.
If $r > 1$, then $y \to -\infty$ as $x \to 0$ and the boat will not land.
If $r = 1$, then $y \to -\frac{1}{2}c$ as $x \to 0$ and the boat will land at $(0, -\frac{1}{2}c)$.
If $r < 1$, then $y \to 0$ as $x \to 0$ and the boat will land at the origin.

27. $r = e^{\theta/\sqrt{3}}$; $0 \le \theta < 2\pi$.

CHAPTER 4

Page 79

1. (a) $y(x) = 1 + \int_1^x t\, dt$. (b) $y(x) = 1 + \int_0^x y(t)\, dt$.
(c) $y(x) = -1 + \int_1^x [2ty(t) + 1]\, dt$. (d) $y(x) = \int_0^x [1 + y^2(t)]\, dt$.

2. (a) $y' = xy + y^2$; $y(1) = -2$. (b) $y' = 2y^{1/2}$; $y(-1) = 0$.
(c) $y' = 2x - xy^2$; $y(1) = 2$. (d) $y' = -x^2y$; $y(0) = 2$.

Page 87

1. (a) 4. (b) 4.
(c) 1. (d) $1/\epsilon^2$.
(e) $1/\epsilon$. (f) $1/\sqrt{\epsilon}$.

2. (a) Any positive constant M will serve as the Lipschitz constant on any domain D.
(b) Any positive constant M will serve as the Lipschitz constant on any domain D on which $g(x)$ is defined.
(c) Lipschitz constant $M \ge 1$ on any domain D.
(d) $M \ge 1$ on any domain D where $g(x)$ is defined.
(e) $M = \sup_{(x,y)\in D} |x|$ on any domain D with bounded abscissa.
(f) $M = \sup_{(x,y)\in D} |1/x|$ on any domain D with abscissa bounded away from 0.
(g) $M = \sup_{(x,y)\in D} |g(x)|$ on any domain D on which $g(x)$ is bounded.
(h) $M = \sup_{(x,y)\in D} |x \cos xy|$ on any convex domain D with bounded abscissa.
(i) $M = \sup_{(x,y)\in D} |2y|$ on any convex domain with bounded ordinate.
(j) $M = \sup_{(x,y)\in D} |\frac{1}{2}y^{-1/2}| = \dfrac{1}{2\sqrt{\epsilon}}$ on any convex domain D with $y > \epsilon > 0$.

(k) $M = \sup_{(x,y)\in D} |1/xy^2|$ on any convex domain D which is bounded away from the coordinate axes.

(l) $M = \sup_{(x,y)\in D} |\frac{1}{2}\sqrt{x/y}|$ on any convex domain D with $xy > \epsilon > 0$.

3. (a) The xy-plane.
(b) The xy-plane with the lines $x = 0$ and $x = (2n + 1)\pi/2$, where n is an integer deleted.
(c) Any domain on which $a(x)$ and $b(x)$ are defined and continuous.
(d) The xy-plane.
(e) The xy-plane with the line $y = 0$ deleted.
(f) The xy-plane with the origin deleted.

4. $(-\infty, 0)$; $(-\infty, \infty)$; for $c_0 > 0$, $(-\infty, x_0 + 1/c_0)$ and for $c_0 < 0$, $(x_0 + 1/c_0, \infty)$.

5. $(-\pi/2, \pi/2)$.

6. (a) $y_0 = 1$; $y_1 = 1 + x^2/2$; $y_2 = 1 + x^2/2$; $y_3 = 1 + x^2/2$; $y_n \to 1 + x^2/2$.
(b) $y_0 = 1$; $y_1 = 1 + x$; $y_2 = 1 + x + x^2/2$; $y_3 = 1 + x + x^2/2 + x^3/3!$; $y_n \to e^x$.
(c) $y_0 = -1$; $y_1 = -1 + x - x^2$; $y_2 = -7/6 + x - x^2 + 2x^3/3 - x^4/2$; $y_3 = -11/10 + x - 7x^2/6 + 2x^3/3 - x^4/2 + 4x^5/15 - x^6/6$.
(d) $y_0 = 0$; $y_1 = x$; $y_2 = x + x^3/3$; $y_3 = x + x^3/3 + 2x^5/15 + x^7/63$; $y_n \to \tan x$.

Page 96

1. 5. **2.** 3.

3. 6.

4. $|y_1(x) - y_0(x)| \leq .5 + x$; .75; 1.; 1.25.

5. $|y_1(x) - y_0(x)| \leq .25 + \pi^2/72 + x$.

Page 103

1. $|y_1(\frac{1}{2}) - y_0(\frac{1}{2})| \leq e^{1/2}$; $|y_1(1) - y_0(1)| \leq e^2$.

2. $|y_1(\frac{1}{2}) - y_0(\frac{1}{2})| \leq (e^{1/2} - 1)/4$; $|y_1(1) - y_0(1)| \leq (e^2 - 1)/2$.

3. $|y_1(\frac{1}{2}) - y_0(\frac{1}{2})| \leq .02e^{.41}$; $|y_1(1) - y_0(1)| \leq .02e^{1.81}$.

4. $|y_1(x) - y_0(x)| \leq (.6 + .02e^{.01})e^{2|x-.1|} - .5$.

Page 120

1. (a) $y_{n+1} = h_n^3/3 + (1 - h_n + h_n^2/2 - h_n^3/6)y_n + (h_n^2 - h_n^3/3)x_n + (h_n - h_n^2/2 + h_n^3/6)x_n^2$.

(b)

x_n	y_n
.0	1.000 000
.1	.905 167
.2	.821 277
.3	.749 192
.4	.689 692
.5	.643 483
.6	.611 203
.7	.593 430
.8	.590 687
.9	.603 447
1.0	.632 137

(c) $E_n \leq .000\ 004\ 167$.
(d) $h \leq .059$.

2. (a)

x_n	y_n
.0	1.
.1	.9
.2	.811
.3	.733 9
.4	.669 51
.5	.618 559
.6	.581 703
.7	.559 533
.8	.552 579
.9	.561 321
1.0	.586 189

(b) $|y_{10} - y(1)| \leq .081\ 606(e - 1) < .140\ 222$.
(c) $h \leq 10^{-3}/\sqrt{1.632\ 121} \doteq .000\ 783$.
(d) $h \leq 10^{-6}/1.632\ 121(e - 1) \doteq .000\ 000\ 357$.

4. (a) $y = x^2 - 2x + 2 - e^{-x}$.

(b)

	Taylor series order 3	*Euler's method*	*Improved Euler*	*Modified Euler*	*Fourth order Runge-Kutta*	*Actual solution*
x_n	y_n	y_n	y_n	y_n	y_n	y_n
.0	1.000 000	1.000 000	1.000 000	1.000 000	1.000 000	1.000 000
.1	.905 167	.900 000	.905 500	.905 250	.905 163	.905 163
.2	.821 277	.811 000	.821 928	.821 451	.821 270	.821 269
.3	.749 192	.733 900	.750 145	.749 464	.749 182	.749 182
.4	.689 692	.669 510	.690 931	.690 064	.689 681	.689 680
.5	.643 483	.618 559	.644 992	.643 958	.643 470	.643 469
.6	.611 203	.581 703	.612 968	.611 782	.611 189	.611 188
.7	.593 430	.559 533	.595 436	.594 113	.593 416	.593 415
.8	.590 687	.552 579	.592 920	.591 472	.590 672	.590 671
.9	.603 447	.561 321	.605 892	.604 332	.603 431	.603 430
1.0	.632 137	.586 189	.634 782	.633 120	.632 121	.632 121

5. $a = \frac{1}{4}$; $b = \frac{3}{4}$; $c = d = \frac{2}{3}$.

Page 126

1. (a)

x_n	y_n
.0	1.000 000
.1	.905 163
.2	.820 888
.3	.748 513
.4	.688 781
.5	.642 389
.6	.609 970
.7	.592 094
.8	.589 278
.9	.601 991
1.0	.630 656

(b) .000 417.
(c) .010 626.

2. (a)

x_n	y_n
.0	1.000 000
.1	.905 163
.2	.820 968
.3	.748 969
.4	.689 174
.5	.643 134
.6	.610 547
.7	.593 025
.8	.589 942
.9	.603 036
1.0	.631 334

(b) .000 167.

3. .000 003.

4. (a) $y_{n+1} = [y_n(2 - h) + h(x_n^2 + x_{n+1}^2)]/(2 + h)$.

(b)

x_n	y_n
.0	1.000 000
.1	.905 238
.2	.821 406
.3	.749 367
.4	.689 904
.5	.643 722
.6	.611 463
.7	.593 705
.8	.590 971
.9	.603 736
1.0	.632 427

5. $y_{n+1} = [y_{n-1}(3 - h) + h(x_{n-1}^2 + 4x_n^2 + x_{n+1}^2 - 4y_n)]/(3 + h)$.

Page 137

1. (i) $y = x^{2/3}$ for $x \in [-1, 0)$. No solution exists for $x \geq 0$.
(ii) $y = x^{3/2}$ for $x \in [-1, 0)$. No solution exists for $x \geq 0$.
(iii) $y = x^2$ for $x \in [-1, 0]$. The solution is not unique for $x > 0$.

CHAPTER 5

Page 145

3. (a) 4. (b) 2.

4. (b) 20 000. (c) $(0, \infty)$.

5. (a) $y_1' = y_2;\ y_2' = -xy_2 - x^2y_1 + x^3$.
(b) $y_1' = y_2;\ y_2' = y_3;\ y_3' = -xy_2y_3 + xy_2^2 + 2xy_1$.

6. (a) $(0, 2)$. (b) $(0, \infty)$.

CHAPTER 6

Page 154

6. (a) $c_1 = 4;\ c_2 = -3;\ c_3 = 1$.
(b) $c_1 = 1;\ c_2 = c_3 = -1$.
(c) $c_1 = 0;\ c_2 = -2;\ c_3 = -1;\ c_4 = 1$.

Page 160

1. $y = c_1e^{-2x} + c_2e^{2x}$.

2. $y = c_1 \cos 2x + c_2 \sin 2x$.

3. $y = (c_1 + c_2x)e^{-2x}$.

4. $y = c_1e^{(-2+\sqrt{2})x} + c_2e^{(-2-\sqrt{2})x}$.

5. $y = c_1e^{-3x/2} + c_2e^{x/2}$.

6. $y = c_1e^x \cos x + c_2e^x \sin x$.

7. $y = c_1e^{2x} + c_2e^x \cos x + c_3e^x \sin x$.

8. $y = c_1 + c_2x + (c_3 + c_4x)e^x$.

9. $y = c_1e^{2x} + c_2e^{-2x} + c_3 \cos 2x + c_4 \sin 2x$.

10. $y = c_1e^{\sqrt{2}x} \cos \sqrt{2}\,x + c_2e^{\sqrt{2}x} \sin \sqrt{2}\,x + c_3e^{-\sqrt{2}x} \cos \sqrt{2}\,x$
$+ c_4e^{\sqrt{2}x} \sin \sqrt{2}\,x$.

11. $y = (c_1 + c_2x)e^x \cos x + (c_3 + c_4x)e^x \sin x$.

12. $y = e^{-3x} + 2e^x$.

13. $y = e^x \cos \sqrt{2}\,x + e^x \sin \sqrt{2}\,x$.

14. $y = 3 - 2x - \cos x + \sin x$.

Page 166

1. $y = c_1e^{-2x} + c_2e^{2x} - x/2 - e^{-x}/3 + 3(\cos x)/5$.

2. $y = c_1e^{-2x} + c_2e^{2x} + 3x^2e^{2x}/8 - 3xe^{2x}/16$.

3. $y = c_1 \cos 2x + c_2 \sin 2x - \frac{3}{4}x \cos 2x - \frac{3}{8} + x^2/4$.

4. $y = c_1 + c_2x + (c_3 + c_4x)e^x + 11x^2 + 7x^3/3 + x^4/4 - x^2e^x/2$.

5. $y = c_1e^{2x} + c_2e^x \cos x + c_3e^x \sin x - x/4 - \frac{3}{8} - xe^x(\cos x)/4 - xe^x(\sin x)/4.$

6. $y = (3e^{-2x} + e^{2x})/4.$

7. $y = -e^{-3x} + x^2 + x + 1.$

8. $y = 3e^x/4 + xe^x/2 + x^3e^x/6 + e^{-x}/4.$

9. $y = 2 \sin 3x - \cos 3x + x/9 - x(\sin 3x)/3.$

10. $y = 2e^{-2x} \sin x - 3e^{-2x} \cos x - x + 1 + 2 \cos x.$

Page 173

1. (a) $s/(s^2 + b^2),\ s > b.$ (b) $s/(s^2 - b^2),\ s > |b|.$
(c) $2bs/(s^2 + b^2)^2,\ s > 0.$ (d) $(s^2 - b^2)/(s^2 + b^2)^2,\ s > 0.$

2. $L[f] = e^{-3s}/s.$

3. $L[g] = 1/s + (2e^{-s} - 1)/s^2.$

4. $L[h] = (e^{-4s} - 2e^{-2s} + 1)/s^2.$

5. $L[k] = (e^{-s} - e^{-2s})/s.$

6. (a) $5/s.$ (b) $e/s.$
(c) $3/s^2 - 2/s.$ (d) $2/(s + 1)^2 + 1/(s + 1).$
(e) $2/(s^2 - 4s + 13) - 2s/(s^2 + 1).$ (f) $e^2/(s - 3).$

9. (a) $3x^2/2.$ (b) $4xe^{-2x}.$
(c) $-2 \cos \sqrt{3}\, x.$ (d) $e^{-x} + x - 1.$
(e) $e^x \cos 2x.$ (f) $e^x \sin 2x.$
(g) $4(\cos x - 1).$ (h) $2e^{-x} \cos x + 3e^{-x} \sin x.$
(i) $x + 2 \sinh x.$ (j) $3e^{2x} \cosh x.$

Page 177

1. $y = Ae^x.$

2. $y = Ae^x \cos 2x + Be^x \sin 2x.$

3. $y = Ae^{-2x} + 2.$

4. $y = Ae^{-3x} + Be^{3x} - \frac{1}{9} \sin 3x.$

5. $y = A \cos 3x + B \sin 3x - \frac{1}{3}x \cos 3x.$

6. $y = Ae^{-2x} + Be^x - xe^x/9 + x^2e^x/6 + 3x^2/2 + 3x/2 + 9/4.$

7. $y = Ae^x + Bxe^x + C + Dx - x^2e^x + x^3e^x/6 - 9x^2 - 2x^3 - x^4/4.$

8. $y = e^x - 2.$

9. $y = e^x + 2xe^x.$

10. $y = (-31e^{-3x} - 11e^{3x} - 6x - 12)/54.$

11. $y = (-33 \cos 3x + 8 \sin 3x + 3x + 6)/27.$

12. $y = (9e^{-2x} - 4e^{3x} - 5 \cos 3x + 5 \sin 3x)/15.$

13. $y = e^x \cos x - x^2/2 - x.$

14. $y = (17e^{-2x} - 20e^{-x} + 55 + 10x^2 - 30x - 12 \cos x + 4 \sin x)/40.$

Page 181

1. $(1 - \cos 3x)/9$.

2. $(e^{2x} - e^{-x})/3$.

3. $[(3x - 1)e^{2x} + e^{-x}]/9$.

4. $[(\sin 2x - 2\cos 2x)e^{x} + 2]/10$.

5. $(2\sin 2x - \cos 2x + e^{x})/5$.

6. $(-2x + \sinh 2x)/8$.

7. $y = -3 + 5e^{2x}$.

8. $y = 2e^{-x} + e^{x}/2$.

9. $y = (1 - \cos 3x)/9$.

10. $y = \sin 3x - 2\cos 3x + e^{3x}$.

11. $y = e^{2x} - e^{-x}$.

12. $y = (28e^{-x} + 14e^{2x} - 6x^2 + 6x - 9)/12$.

13. $y = 3xe^{x} - 3e^{x} + \cos x$.

14. $y = 2\sin 2x - \cos 2x + e^{x}$.

Page 187

1. (a) $f_1(x) = 2u(x) - u(x-1)$.
(b) $f_2(x) = u(x-2) - u(x-4)$.
(c) $f_3(x) = u(x-1)(x-1)^2$.
(d) $f_4(x) = u(x-1)(x^2 - 2x + 3)$.
(e) $f_5(x) = u(x-\pi)\sin 3(x-\pi)$.
(f) $f_6(x) = u(x)x + u(x-1)(x-1)$.
(g) $f_7(x) = [u(x) - u(x-1)]x$.

2. (a) $2/s - e^{-s}/s$.
(b) $(e^{-2s} - e^{-4s})/s$.
(c) $2e^{-s}/s^3$.
(d) $2e^{-s}/s^3 + 2e^{-s}/s$.
(e) $3e^{-\pi}/(s^2 + 9)$.
(f) $(1 + e^{-s})/s^2$.
(g) $[1 - (s+1)e^{-s}]/s^2$.

3. (a) $u(x-1)e^{-2(x-1)}$.
(b) $x - u(x-2)(x-2)$.
(c) $u(x-\pi)\cos 3(x-\pi)$.
(d) $u(x-\pi)\cosh 3(x-\pi)$.
(e) $u(x-2)e^{-(x-2)}\sin(x-2)$.
(f) $u(x-3)e^{-(x-3)}[\cos(x-3) - \sin(x-3)]$.
(g) $u(x-3)(e^{x-3} - e^{-3(x-3)})/4$.
(h) $u(x-1)(x-1)e^{x-1}$.

4. (a) $y = 1 + u(x-1)(e^{-2(x-1)} - 1)/2$.
(b) $y = [2e^{2x} - 2e^{-2x} + u(x-2)(-3 + 2e^{x-2} + e^{2(x-2)}) + u(x-4)(3 - 2e^{x-4} - e^{2(x-4)})]/6$.
(c) $y = u(x-1)[-\frac{1}{8} - \frac{1}{4}(x-1) - \frac{1}{4}(x-1)^2 - \frac{1}{6}(x-1)^3 + \frac{1}{8}e^{2(x-1)}] + 1$.
(d) $y = u(x-1)(8 + 4(x-1) + (x-1)^2 - 8e^{x-1} + 4(x-1)e^{x-1}) + xe^{x}$.
(e) $y = u(x-\pi)\{3\sin[2(x-\pi)] - 2\sin[3(x-\pi)]\}/10 + \cos 2x + (\sin 2x)/2$.
(f) $y = \{u(x-1)[-4(x-1) - e^{-2(x-1)} + e^{2(x-1)}] - 4x - e^{-2x} + e^{2x}\}/16$.
(g) $y = \{-u(x-1)[9 + 5(x-1) - 9e^{2(x-1)}\cos(x-1) + 13e^{2(x-1)}\sin(x-1)] + 4 + 5x + 21e^{2x}\cos x - 47e^{2x}\sin x\}/25$.

Page 190

1. $y = u(x-2)e^{-3(x-2)}$.

2. $y = u(x-1)e^{3(x-1)} + 2u(x-2)(e^{3(x-2)} - 1)/3$.

3. $y = [u(x - \pi)\sin 3(x - \pi) - u(x - 3\pi)\sin 3(x - 3\pi)]/3.$

4. $y = 2u(x - 1)(x - 1)e^{x-1} + xe^x.$

5. $y = [10u(x - \pi)e^{x-\pi}\sin 2(x - \pi) + 4\cos x - 2\sin x$
$+ 16e^x\cos 2x - 7e^x\sin 2x]/20.$

6. $y = [\sin 2x - u(x - \pi)\sin 2(x - \pi)]/2.$

7. $y = [f(\pi)u(x - \pi)\sin ax]/a.$

Page 194

1. $y = x^{-1}$; $(-\infty, 0)$ or $(0, \infty)$.
2. $y = x^{-2}$; $(-\infty, 0)$ or $(0, \infty)$.
3. $y = x\ln x$; $(0, \infty)$.
4. $y = \sin(\ln x)$; $(0, \infty)$.
5. $y = x$; $(-\infty, \infty)$.
6. $y = x + 1$; $(-\infty, \infty)$.
7. $y = x^{-1/2}\sin x$; $(0, \infty)$.
8. $y = \dfrac{x}{2}\ln\left(\dfrac{x - 1}{x + 1}\right) + 1$; $(-\infty, -1)$, $(1, \infty)$.
9. $y = (A + B\sin x + C\cos x)x^2$; $(-\infty, \infty)$.

Page 201

1. $y_p = x\sin x + \cos x\ln|\cos x|.$
2. $y_p = -e^{2x}\cos e^{-x}.$
3. $y_p = e^x\ln(1 + e^x).$
4. $y_p = -e^{-2x}\ln x.$
5. $y_p = -x^2/2.$
6. $y_p = e^{4x}/6.$
7. $y_p = -(x\sin x)/4.$
8. $y_p = (\cos x\ln|\cos x|)/2.$
9. $y_p = x^4e^x/24.$
10. $y = c_1x + c_2(x^2 - 1) + x^4/6 + x^2/2.$
11. $y = c_1x + c_2x^2 + x^3/3.$
12. $y = (c_1 + c_2e^x + e^{2x})x^2.$
13. $y = 3x^3/4 + x^2/2 + x/2 + 1/2 - 1/4x.$

CHAPTER 7

Page 209

1. $y = 5\cos 7t$; $C = 5$ cm; $P = \frac{1}{7}$ s; $\omega = 7$ cycles/s.
2. $y = -5\cos 7t.$
3. $y = -\sqrt{7}\sin 2\sqrt{7}\,t$; $C = \sqrt{7}$ cm; $P = \dfrac{1}{2\sqrt{7}}$ s; $\omega = 2\sqrt{7}$ cycles/s.

4. $y = \sqrt{35}\sin\sqrt{35}\,t + 8\cos\sqrt{35}\,t$; $C = 3\sqrt{11}$ cm; $P = \dfrac{1}{\sqrt{35}}$ s;

$\omega = \sqrt{35}$ cycles/s; $\varphi = \text{Arcsin}\,(8\sqrt{11}/33)$ and $\varphi = \text{Arccos}\,(\sqrt{385}/33)$.

5. $y = -\sqrt{35}\sin\sqrt{35}\,t + 8\cos\sqrt{35}\,t$; $C = 3\sqrt{11}$ cm; $P = \dfrac{1}{\sqrt{35}}$ s;

$\omega = \sqrt{35}$ cycles/s; $\varphi = \text{Arcsin}\,(8\sqrt{11}/33)$ and $\varphi = \text{Arccos}\,(-\sqrt{385}/33)$.

6. (a) 0. (b) equilibrium position.

7. $m = 200$ g.

8. $X = 4$ cm; $k = 128\ \text{kg/s}^2$.

9. $m = 4$ kg.

10. $(245/\pi^2)$ cm $\doteq 24.8$ cm.

11. $\theta = \sqrt{2}\,[\sin(5\sqrt{2}\,t + 3\pi/4)]/32$; $C = .866$ cm; $P = (1/5\sqrt{2})$ s;
$\omega = 5\sqrt{2}$ cycles/s; $\varphi = 3\pi/4$.

Page 215

1. (a) $y = (5 + 77t)e^{-14t}$; critically damped motion.

(b) $y = \left(\dfrac{7 + 10\sqrt{13}}{12}\sin 12t + 5\cos 12t\right)e^{-2\sqrt{13}t}$; damped oscillatory motion.

(c) $y = 7e^{-7t} - 2e^{-28t}$; overdamped motion.

2. $\sqrt{4949 + 140\sqrt{13}}\,e^{-2\sqrt{13}t}/12$; $P = (\pi/6)$ s.

3. oscillatory motion, $c < 2000\sqrt{5}$ g/s; critically damped motion, $c = 2000\sqrt{5}$ g/s; overdamped motion, $c > 2000\sqrt{5}$ g/s.

4. $k = 600\ \text{g/s}^2$; $c = 600\sqrt{3}\,\pi$ g/s.

Page 220

1. (a) 7 cycles/s. (b) $y = -\frac{10}{7}\sin 7t + 4\cos 7t + 2\sin 5t$.

2. $y = \left[\dfrac{c_1}{\omega} - \dfrac{E}{\omega(c - a\bar{\omega}^2)}\right]\sin\omega t + c_0\cos\omega t + \dfrac{E}{c - a\bar{\omega}^2}\sin\bar{\omega}t.$

3. $(\sqrt{41}/2\pi)$ cycles/s.

4. $y = e^{-bt/a}(A\sin\omega t + B\cos\omega t) + \dfrac{E}{H(\bar{\omega})}[(c - a\bar{\omega}^2)\sin\bar{\omega}t - 2b\bar{\omega}\cos\bar{\omega}t],$

where $\omega = \sqrt{ac - b^2}/|a|$, $H(\bar{\omega}) = (c - a\bar{\omega}^2)^2 + 4b^2\bar{\omega}^2$, $A = c_1/\omega + bc_0/a\omega + E\bar{\omega}(2b^2/a - c + a\bar{\omega}^2)/H(\bar{\omega})\omega$, and $B = c_0 + 2b\bar{\omega}E/H(\bar{\omega})$.

5. (a) $K = \sqrt{24 - 2\bar{\omega}^2}$. (b) $\bar{\omega} = 2$ cycles/s.

Page 225

1. (a) $i_0 = \dfrac{E(0)}{R} - \dfrac{q_0}{RC}$.

(b) (i) $q(t) = q_0e^{-t/RC} + EC(1 - e^{-t/RC})$; $i(t) = i_0e^{-t/RC}$; $V(t) = q(t)/C$.

(ii) $$q(t) = \left(q_0 - \frac{EC}{1 + R^2C^2\bar{\omega}^2}\right)e^{-t/RC} + \frac{EC}{1 + R^2C^2\bar{\omega}^2}(RC\bar{\omega}\sin\bar{\omega}t + \cos\bar{\omega}t);$$

$i(t) = dq/dt$; $V(t) = q(t)/C$.

2. (a) $i_0' = E_0/L - q_0/LC$.

(b) (i) $i(t) = \dfrac{E_0C - q_0}{\sqrt{LC}}\sin\dfrac{t}{\sqrt{LC}} + i_0\cos\dfrac{t}{\sqrt{LC}}$.

(ii) $$i(t) = \sqrt{LC}\left(\frac{E_0C - q_0}{LC} + \frac{EC\bar{\omega}^2}{1 - LC\bar{\omega}^2}\right)\sin\frac{t}{\sqrt{LC}} + i_0\cos\frac{t}{\sqrt{LC}} + \frac{EC}{1 - LC\bar{\omega}^2}\sin\bar{\omega}t.$$

(iii) $i(t) = \sqrt{\dfrac{C}{L}}\left(\dfrac{E_0C - q_0}{C} - \dfrac{E}{2}\right)\sin\dfrac{t}{\sqrt{LC}} + \left(i_0 + \dfrac{Et}{2L}\right)\cos\dfrac{t}{\sqrt{LC}}$.

3. (a) $i(t) = i_0e^{-Rt/L} + \dfrac{E}{R}(1 - e^{-Rt/L})$.

(b) $i(t) = \left(i_0 - \dfrac{ER}{R^2 + L^2\bar{\omega}^2}\right)e^{-Rt/L} + \dfrac{E}{R^2 + L^2\bar{\omega}^2}(L\bar{\omega}\sin\bar{\omega}t + R\cos\bar{\omega}t)$.

4. $$i(t) = \frac{RE\bar{\omega}^2\sin\bar{\omega}t + E\bar{\omega}\left(\frac{1}{C} - L\bar{\omega}^2\right)\cos\bar{\omega}t}{\left(\frac{1}{C} - L\bar{\omega}^2\right)^2 + R^2\bar{\omega}^2}, \quad \bar{\omega} = 1/\sqrt{LC}.$$

CHAPTER 8

Page 232

1. (a) $(-1, 1)$. (b) $(-1, 1)$.
(c) $(-\frac{1}{2}, \frac{1}{2})$. (d) $x = 0$.
(e) $(-\frac{7}{3}, -\frac{5}{3})$. (f) $(-3, 3)$.
(g) $(-\infty, \infty)$. (h) $(0, 1)$.

2. (a) $\sum_{n=0}^{\infty} x^n/n!$; $(-\infty, \infty)$. (b) $\sum_{n=0}^{\infty} (-1)^n x^{2n}/(2n)!$; $(-\infty, \infty)$.

(c) $1 + 2(x - 1) + (x - 1)^2$; $(-\infty, \infty)$.

(d) $\sum_{n=1}^{\infty} (-1)^{n+1}(x-1)^n/n$; $(0, 2)$.

(e) $\sum_{n=0}^{\infty} (-1)^n(n+1)(x-1)^n$; $(0, 2)$.

(f) $\sum_{n=0}^{\infty} (-1)^n x^n$; $(-1, 1)$.

(g) $-\sum_{n=0}^{\infty} (x+3)^n/2^{n+1}$; $(-5, -1)$

(h) $\sum_{n=0}^{\infty} (-1)^n x^{2n}$; $(-1, 1)$.

3. (a) $(-\infty, -1) \cup (-1, 1) \cup (1, \infty)$. (b) $(0, 1)$.
   (c) $(1, 5)$.

Page 237

1. $\sum_{n=0}^{\infty} x^n/n!$.

2. $\sum_{n=0}^{\infty} x^{2n}/2^n n!$.

3. $\sum_{n=0}^{\infty} (-1)^n x^{2n+1}/(2n+1)!$.

4. $\sum_{n=0}^{\infty} (n+1)x^n/n!$.

5. $\sum_{n=0}^{\infty} (-1)^{n+1}(n-1)x^{2n+1}/(2n+1)!$.

6. $y(x) = 1 + x + x^2/2 + 2x^3/3 + x^4/2 + \cdots$.

7. $y = 1 + ex + (1 + e^2)x^2/2 + e(1 + 2e^2)x^3/6$
   $+ e^2(1 + e + 2e^2 + 2e^3)x^4/12 + \cdots$.

8. $y = 1 + 2x + 5x^2/2 + 11x^3/3 + 119x^4/24 + \cdots$.

9. $y = 1 - x^2/2 + x^4/6 - 7x^6/360 + 37x^8/100\,80 - \cdots$.

10. $y = 1 + x/2 - 3x^2/8 + 7x^3/16 - 77x^4/128 + \cdots$.

11. $y = x + x^2 + x^4/24 + x^5/15 + x^6/40 + \cdots$.

12. $y = \pi/60 + \pi x/2 - g \sin(\pi/60)x^2/2L - g\pi \cos(\pi/60)x^3/12L$
    $+ [gL\pi^2 + 4g^2 \cos(\pi/60)] \sin(\pi/60)x^4/96L^2 + \cdots$.

Page 244

1.

	Singular points	
(a)	None;	∞, ∞.
(b)	$-2, 1$;	1, 1, 2.
(c)	$-2 \pm i$;	$1, \sqrt{5}, \sqrt{17}$.
(d)	$(2n+1)\pi/2$, n an integer;	$\pi/2$.
(e)	None;	∞.
(f)	$(2n+1)\pi i$, n an integer;	$\pi, \sqrt{1+\pi^2}$.

2. (a) 1.
(b) $y_1 = x + 2x^3/3$.
$y_2 = 1 + 3x^2/2 + 3x^4/8 - x^6/16 + x^8/32 - \cdots$.
(c) y_1 converges for all x. y_2 converges for $|x| < 1$.

3. (a) 1. (b) $y_1 = 1 - 3x^2$; $y_2 = x - x^3/3$.
(c) y_1 and y_2 both converge for all x.

4. (a) $y = 1 - x + x^3/6 + x^4/12 - 3x^5/40 + \cdots$.
(b) $y = 2 + 3(x-1) - 2(x-1)^2 + 5(x-1)^3/2 - 47(x-1)^4/12 + 19(x-1)^5/8 + \cdots$.
(c) $y = -1 + (x+1) + (x+1)^2 + 2(x+1)^3/3 + 5(x+1)^4/12 + (x+1)^5/4 + \cdots$.

5. (a) $$y_1 = 1 - \frac{2\lambda}{2!}x^2 + \frac{2^2\lambda(\lambda-2)}{4!}x^4 - \frac{2^3\lambda(\lambda-2)(\lambda-4)}{6!}x^6 + \cdots.$$
$$y_2 = x - \frac{2(\lambda-1)}{3!}x^3 + \frac{2^2(\lambda-1)(\lambda-3)}{7!}x^5 - \frac{2^3(\lambda-1)(\lambda-3)(\lambda-5)}{7!}x^7 + \cdots.$$
(b) 1; x; $1 - 2x^2$; $x - 2x^3/3$.

6. (a) $$y_1 = 1 - \frac{\lambda(\lambda+1)}{2!}x^2 + \frac{\lambda(\lambda-2)(\lambda+1)(\lambda+3)}{4!}x^4 - \frac{\lambda(\lambda-2)(\lambda-4)(\lambda+1)(\lambda+3)(\lambda+5)}{6!}x^6 + \cdots;$$
$$y_2 = x - \frac{(\lambda-1)(\lambda+2)}{3!}x^3 + \frac{(\lambda-1)(\lambda-3)(\lambda+2)(\lambda+4)}{5!}x^5 - \frac{(\lambda-1)(\lambda-3)(\lambda-5)(\lambda+2)(\lambda+4)(\lambda+6)}{7!}x^7 + \cdots.$$
(b) 1; x; $1 - 3x^2$; $x - 5x^3/3$.
(c) 1; x; $-1/2 + 3x^2/2$; $-3x/2 + 5x^3/2$.

7. (a) $$y_1 = 1 - \frac{\lambda^2}{2!}x^2 - \frac{(2^2-\lambda^2)\lambda^2}{4!}x^4 - \frac{(4^2-\lambda^2)(2^2-\lambda^2)\lambda^2}{6!}x^6 - \cdots;$$
$$y_2 = x + \frac{1-\lambda^2}{3!}x^3 + \frac{(3^2-\lambda^2)(1-\lambda^2)}{5!}x^5 + \frac{(5^2-\lambda^2)(3^2-\lambda^2)(1-\lambda^2)}{7!}x^7 + \cdots.$$
(b) 1; x; $1 - 2x^2$; $x - 4x^3/3$.

Page 261

1. (a) R.S.P.: $x = 0$, I.S.P.: $x = 2$.
(b) R.S.P.: $x = 2$, I.S.P.: $x = 0$.

(c) None. (d) R.S.P.: $x = 0$.
(e) I.S.P.: $x = 0$. (f) R.S.P.: $x = \pm 2i$.

2. (a) (i) $y_1 = |x|^{1/2}$; $y_2 = |x|$.
(ii) $y_1 = x^2$; $y_2 = |x|^3$.

3. (a) (i) $y_1 = 1$; $y_2 = \ln |x|$.
(ii) $y_1 = x^2$; $y_2 = x^2 \ln |x|$.

4. (a) $y_1 = |x|^{-1}\left[\sum_{n=0}^{\infty} \frac{(-1)^n x^n}{(n!)^2}\right]$; choosing $b_0 = 0$,

$$y_2 = y_1 \ln |x| - 2|x|^{-1} \sum_{n=1}^{\infty} \frac{(-1)^n\left(1 + \frac{1}{2} + \cdots + \frac{1}{n}\right)x^n}{(n!)^2}.$$

Both functions converge for $0 < |x|$.

(b) $y_1 = |x| \sum_{n=0}^{\infty} (-1)^n x^n$; $y_2 = |x|^{1/2} \sum_{n=0}^{\infty} (-1)^n x^n$.

Both functions converge for $0 < |x| < 1$.

(c) $y_1 = \sum_{n=2}^{\infty} \frac{(-1)^n 2^{n-1} x^{n-2}}{n(n-2)!}$; $y_2 = x^{-2}$.

y_1 converges to all x and y_2 is defined for $x \neq 0$.

(d) $y_1 = |x| \sum_{n=0}^{\infty} \frac{(-1)^n x^n}{n!\,(n+1)!}$; choosing $b_0 = b_1 = 1$,

$$y_2 = -\frac{|x|}{x} y_1 \ln |x| + \left[1 - \sum_{n=1}^{\infty} \frac{(-1)^n\left(2\left(1 + \frac{1}{2} + \cdots + \frac{1}{n}\right) - \frac{1}{n}\right)x^n}{(n-1)!\,n!}\right].$$

Both functions converge for $0 < |x|$.

(e) $y_1 = \sum_{n=0}^{\infty} \frac{x^n}{1\cdot 3\cdot 5\cdots(2n+1)}$; $y_2 = |x|^{-1/2} \sum_{n=0}^{\infty} \frac{x^n}{2^n n!}$.

y_1 converges for all x and y_2 converges for $x \neq 0$.

(f) $y_1 = |x| \sum_{n=0}^{\infty} \frac{(-1)^n x^{2n}}{2^{2n}(n!)^2}$; choosing $b_0 = 0$,

$$y_2 = y_1 \ln |x| + |x| \sum_{n=1}^{\infty} \frac{(-1)^{n+1}\left(1 + \frac{1}{2} + \cdots + \frac{1}{n}\right)x^{2n}}{2^{2n}(n!)^2}.$$

Both functions converge for $0 < |x|$.

(g) $y_1 = x^2\left[1 + \sum_{n=1}^{\infty} \frac{(-1)^n x^{2n}}{5\cdot 7\cdot 9\cdots(2n+3)}\right]$; $y_1 = |x|^{-1} \sum_{n=0}^{\infty} \frac{(-1)^n x^{2n}}{2^n n!}$.

y_1 converges for all x and y_2 converges for $x \neq 0$.

(h) $y_1 = \sum_{n=0}^{\infty} \frac{(-1)^n x^n}{n!\,(n+1)!}$; choosing $b_0 = b_1 = 1$,

$$y_2 = -\frac{|x|}{x} y_1 \ln|x| + |x|^{-1}\left[1 - \sum_{n=1}^{\infty} \frac{(-1)^n\left(2\left(1 + \frac{1}{2} + \cdots + \frac{1}{n}\right) - \frac{1}{n}\right)x^n}{(n-1)!\,n!}\right]$$

y_1 converges for all x and y_2 converges for $x \neq 0$.

5. (b) $r_1 = r_2 = 0$.

(c) $y_1 = 1 + \sum_{n=1}^{\infty} \frac{(-\lambda)(1-\lambda)\cdots(n-1-\lambda)x^n}{(n!)^2}$.

(d) 1; $1 - x$; $1 - 2x + x^2/2$; $1 - 3x + 3x^2/2 - x^3/6$.

(e) $y_2 = y_1 \ln|x| + \sum_{n=0}^{\infty} b_n x^n$.

6. (b) $r_1 = r_2 = 0$.

(c) $y_1 = 1 + \sum_{n=1}^{\infty} \frac{(\lambda+n)(\lambda+n-1)\cdots(\lambda+1)\lambda(\lambda-1)\cdots(\lambda-n+2)(\lambda-n+1)(x-1)^n}{2^n(n!)^2}$.

(d) $|x-1| < 2$.

Page 281

1. (a) $a_{n+1} = 2a_n/(n+1)$, $n = 0, 1, 2, \ldots$.
(c) $a_n = 2^n/n!$, $N \geq 38$.

2. (a) $y = \frac{2}{\sqrt{\pi}} \sum_{n=0}^{\infty} \frac{(-1)^n x^{2n+1}}{(2n+1)n!}$.

(b) $N \geq 48$.

(c) $a_{2n} = 0$, $n = 0, 1, 2, \ldots$; $a_1 = 2/\sqrt{\pi}$;
$a_{2n+3} = -(2n+1)a_{2n+1}/(2n+3)(n+1)$, $n = 0, 1, 2, \ldots$.

3. (b) $(w+1)\,dy/dw = 1$; $y(0) = 0$.
(c) $w = -1$.
(d) $a_0 = 0$; $a_1 = 1$; $a_{n+1} = -na_n/(n+1)$, $n = 1, 2, \ldots$.
(e) 1.
(f) $a_0 = 0$; $a_n = (-1)^{n+1}/n$, $n = 1, 2, \ldots$.
(g) $N \geq 30$.

4. (a) $a_0 = y(0)$; $a_1 = y'(0)$; $a_{n+2} = -a_n/(n+2)(n+1)$, $n = 0, 1, 2, \ldots$.

CHAPTER 9

Page 293

2. (a) $c_1 = 3$; $c_2 = 1$.
(b) $c_1 = -2$; $c_2 = 1$.
(c) $c_1 = 2$; $c_2 = -1$; $c_3 = 3$.
(d) $c_1 = 3$; $c_2 = 1$; $c_3 = -2$.

4. $\mathbf{y} = k_1\begin{pmatrix}1\\2\end{pmatrix}e^x + k_2\begin{pmatrix}1\\-2\end{pmatrix}e^{-x} + \begin{pmatrix}1 - e^x\\2x\end{pmatrix}$.

Page 305

5. $$\mathbf{y} = k_1\begin{pmatrix}1\\-2\end{pmatrix}e^{-x} + k_2\begin{pmatrix}1\\1\end{pmatrix}e^{2x}.$$

6. $$\mathbf{y} = k_1\begin{pmatrix}1\\-1\end{pmatrix}e^{2x} + k_2\left[\begin{pmatrix}1\\-1\end{pmatrix}xe^{2x} + \begin{pmatrix}1\\0\end{pmatrix}e^{2x}\right].$$

7. $$\mathbf{y} = k_1\begin{pmatrix}\cos 2x + 2\sin 2x\\ \cos 2x\end{pmatrix} + k_2\begin{pmatrix}\sin 2x - 2\cos 2x\\ \sin 2x\end{pmatrix}.$$

8. $$\mathbf{y} = k_1\begin{pmatrix}1\\1\\-1\end{pmatrix}e^{-x} + k_2\begin{pmatrix}1\\-2\\-1\end{pmatrix}e^{2x} + k_3\left[\begin{pmatrix}1\\-2\\-1\end{pmatrix}xe^{2x} + \begin{pmatrix}1\\-4\\-2\end{pmatrix}e^{2x}\right].$$

9. $$\mathbf{y} = k_1\begin{pmatrix}1\\1\\-1\end{pmatrix}e^{-x} + k_2\begin{pmatrix}1\\0\\1\end{pmatrix}e^{2x} + k_3\begin{pmatrix}0\\1\\1\end{pmatrix}e^{2x}.$$

10. $$\mathbf{y} = k_1\begin{pmatrix}2\\7\\4\end{pmatrix}e^{2x} + k_2\begin{pmatrix}\cos 3x - 3\sin 3x\\ -\cos 3x - 3\sin 3x\\ 2\cos 3x\end{pmatrix}e^{2x}$$
$$+ k_3\begin{pmatrix}3\cos 3x + \sin 3x\\ 3\cos 3x - \sin 3x\\ 2\sin 3x\end{pmatrix}e^{2x}.$$

11. $$\mathbf{y}_1 = \begin{pmatrix}1\\0\\1\end{pmatrix}e^{3x};\ \mathbf{y}_2 = \begin{pmatrix}1\\0\\1\end{pmatrix}xe^{3x} + \begin{pmatrix}1\\1\\1\end{pmatrix}e^{3x};$$
$$\mathbf{y}_3 = \begin{pmatrix}1\\0\\1\end{pmatrix}x^2e^{3x} + \begin{pmatrix}1\\2\\1\end{pmatrix}xe^{3x} + \begin{pmatrix}1\\2\\2\end{pmatrix}e^{3x}.$$

Page 309

1. $$\mathbf{y}_p = \begin{pmatrix}2\sin x\\ \cos x\end{pmatrix}.$$

2. $$\mathbf{y}_p = \begin{pmatrix}-1\\1\\2\end{pmatrix}e^{2x}.$$

3. $$\mathbf{y}_p = \begin{pmatrix}1\\2\\-1\end{pmatrix}e^{x}.$$

4. $$\mathbf{y}_p = \begin{pmatrix}1\\2\\-1\end{pmatrix}e^{x}.$$

5. $\mathbf{y}_p = \begin{pmatrix} \sin 2x + \cos x \\ \cos 2x + \sin x \end{pmatrix}.$

6. $\mathbf{y}_p = \begin{pmatrix} 1 \\ 0 \end{pmatrix} xe^{2x}.$

7. $\mathbf{y}_p = \begin{pmatrix} 4x^2 + 3e^x \\ 2x - 4 - e^x \end{pmatrix}.$

8. $\mathbf{y}_p = \begin{pmatrix} -3x - 1 \\ -6x^2 - x - 1 \end{pmatrix}.$

9. $\mathbf{y}_p = \begin{pmatrix} 2x^2 \\ -x^2 \end{pmatrix}.$

10. $\mathbf{y}_p = \begin{pmatrix} 1 \\ -1 \\ 1 \end{pmatrix} xe^{2x}.$

Page 314

1. $\begin{pmatrix} i_1 \\ i_2 \end{pmatrix} = k_1 \begin{pmatrix} 100 \\ 100 \end{pmatrix} e^{-200t} + k_2 \left[\begin{pmatrix} 100 \\ 100 \end{pmatrix} te^{-200t} + \begin{pmatrix} 2 \\ 1 \end{pmatrix} e^{-200t} \right].$

2. $\begin{pmatrix} i_1 \\ i_2 \end{pmatrix} = k_1 \begin{pmatrix} -\sin 100t \\ \cos 100t \end{pmatrix} e^{-100t} + k_2 \begin{pmatrix} \cos 100t \\ \sin 100t \end{pmatrix} e^{-100t}.$

3. $\begin{pmatrix} i_1 \\ i_2 \end{pmatrix}_p = \frac{1}{25} \begin{pmatrix} 36 \cos 100t + 48 \sin 100t \\ -9 \cos 100t - 12 \sin 100t \end{pmatrix}.$

4. (a)

$$\begin{pmatrix} i_1' \\ i_2' \end{pmatrix} = \begin{pmatrix} -\frac{1}{R_1 C_1} - \frac{1}{R_2 C_1} & \frac{1}{R_2 C_2} \\ \frac{1}{R_2 C_1} & -\frac{1}{R_2 C_2} \end{pmatrix} \begin{pmatrix} i_1 \\ i_2 \end{pmatrix} + \begin{pmatrix} \frac{1}{R_1} \frac{dE}{dt} \\ 0 \end{pmatrix}.$$

(b) $\begin{pmatrix} i_1 \\ i_2 \end{pmatrix} = \begin{pmatrix} 4 \\ -2 \end{pmatrix} e^{-1000t} + \begin{pmatrix} 2 \\ 2 \end{pmatrix} e^{-250t}.$

(c) $q_1(0) = 6 \times 10^{-3}$ C; $q_2(0) = 3 \times 10^{-3}$ C.

5. (a)

$$\begin{pmatrix} i_1' \\ i_2' \end{pmatrix} = \begin{pmatrix} -\frac{R_1}{L_1} & -\frac{R_1}{L_1} \\ -\frac{R_1}{L_2} & -\frac{R_1}{L_2} - \frac{R_2}{L_2} \end{pmatrix} \begin{pmatrix} i_1 \\ i_2 \end{pmatrix} + \begin{pmatrix} \frac{E(t)}{L_1} \\ \frac{E(t)}{L_2} \end{pmatrix}.$$

(b) $\begin{pmatrix} i_1 \\ i_2 \end{pmatrix} = \begin{pmatrix} (\sqrt{3} - 3)/4 \\ -\sqrt{3}/2 \end{pmatrix} e^{-(200+100\sqrt{3})t} + \begin{pmatrix} -(\sqrt{3} + 3)/4 \\ \sqrt{3}/2 \end{pmatrix} e^{-(200-100\sqrt{3}t)}.$

6. (a)

$$\begin{pmatrix} i_1' \\ i_2' \end{pmatrix} = \begin{pmatrix} -\frac{1}{RC} + \frac{R}{L} & \frac{R}{L} \\ -\frac{R}{L} & -\frac{R}{L} \end{pmatrix} \begin{pmatrix} i_1 \\ i_2 \end{pmatrix} + \begin{pmatrix} \frac{1}{R} \frac{dE}{dt} - \frac{E}{L} \\ \frac{E}{L} \end{pmatrix}.$$

(b)

$$\begin{pmatrix} i_1 \\ i_2 \end{pmatrix} = \begin{pmatrix} -\frac{12\sqrt{7}}{7} \sin 50\sqrt{7}\, t \\ -3 \cos 50\sqrt{7}\, t + \frac{9\sqrt{7}}{7} \sin 50\sqrt{7}\, t + 3 \end{pmatrix}.$$

Page 319

1.
$$\begin{pmatrix} Q_1 \\ Q_2 \end{pmatrix} = \begin{pmatrix} -95 - 145\sqrt{3}/3 \\ -50 - 140\sqrt{3}/3 \end{pmatrix} e^{-(3-\sqrt{3})t/20} + \begin{pmatrix} -95 + 145\sqrt{3}/3 \\ -50 + 140\sqrt{3}/3 \end{pmatrix} e^{-(3+\sqrt{3})t/20} + \begin{pmatrix} 200 \\ 100 \end{pmatrix}.$$

2. (a) $\mathbf{v} = \dfrac{2C}{5}\mathbf{v}_1 + \dfrac{3C}{5}\mathbf{v}_2.$

(b) $\mathbf{v} = \dfrac{3C}{5}\mathbf{v}_1 + \dfrac{2C}{5}\mathbf{v}_2.$

(c) $\mathbf{v} = -\dfrac{3C}{5}\mathbf{v}_1 - \dfrac{C}{15}\mathbf{v}_3 + \dfrac{C}{10}\mathbf{v}_4.$

(d) $\mathbf{v} = \dfrac{90C - 30D}{150}\mathbf{v}_1 + \dfrac{60C + 30D}{150}\mathbf{v}_2$, where

$$\mathbf{v}_1 = \begin{pmatrix} \cos 3t \\ -3\sin 3t \\ -2\cos 3t \\ 6\sin 3t \end{pmatrix}, \quad \mathbf{v}_2 = \begin{pmatrix} \cos 2t \\ -2\sin 2t \\ 3\cos 2t \\ -6\sin 2t \end{pmatrix},$$

$$\mathbf{v}_3 = \begin{pmatrix} \sin 3t \\ 3\cos 3t \\ -2\sin 3t \\ -6\cos 3t \end{pmatrix}, \quad \mathbf{v}_4 = \begin{pmatrix} \sin 2t \\ 2\cos 2t \\ 3\sin 2t \\ 6\cos 2t \end{pmatrix}.$$

CHAPTER 10
Page 362

1.

x	*Exact solution* $y = 2\left(\dfrac{e^x - e^{-x}}{e - e^{-1}}\right) - x$	*Approximate solution*
.0	.000 000	.000 000
.1	.070 468	.705 625
.2	.142 641	.142 812
.3	.218 244	.218 470
.4	.299 034	.299 297
.5	.386 819	.387 099
.6	.483 480	.483 756
.7	.590 985	.591 234
.8	.711 411	.711 607
.9	.846 962	.847 077
1.0	1.000 000	1.000 000

2.

x	*Exact solution* $y = x^2 + 1/x^2$	*Approximate solution*
1.0	2.000 00	2.000 00
1.1	2.036 45	2.036 31
1.2	2.134 44	2.134 28
1.3	2.281 71	2.281 58
1.4	2.470 20	2.470 11
1.5	2.694 44	2.694 40
1.6	2.950 61	2.950 63
1.7	3.236 01	3.236 05
1.8	3.548 63	3.548 68
1.9	3.886 99	3.887 04
2.0	4.250 00	4.250 00

4.

x	*Exact solution* $y = \sin x - \cos x + 2$	*Approximate solution*
.0	1.000 00	1.000 00
.1	1.104 83	1.104 65
.2	1.218 60	1.218 26
.3	1.340 18	1.339 71
.4	1.468 36	1.467 77
.5	1.601 84	1.601 17
.6	1.739 31	1.738 58
.7	1.879 38	1.878 63
.8	2.020 65	2.019 91
.9	2.161 72	2.161 01
1.0	2.301 17	2.300 53

6.

x	*Exact solution* $y = x + 1$	*Approximate solution*
.0	1.000 00	0.999 97
.1	1.100 00	1.099 97
.2	1.200 00	1.199 98
.3	1.300 00	1.299 98
.4	1.400 00	1.399 98
.5	1.500 00	1.499 98
.6	1.600 00	1.599 98
.7	1.700 00	1.699 98
.8	1.800 00	1.799 98
.9	1.900 00	1.899 98
1.0	2.000 00	1.999 97

8.

x	*Exact solution* $y = 1/x$	*Approximate solution*
1.0	1.000 000	1.000 000
1.1	.909 091	.908 933
1.2	.833 334	.833 080
1.3	.769 232	.768 921
1.4	.714 287	.713 944
1.5	.666 668	.666 309
1.6	.625 001	.624 637
1.7	.588 237	.587 874
1.8	.555 557	.555 201
1.9	.526 317	.525 971
2.0	.500 001	.499 667

9.

x	*Exact solution* $y = 2x + 2$	*Approximate solution*
.0	1.414 21	1.413 56
.1	1.483 24	1.482 67
.2	1.549 19	1.548 69
.3	1.612 45	1.612 01
.4	1.673 32	1.672 92
.5	1.732 05	1.731 69
.6	1.788 85	1.788 53
.7	1.843 91	1.843 62
.8	1.897 37	1.897 10
.9	1.949 36	1.949 12
1.0	2.000 00	1.999 78

INDEX